262-2713

612/654-5161

Energy Methods
in Applied Mechanics

Energy
Methods
in Applied
Mechanics

Henry L. Langhaar

Department of Theoretical
and Applied Mechanics
University of Illinois

John Wiley and
Sons, Inc.

New York
London

ISBN 0 471 51711 9

Library of Congress Catalog Card Number: 62-10925

Printed in the United States of America

Preface

Students of engineering usually receive only fragmentary instruction in the important principles of classical mechanics, stemming from the works of Huygens, Leibniz, Bernoulli, and Lagrange, which assign a central role to the concepts of work, potential energy, and kinetic energy. These laws, designated as "energy principles of mechanics," are sufficiently general to allow Newton's second law to be deduced from them. An integrated and modern treatment of energy principles of mechanics, with applications to dynamics of rigid bodies, analyses of elastic frames, general elasticity theory, the theories of plates and shells, the theory of buckling, and the theory of vibrations, is undertaken in this work.

The book has been planned for a two-semester course in advanced mechanics for graduate students. In addition, it may be used for independent study and for reference purposes. A sound course in advanced calculus, the rudiments of Gibb's vector theory, some knowledge of advanced strength of materials, and the fundamentals of elasticity theory are prerequisites. The student who has mastered these subjects will not find the mathematics too difficult. There are many abstractions, however, that must be digested slowly. The instructor should use many simple illustrations. Chapter 1, which serves as the groundwork for what follows, need not be comprehended fully at the first reading; the student should refer back to that chapter frequently.

The core of the theory is Jean Bernoulli's principle of virtual work. This law leads immediately to the more special principle of stationary potential energy. By means of the Legendre transformation the principle of virtual work yields the principle of complementary energy, which is a generalization of Castigliano's theorem. Hamilton's principle is derived from the principle of virtual work with the aid of the concept of virtual work of inertial forces. Hamilton's principle yields Lagrange's equations of motion directly. The theory of dynamics of rigid bodies and the theory of vibrations of systems with many degrees of freedom are developed with

v

the aid of the equations of Lagrange and Hamilton. Applications of the theory of real quadratic forms arise in the criterion for minimum potential energy, in the existence proofs for principal axes of stress, strain, and inertia, and in the theory of normal coordinates of vibrating systems. Accordingly, an appendix is devoted to this algebraic topic. Useful approximation methods based on variational principles of mechanics were devised by Rayleigh, Ritz, Galerkin, and others, but, in some environments, they have stigmatized energy principles with a reputation for inexactitude. Actually, the vectorial principles and the energy principles of mechanics are equally precise. Variational methods of approximation have been treated comprehensively in several treatises; these methods are discussed only briefly in this book.

Energy Methods in Applied Mechanics was developed from class notes used in a graduate course at the University of Illinois for about ten years. I have sustained an interest in energy principles of mechanics for more than twenty years through my research efforts, teaching activities, and employment as a structural research engineer. I am pleased to acknowledge as a notable source of information the inspiring lectures on variational principles of mechanics by C. Lanczos at Purdue University during 1940 and 1941.

H. L. LANGHAAR

Urbana, Illinois
May 1962

Contents

Notations

The following incomplete list of notations summarizes the main uses of certain symbols. In discussions of special topics other meanings may be ascribed to the symbols.

A dot over a letter denotes the time derivative. Primes are used occasionally to denote derivatives with respect to other variables. Bold face type denotes a vector or a configuration.

$\mathbf{X}$ a variable point in configuration space

$\mathbf{X}_0, \mathbf{X}_1, \cdots$ fixed points in configuration space

s distance in configuration space; Also, distance along the axis of a beam or strut

x, y, z rectangular coordinates or orthogonal curvilinear coordinates

$x_1, x_2, \cdots, x_n$ generalized coordinates of a mechanical system.

$\mathbf{r}$ position vector

$\mathbf{F}$ force vector

T kinetic energy of a mechanical system; also torque

U total internal energy or potential energy of internal forces (strain energy) of a mechanical system

W_e', W_i', W' respective virtual work of external forces, internal forces, or all forces, corresponding to any virtual displacement of a mechanical system; prime indicates that the quantities depend on the path in configuration space, $W' = W_e' + W_i'$

W_e, W_i, W least upper bounds of W_e', W_i', W', respectively, for all paths in configuration space that connect given terminal points, $W = W_e + W_i$; also, in discussions of real motions rather than virtual displacements W_e, W_i, W denote the actual work of external forces, internal forces, or all forces, respectively

$\delta W, \delta^2 W$ first and second variations of W [Eq. (1-5)]

xi

$Q_1, Q_2, \cdots, Q_n$ components of generalized force [Eqs. (1-6) and (1-8)]

$P_1, P_2, \cdots, P_n$ components of generalized external force [Eqs. (1-7) and (1-9)]

$R_1, R_2, \cdots, R_n$ components of nonconservative generalized force [Eq. (7-11)]

Ω potential energy of external forces that act on a mechanical system

V total potential energy of a mechanical system, $V = U + \Omega$, where U is the potential energy of internal forces

N tension in a bar, beam, or strut

E Young's modulus

ν Poisson's ratio; also, any positive integer

G shear modulus

L length of a bar, beam, or strut; also, $L = T - V$, the Lagrangian function

l length of a bar, beam, or strut

A cross-sectional area of a bar, beam, or strut; also, A denotes action

e extension of a bar; also volumetric strain [Eq. (4-10)]

e' initial or residual value of extension

P magnitude of a force

F magnitude of a force; an integrand; Airy stress function

$h_1, h_2, \cdots, h_n$ arbitrary increments of the generalized coordinates

λ Lagrange multiplier; shear rigidity of a beam [Eq. (2-13)] Lamé constant for a Hookean material [Eq. (4-47)]

σ tensile stress

ϵ tensile strain of a bar or a fiber; also an arbitrary infinitesimal

M bending moment

I moment of inertia of the cross section of a beam

τ shearing stress

S shearing force on the cross section of a beam

κ shape factor in formula for strain energy of shear [Eq. (2-10)]

β slope of a beam due to shear deformation

θ temperature; an angle or an angular coordinate

GJ torsional stiffness

ρ mass density

a radius of a circle; a length

z ordinate in a cross section of a curved beam, shell, or plate

$\mathbf{q}$ displacement vector of a deformable body

u, v, w components of the displacement vector of a deformable body in the x, y, z directions

$\bar{u}, \bar{v}, \bar{w}$ values of u, v, w at the middle surface or centroidal axis of a shell, plate, or curved beam

Z Winkler constant in the theory of curved beams [Eq. (2-22)]

ϕ angular displacement of a bar [Eq. (2-31)]; an Euler angle; a function symbol; in theory of strain $\phi = \epsilon + \frac{1}{2}\epsilon^2$; in theory of shells of revolution ϕ is angle between normal to the middle surface and the axis of symmetry

K stiffness factor of a beam ($K = 2EI/L$); also Gaussian curvature.

$\epsilon\eta(x)$ variation of a function.

Γ specified class of functions or configurations

$\epsilon_x, \epsilon_y, \epsilon_z, \gamma_{yz}, \gamma_{zx}, \gamma_{xy}$ strain components [Eqs. (4-3) and (4-28)]

l, m, n direction cosines

I_1, I_2, I_3 invariants of the strain tensor [Eq. (4-7)]

$\sigma_x, \sigma_y, \sigma_z, \tau_{yz}, \tau_{zx}, \tau_{xy}$ stress components

$\mathbf{p} = (p_x, p_y, p_z)$ stress vector on an oblique plane

p_n normal component of vector p

J_1, J_2, J_3 invariants of stress tensor [Eq. (4-21)]

F_x, F_y, F_z components of body force per unit mass

α, β, γ Lamé coefficients [Eqs. (4-26) and (4-27)]

U_0 internal energy density or strain energy density

$\epsilon_1, \epsilon_2, \cdots, \epsilon_6;$ alternative notations for components of strain and $\sigma_1, \sigma_2, \cdots, \sigma_6$ stress [Eq. (4-36)]; also, $\epsilon_1, \epsilon_2, \epsilon_3$ and $\sigma_1, \sigma_2, \sigma_3$ denote principal strains and stresses, respectively

Υ_0 complementary energy density (Sec. 4-6)

k coefficient of thermal expansion

Υ complementary energy [Eq. (4-58)]

m bending moment caused by a unit load

n tension caused by a unit load

t time; an arbitrary parameter; web thickness; torque caused by a unit load

θ_0, θ_1 temperature terms in theory of plates and shells [Eq. (5-8)]

e_x, e_y, e_{xy} values of ϵ_x, ϵ_y, γ_{xy} on the middle surface of a plate or shell

U_m strain energy due to stretching of the middle surface of a plate or shell

U_b strain energy due to bending of a plate or shell

U_θ strain energy due to heating of a plate or shell

h thickness of a plate or shell; depth of web of a beam

p a load parameter; lateral load per unit area of a plate or shell

p_{cr} critical value of load parameter

D flexural rigidity of a plate or shell [Eq. (5-10)]

N_x, N_y, N_{xy}, N_{yx} intensities of normal and shearing tractive forces in a plate or shell [Eqs. (5-11) and (5-59)]

Q_x, Q_y intensities of transverse shearing forces in a plate or shell [Eqs. (5-22) and (5-59)]

M_x, M_y, M_{xy}, M_{yx} intensities of bending moments and twisting moments in a plate or shell [Eqs. (5-22) and (5-59)]

A, B Lamé coefficients for the middle surface of a shell or plate [Eqs. (5-48) and (5-49)]

$\mathbf{n}$ unit normal vector of a surface

e, f, g coefficients of the second fundamental form of a surface [Eq. (5-51)]

r_1, r_2 principal radii of curvature of a surface [Eq. (5-53)]

$\mathbf{P} = (P_x, P_y, P_z)$ intensity of external force acting on a shell (Sec. 5-7)

$\mathbf{R} = (R_x, R_y, R_z)$ intensity of external couple acting on a shell (Sec. 5-7)

$\kappa_x, \kappa_y, \kappa_{xy}$ incremental curvatures and twist due to bending of a shell [Eqs. (5-69) and (5-71)]

r radius of cross section of a surface of revolution; radial coordinate for polar coordinates; rank of a matrix

H Hamiltonian function

$p_1, p_2, \cdots, p_n$ components of generalized momentum

ξ, η, ζ body-centered rectangular coordinates (Sec. 7-8)

θ, ϕ, ψ Euler angles (Sec. 7-8)

$\boldsymbol{\omega}$ angular velocity vector

$\omega_1, \omega_2, \omega_3$ components of $\boldsymbol{\omega}$ on (ξ, η, ζ) axes

A, B, C principal moments of inertia of a rigid body

1 General concepts and principles of mechanics

If any system of bodies or particles, whatsoever, each being acted upon by arbitrary forces, is in equilibrium, and if one gives to this system a small arbitrary movement, by virtue of which each particle traverses an infinitely small space, which will determine its virtual velocity, then the sum of the forces, each multiplied by the distance through which the corresponding particle moves in the direction of the same force, will always equal zero, where we regard as positive the small distances traveled in the directions of the forces, and as negative the distances traveled in the opposite directions.

J. L. LAGRANGE

Classical mechanics developed along two different lines. One path, known as "vectorial mechanics," issues directly from Newton's laws. It exploits the idea that any mechanical system is composed of particles that move in accordance with the vector equation $\mathbf{F} = m\mathbf{a}$. The other path originates in a law of statics that was known formerly as the "principle of virtual velocities." It is now called the "principle of virtual work" or "virtual displacements." In a rudimentary form this law was known (31)* to Leonardo da Vinci (1452–1519), and its origin may be traced to the ancients. Galileo (1564–1642) recognized the principle of virtual work as a general law that could be applied to simple machines (21).

Jean Bernoulli (1667–1748) first gave the principle of virtual work a general formulation that applies to nearly all mechanical systems. Also, in the eighteenth century the mathematical tools for applying the principle of virtual work were developed extensively. The main contributor to this

---

* Numbers in parentheses designate references at the end of the book.

theory was the French mathematician J. L. Lagrange (1736–1813), whose treatise *Mécanique analytique* (46) is a classic on energy principles of mechanics. Through the works of D'Alembert, Lagrange, and Hamilton (21), the principle of virtual work was extended to kinetics, where it has proved to be very fruitful.

The principle of virtual work is only one of several laws of classical physics that relate to the concept of work. The equivalence of heat and work, discovered by Mayer and Joule in the first half of the nineteenth century, is the foundation of thermodynamics. An equally general law, the principle of kinetic energy or "vis viva," propounded by G. W. Leibniz (1646–1716), formed the basis of the theory of hydraulics in the eighteenth century.

The present chapter is devoted to general energy principles of mechanics and some allied concepts.

1-1. MECHANICAL SYSTEMS.   Some kinematical concepts that apply to all mechanical systems are discussed in this article. A mechanical system is defined as anything that is composed of matter. The first step in an analysis of a mechanical system should be a precise and definitive description of the system under consideration. Since the modern theories of the constitution of matter will not be considered, the particles that compose a mechanical system are regarded as mathematical abstractions; they are more properly called "material points." The simultaneous positions of all the material points of a mechanical system are called the "configuration" of the system. For example, the displacement vector field of a deformable body defines a configuration of the body. To define the configuration of a mechanical system, we require a coordinate system that is attached to some rigid system, known as a "reference frame." In the theory of kinematics the reference frame is arbitrary.

A general problem of statics is to determine the equilibrium configurations of mechanical systems under prescribed types of loadings and to ascertain which among them are stable. An important general problem of dynamics is to express the configuration of a given mechanical system as a function of time.

A mechanical system is said to experience a displacement if any of its material points are displaced. In other words, any change of the configuration of a mechanical system is a displacement.

Usually the material points of a mechanical system cannot be displaced independently. Geometrical restrictions on the displacements of the material points of a system are called "constraints." For example, the constraints in a rigid body are such that any two particles of the body remain at a constant distance from each other. The constraints in an incompressible fluid are such that the volume of any part of the fluid remains constant. The constraints of an ideal cantilever beam are such that the displacement vector vanishes at the clamped end.

Geometric Terminology. Geometric terminology is used in mechanics as an aid to the imagination and as a convenient and concise mode of expression of ideas. There are remarkable analogies between the familiar concepts of space relations and certain general ideas in kinematics. The theory of point sets is the logical basis for such analogies.

A set is defined as any collection of things. The individual things that constitute a set are called the elements of the set. The most important sets, from a theoretical standpoint, contain an infinite number of elements. Mechanics deals particularly with the set of all configurations that a mechanical system can assume. This set is called the configuration space of the system. Any configuration is called a point in configuration space. If the set of all configurations is such that any configuration can be reached from any other configuration, the configuration space is said to be connected. The set of all conceivable configurations of two rings is an example of a disconnected configuration space, since configurations in which the rings are linked are inaccessible from those in which they are not linked. A more significant example is a unidirectional mechanism such as a ratchet. A device of this type may possess two configurations (say A and B) such that the passage from A to B is possible but the passage from B to A is impossible. Disconnected configuration spaces are not considered in this book.

A variable point in the configuration space of a mechanical system is denoted by the symbol $\mathbf{X}$. Numerical subscripts are appended to $\mathbf{X}$ to denote fixed points. The notation $\mathbf{X}_1 - \mathbf{X}_0$ designates the displacement from $\mathbf{X}_0$ to $\mathbf{X}_1$. If the system is a single particle, $\mathbf{X}_0$ and $\mathbf{X}_1$ may be regarded as vectors from a fixed origin O to the particle (Fig. 1-1). Then $\mathbf{X}_1 - \mathbf{X}_0$ denotes the difference of the vectors. In an abstract sense this picture remains applicable even though the system is not a particle.* H. Hertz (33) conceived this idea when he wrote:

A vector quantity with regard to a system is any quantity which bears a relation to the system and which has the same kind of mathematical manifold as a conceivable displacement of the system. A displacement of a system is itself a vector quantity with regard to the system. A vector quantity with regard to a single material point is a vector in the ordinary sense of the word.

Distance in Configuration Space. A displacement of a particle is a vector quantity, but, in the present context, the term displacement denotes

---

* J. L. Synge has employed analogous vector representations of various physical states to develop a general approximation procedure for solving problems of applied mathematics (80).

the magnitude of the displacement vector. The displacements of the particles of a mechanical system constitute a set of non-negative real numbers that is bounded above.* Accordingly, the following definition is admissible:

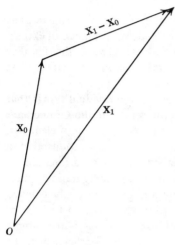

Fig. 1-1

> *The magnitude s of a displacement of a mechanical system is the least upper bound of the displacements of the material points of the system.*

In other words, s is the smallest number that is not exceeded by the displacement of some material point of the system. Consequently, s is the maximum of the displacements of the individual particles of the system, provided that this maximum exists. The distance s between two points $\mathbf{X}_0$ and $\mathbf{X}_1$ in configuration space is understood to be the magnitude of the displacement that the system receives when it is transferred from $\mathbf{X}_0$ to $\mathbf{X}_1$. It is denoted by $s = |\mathbf{X}_1 - \mathbf{X}_0|$.

Distance in configuration space has the following properties in common with distance in Euclidean space:

(a) $|\mathbf{X}_0 - \mathbf{X}_0| = 0$.

(b) If $\mathbf{X}_1 \neq \mathbf{X}_0$, $|\mathbf{X}_1 - \mathbf{X}_0| = |\mathbf{X}_0 - \mathbf{X}_1| > 0$.

(c) If $\mathbf{X}_0$, $\mathbf{X}_1$, $\mathbf{X}_2$ are any three points, $|\mathbf{X}_0 - \mathbf{X}_1| + |\mathbf{X}_1 - \mathbf{X}_2| \geq |\mathbf{X}_0 - \mathbf{X}_2|$.

Condition (c) is known as the "triangle law," since it is analogous to the Euclidean theorem that the sum of two sides of a triangle is no less than the third side.

A "neighborhood" of a point $\mathbf{X}_0$ in configuration space is defined as the

---

* A number c is said to be an upper bound of a set S of real numbers if there is no number in S which is greater than c. If a number c with this property exists, set S is said to be bounded above. If c is an upper bound of S, so is any number greater than c. It is shown in real variable theory (36) that the set of all upper bounds of a set S always contains a minimum number b, called the "least upper bound" of S. The number b is the maximum number in S if b belongs to S. However, not every set of real numbers that is bounded above contains a maximum. For example, let S be the set of all real numbers less than 3. Then, if x is any given number in S, there are infinitely many numbers in S that are greater than x; hence S contains no maximum. In this case the least upper bound of S is 3, but 3 is not a number belonging to S.

Similarly, if a set S of real numbers is bounded below, it possesses a "greatest lower bound" which coincides with the minimum number in S, whenever a minimum exists.

set of points X that satisfies the inequality $|X - X_0| < \rho$, where ρ is any given positive number. A deleted neighborhood of point X_0 is defined as the set of points X that satisfies the inequality $0 < |X - X_0| < \rho$.

Paths in Configuration Space. The configuration X of a mechanical system is said to be a function of a real variable t in the interval $a \leq t \leq b$ if, to each value of t in this interval, there corresponds a single configuration X. This relationship is indicated by the conventional function notation, $X = X(t)$, $a \leq t \leq b$. For example, t may denote time. Then the relation $X = X(t)$ represents the motion of the system. The function $X(t)$ is then continuous, since the system does not leap a finite distance in zero time. With the preceding definition of distance in configuration space, continuity of the function $X(t)$ is defined as in real variable theory; that is, the function $X(t)$ is continuous at the point $X_0 = X(t_0)$ if to each positive number ϵ there corresponds a positive number δ such that $|X - X_0| < \epsilon$ for all t in the interval $|t - t_0| < \delta$. The function is said to be continuous at the end point $t = a$ (or $t = b$) if $|X - X_0| < \epsilon$ for all t in the interval $0 < t - a < \delta$ (or $0 < b - t < \delta$).

If the coordinates of a particle in Euclidean space are continuous functions of a real variable t, the locus of the particle is a continuous curve. Analogously, a function $X = X(t)$ that is defined and that is continuous at all points of the interval $a \leq t \leq b$ is said to represent a continuous curve in configuration space. The points $X(a)$ and $X(b)$ are called the "end points" of the curve. The curve is said to have no double points (in other words, the curve does not cross itself) if any pair of distinct values of t in the interval $a < t < b$ (say t_1 and t_2) correspond to distinct points X_1 and X_2. In real variable theory a continuous curve of this type is called a "Jordan curve." A Jordan curve in configuration space is called a "path." If $X(a) = X(b)$, the path is said to be "closed."

Nonholonomic Systems. A system may be constrained to follow certain paths in configuration space, even though the space is connected. In other words, although the system can pass from any point to any other point in its configuration space, it may be constrained to follow routes that coincide with a certain dense network of paths. These paths may be regarded as tracks that guide the system. If such constraints exist, the system is said to be "nonholonomic." Since nonholonomic constraints exclude certain paths, a path that the system can follow is said to be "admissible."

The word "holonomic" was introduced by H. Hertz, (33), who stated:

"A material system, between whose possible positions all conceivable continuous motions are also possible motions, is called a holonomic system."

A mathematical example of a nonholonomic system is provided by a single particle that is located by rectangular coordinates (x, y, z). Suppose there is a mechanism that constrains the particle so that any infinitesimal displacement (dx, dy, dz) conforms to the relationship

$$dx - z\,dy = 0 \tag{a}$$

If there were an integrating factor for this differential relation, it would lead to a finite equation among the coordinates, which would imply that the particle were constrained to lie on a definite surface. However, this interpretation is incorrect, since Eq. (a) does not possess an integrating factor. Any point in space is accessible to the particle, even though Eq. (a) is satisfied. In other words, if (x_0, y_0, z_0) and (x_1, y_1, z_1) are any two points, the particle can travel from (x_0, y_0, z_0) to (x_1, y_1, z_1) by paths that satisfy Eq. (a).* Since the particle is not free to follow every path between these points, the constraint represented by Eq. (a) is nonholonomic (also said to be nonintegrable). A similar two-dimensional example does not exist, since a linear differential expression in two variables always possesses an integrating factor. Nonholonomic constraints are usually defined by nonintegrable differential relations, such as Eq. (a). Perhaps the simplest practical example of a nonholonomic system is a sphere that rolls on a plane. There are two nonintegrable differential equations which specify that any infinitesimal displacement is effected without slipping (see Prob. 30, Chap. 7).

Rolling motion might provide the only real mechanical illustration of nonholonomic constraints. Furthermore, a rolling body such as a railroad car wheel, whose configuration is determined completely by the angle through which the body has turned, is holonomic. Consequently, little attention is given to nonholonomic systems in this book, although general laws are phrased to apply to nonholonomic cases unless the contrary is stated. Treatises on theoretical mechanics (48, 90) present special methods for analyzing nonholonomic systems.

1-2. GENERALIZED COORDINATES. If a mechanical system consists of a finite number of material points or rigid bodies, its configuration

* If $x_0 = x_1$ and $y_0 = y_1$, the particle may follow the path $x = x_0$, $y = y_0$. If $y_0 \neq y_1$, let it follow the line $x = x_0$, $y = y_0$ from $(x_0, y_0. z_0)$ to (x_0, y_0, z'), where $z' = (x_1 - x_0)/(y_1 - y_0)$; then let it follow a straight line from (x_0, y_0, z') to (x_1, y_1, z'). Finally, let it follow the path $x = x_1$, $y = y_1$ from (x_1, y_1, z') to (x_1, y_1, z_1). In the special case for which $y_1 = y_0$ the preceding method may be used to transfer the particle from (x_0, y_0, z_0) to (x_2, y_2, z_2), where (x_2, y_2, z_2) is any point such that $y_2 \neq y_0$. Then, by the same method, the particle may be transferred from (x_2, y_2, z_2) to (x_1, y_1, z_1). Thus, in all cases the particle may be transferred from (x_0, y_0, z_0) to (x_1, y_1, z_1) by paths that satisfy the equation $dx - z\,dy = 0$ and indeed by paths consisting of straight-line segments.

can be specified by a finite number of real variables, called "generalized coordinates" (or simply "coordinates"). For example, the angle θ through which the flywheel of an engine rotates is a coordinate for the engine; it determines the positions of the crankshaft and the pistons if the members are rigid. A free rigid body requires six coordinates; for example, three cartesian coordinates of the center of mass and three angles (Euler angles, Sec. 7-8) that determine the orientation of the body. A flexible beam does not possess a finite set of coordinates, but it may possess an enumerably infinite set; for example, the coefficients $a_1, a_2, \cdots$ of a Fourier series $y = \sum a_n \sin n\pi x/L$ that represents the deflection of the beam, since these coefficients determine the configuration of the beam. Similarly, by means of a double series, an enumerable set of coordinates may be employed for a flexible plate. If the series is approximated by a finite number of terms, the beam or plate is effectively approximated by a system with a finite number of generalized coordinates. Such approximations are used extensively in applied mechanics.

The generalized coordinates of an arbitrary mechanical system with a finite number of movable parts are denoted by $(x_1, x_2, \cdots, x_n)$. As a condition of regularity of the coordinates, we require that a one-to-one continuous correspondence exist between the number n-tuples $(x_1, x_2, \cdots, x_n)$ and the points in configuration space, at least in the region R under consideration. This condition precludes an equation of constraint of the type $f(x_1, x_2, \cdots, x_n) = 0$, since such an equation would exclude some points of region R from the range of possible configurations.

As a second condition of regularity of the coordinates, we require that if any coordinate x_i receives an increment Δx_i the corresponding displacement Δs in configuration space shall be an infinitesimal of the same order of magnitude; that is, the limit of $\Delta s/\Delta x_i$ as Δx_i approaches zero shall exist, and it shall be different from zero.

The need for regular coordinates is illustrated by a simple example. Suppose that a frictionless bead is constrained to a fixed smooth curve (e.g. a rigid wire) that is defined parametrically by differentiable functions $x = x(t)$, $y = y(t)$, $z = z(t)$, where (x, y, z) are rectangular coordinates with the z-axis directed upward. The parameter t is a generalized coordinate for the bead. For t to be a regular coordinate, there must be a one-to-one continuous correspondence between values of t and points on the curve. Furthermore, the derivative of arc length s with respect to t must exist, and it must not vanish in the interval under consideration. It is shown in Sec. 1-9 that the equilibrium points of the bead are those points for which $dz/dt = 0$, provided that t is a regular coordinate. This may be written in the form $(dz/ds)(ds/dt) = 0$. Since $ds/dt \neq 0$, this yields $dz/ds = 0$. Thus it is seen that a sufficient condition for equilibrium is that the bead lie at a point at which the tangent to the curve is horizontal. If t were not required to be a regular coordinate, we could set $t = z$, provided that there is a single value of z for each point on the curve, and vice versa. However, if $t = z$, the equilibrium points are no longer determined by $dz/dt = 0$, since dz/dt is now identically equal to 1. The trouble is that ds/dt becomes infinite at the equilibrium points; that is, the coordinate $t = z$ is irregular at the points where equilibrium occurs.

Degrees of Freedom. For a mechanical system with n regular coordinates $(x_1, x_2, \cdots, x_n)$, the number of degrees of freedom is defined as $n - r$, where r is the number of independent nonintegrable differential equations of constraint. For example, the particle discussed in Sec. 1-1 has a three-dimensional configuration space, but it has two degrees of freedom, since there is a nonholonomic constraint defined by $dx - z\, dy = 0$. A rolling sphere has a five-dimensional configuration space, but it has three degrees of freedom, since there are two nonintegrable differential equations of constraint. For a holonomic system, the number of degrees of freedom equals the number of generalized coordinates required to specify the configuration of the system. If a finite number of coordinates will not serve this purpose, the system is said to have infinitely many degrees of freedom.

1-3. ELEMENTARY PRINCIPLES OF DYNAMICS. To measure displacements, velocities, and accelerations, an observer must first establish a coordinate system with respect to some material bodies known as a "reference frame." For example, latitude and longitude are coordinates attached to the earth; the earth is the reference frame for this system. Newton's laws of mechanics are valid* only for reference frames that do not rotate with respect to the remote stars and that do not accelerate with respect to the center of mass of the solar system. Such reference frames are said to be "Newtonian" (or "Galilean" or "inertial"). The earth is not exactly a Newtonian reference frame, but, for most engineering purposes, the rotation and the acceleration of the earth are negligible. Unless the contrary is specified, the reference frame under consideration is Newtonian.

Let **r** be the position vector of a particle with respect to a Newtonian reference frame. If **r** is a function of time, the velocity of the particle is

---

* According to Einstein's relativity theory, there is no reference frame for which Newton's laws are exactly valid. However, for phenomena in which the particles are large compared to the elementary particles of physics and for which the speeds are small compared to the speed of light, Newton's laws are a close approximation to reality.

$\mathbf{v} = d\mathbf{r}/dt$. The work dW that an applied force $\mathbf{F}$ performs on the particle when the infinitesimal displacement $d\mathbf{r}$ occurs is defined by

$$dW = \mathbf{F} \cdot d\mathbf{r} = \mathbf{F} \cdot \mathbf{v}\, dt \tag{a}$$

Accordingly, the work that force $\mathbf{F}(t)$ performs on the particle during the time interval (t_0, t_1) is

$$W = \int_{t_0}^{t_1} \mathbf{F} \cdot \mathbf{v}\, dt \tag{1-1}$$

Since $\mathbf{v}$ depends on the choice of the reference frame, W also depends on the choice of the reference frame. In other words, work is a relative quantity. In fact, by a suitable selection of a Newtonian reference frame, W may be given any preassigned value.

If a particle belongs to a mechanical system, work W performed on that particle is said to be performed on the system. If several forces act on particles of the system, the work that they perform on the system is understood to be the sum of their individual amounts of work. Frequently a force shifts its point of application from one particle to another as the motion proceeds. The work that the force performs on the system is then understood to be the sum of the amounts of work that it performs on each particle on which it acts. The meaning of this statement is clear, provided that we can enumerate the particles on which the force acts. However, an explanation is necessary if the force shifts its point of application in a continuous manner. Such cases arise frequently. For example, the force that the track exerts on a railroad car wheel moves its point of application with respect to the wheel as the wheel rotates.

To arrive at a suitable general definition of work, we regard continuous shifting of the point of action of a force as the limiting case of discrete jumps. Let a force act successively on particles 1, 2, 3, $\cdots$ of a mechanical system, and let it have the constant value $\mathbf{F}_i$ while it is acting on the ith particle. Let the ith particle experience a displacement $\Delta\mathbf{r}_i$ while the force $\mathbf{F}_i$ acts on it. Then the work that the force performs upon the ith particle is $\mathbf{F}_i \cdot \Delta\mathbf{r}_i$, and the total work that the force performs on the mechanical system is

$$W = \sum \mathbf{F}_i \cdot \Delta\mathbf{r}_i \tag{b}$$

where the sum extends over all particles on which the force acts. Let the displacement $\Delta\mathbf{r}_i$ occur in a time interval Δt_i. Then Eq. (b) may be written as

$$W = \sum \mathbf{F}_i \cdot \frac{\Delta\mathbf{r}_i}{\Delta t_i} \Delta t_i \tag{c}$$

Now $\Delta\mathbf{r}_i/\Delta t_i$ is the mean velocity $\mathbf{v}_i$ of the ith particle during the interval Δt_i. Hence Eq. (c) yields

$$W = \sum \mathbf{F}_i \cdot \mathbf{v}_i \Delta t_i \tag{d}$$

If the increments Δt_i approach zero, the sum approaches a limiting value which is represented by an integral. Accordingly, Eq. (1-1) remains valid for a force that is applied in any continuous way to a mechanical system.

When Eq. (1-1) is applied to a mechanical system, $\mathbf{v}$ must be interpreted as the velocity of the particle on which force $\mathbf{F}$ acts at time t; it must not be regarded as the velocity of the point of action of the force. For example, a rigid ball that rolls on a fixed rigid body receives no work from the reactive force of the body, since the velocity of the particle of the ball that is in contact with the body is zero. Likewise, Eq. (1-1) shows that if a brick slides on a pavement the friction of the pavement performs work on the brick, although the friction of the brick performs no work on the pavement. These examples show that Eq. (1-1) is more general than the equation $W = \int \mathbf{F} \cdot d\mathbf{r}$; the latter equation applies only if the force $\mathbf{F}$ acts continuously on the same particle. This fact was emphasized by Osgood (62).

Newton's third law asserts that if body A exerts a force $\mathbf{F}$ on body B then body B exerts the force $-\mathbf{F}$ on body A. This law signifies that forces may be mated, "action" and "reaction." The reaction of a given force $\mathbf{F}$ is understood to act on the body that causes or exerts force $\mathbf{F}$. If a force $\mathbf{F}$ acts on a mechanical system, its reaction $-\mathbf{F}$ acts on another part of the same system, or it acts on a body outside the system. In the first case, it is called an "internal force;" in the second case, it is called an "external force." Accordingly, all the forces that act on a mechanical system may be classified as internal or external. Hence the work W of all the forces that act on a mechanical system is separated into a sum, $W = W_e + W_i$, where W_e is the work of the external forces and W_i is the work of the internal forces.

Law of Kinetic Energy. Newton's equation for a mass particle is $\mathbf{F} = m\,d\mathbf{v}/dt$. Hence, if $\mathbf{r}$ is a vector from a fixed origin to the particle, $\mathbf{F} \cdot d\mathbf{r} = m(d\mathbf{v}/dt) \cdot d\mathbf{r}$. Now, $\mathbf{F} \cdot d\mathbf{r}$ is the infinitesimal work dW that the force $\mathbf{F}$ performs on the particle. Also, $\mathbf{v} = d\mathbf{r}/dt$. Therefore,

$$dW = m\mathbf{v} \cdot d\mathbf{v} = \frac{d}{dt}(\tfrac{1}{2}mv^2)\,dt \tag{e}$$

where v is the magnitude of vector $\mathbf{v}$. Since, by definition, the kinetic energy of the particle is $T = \tfrac{1}{2}mv^2$, Eq. (e) yields $dW = dT$. Consequently, by integration, $W = \Delta T$, where ΔT is the increment of kinetic energy that results from work W.

This conclusion may be generalized immediately to apply to all finite mechanical systems. The kinetic energy of any mechanical system is defined as the sum of the kinetic energies of its particles. Likewise, the total work that is performed on a mechanical system is the sum of the amounts of work performed on each of its particles. Consequently, by summing the equation $W = \Delta T$ over all the particles of a system, we obtain the following conclusion:

> *The work of all the forces (internal and external) that act on a mechanical system equals the increase of kinetic energy of the system.*

This theorm is a modern statement of Leibniz's law of vis viva; it is called the "law of kinetic energy."

The law of kinetic energy is naturally restricted to Newtonian reference frames. In general, ΔT changes its value if one Newtonian reference frame is substituted for another, since the speed v of any particle relative to its reference frame is changed. However, the law of kinetic energy is not violated, for W changes also. The law of kinetic energy thus illustrates the relativity of work.

1-4. FIRST LAW OF THERMODYNAMICS. It is well known that heat is a form of energy and that 778 ft-lb of work are equivalent to one British thermal unit (or 4.185×10^7 ergs $= 4.185$ joules $= 1$ calorie). In the occasional references to heat in this book, we suppose that heat is measured in mechanical units.

The principle of conservation of energy is based, to some extent, on the idea that any mechanical system possesses internal energy U that is stored partly as kinetic and potential energy of the atoms and partly as energy within the atoms. The internal energy of a system is usually regarded as a function of the state of the system. However, this condition depends on the definition of "state." For example, the internal energy of a given mass of fluid is determined by the pressure and the volume; the internal energy of an elastic body is determined by the strains and the temperature. Generally speaking, however, the internal energy of a system depends on factors other than the configuration and the temperature. For example, by slow extension and compression of an inelastic rod, we may lead the rod around a hysteresis loop (Fig. 1-2). The external work supplied for one cycle is represented by the area enclosed by the loop. If no heat escapes, this work is equal to the increase ΔU of internal energy during the cycle. The temperature may be changed by the cycle, but the heat that must be extracted to restore the initial temperature is not equal to ΔU, since some energy is locked in the rod by residual stresses on the microscopic level. Consequently, by leading the rod around a hysteresis loop and then restoring the initial temperature, we alter the internal energy of the rod without changing the macroscopic strain or the temperature. This conclusion shows that the internal energy of a body may depend on parameters other than strain and temperature.

The total amount of internal energy in a system is generally indeterminate. Only changes of internal energy are measurable. These changes are determined by the first law of thermodynamics, a precise statement of

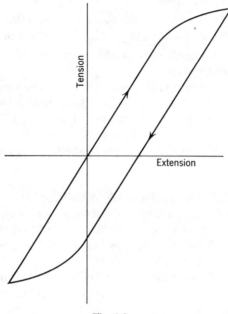

Fig. 1-2

the law of conservation of energy. If electromagnetic effects are disregarded, this law is expressed as follows:

The work that is performed on a mechanical system by external forces plus the heat that flows into the system from the outside equals the increase of kinetic energy plus the increase of internal energy.

In symbols the first law of thermodynamics is expressed by the equation

$$W_e + Q = \Delta T + \Delta U \tag{1-2}$$

Here, W_e is the work performed on the system by *external* forces, Q is the heat that flows into the system, ΔT is the increase of kinetic energy, and ΔU is the increase of internal energy.

On the other hand, the law of kinetic energy (Sec. 1-3) is expressed by the equation

$$W_e + W_i = \Delta T \tag{1-3}$$

Here, W_i is the work performed by internal forces in the system. Equations (1-2) and (1-3) yield.

$$W_i = Q - \Delta U \tag{1-4}$$

Equation (1-4) is useful for determining the work of internal forces. For a rigid body or a frictionless incompressible fluid, $Q = \Delta U$; consequently, the internal forces perform no work.

A process for which $Q = 0$ is said to be "adiabatic." For an adiabatic process, Eq. (1-4) yields the important special case $W_i = -\Delta U$.

1-5. FOURIER'S INEQUALITY. The law of kinetic energy leads immediately to a general principle of statics. When a mechanical system begins to move, it is gaining kinetic energy. Therefore, by the law of kinetic energy, the forces that act on the system (internal and external forces) are performing net positive work. Consequently, the system does not move spontaneously unless it can experience some arbitrarily small displacement for which the net work W' of all the forces is positive. In other words, spontaneous movement is impossible if $W' \leq 0$ for all small displacments. The small displacements considered here are called *virtual displacements*, since they are not necessarily realized. Likewise, the work W' is called *virtual work*. To eliminate a possible dependence of W' on the speed with which the virtual displacement is executed, we suppose that virtual displacements are performed with infinitesimal speed. The forces that act when a system moves with infinitesimal speed do not necessarily coincide with the forces that exist when the system is motionless. For example, if we drag an object with infinitesimal speed, the frictional force is not the same as though the object were at rest.

The conclusion that has been deduced may be expressed as follows:

A motionless mechanical system remains at rest if $W' \leq 0$ for every small virtual displacement that is consistent with the constraints.

This theorem was first stated by the French mathematician J. Fourier (1768–1830); it is known as "Fourier's inequality" (105). It may be generalized immediately to apply to systems that translate at constant velocity, since uniform translation is annulled by a proper change of the reference frame. Such a change does not violate the requirement that the reference frame be Newtonian (49). A mechanical system will be said to be in equilibrium if the conditions are such that it can remain at rest in a Newtonian reference frame.

For example, consider a brick that rests on a table. If the brick rises or tips slightly, gravity performs negative work. If it slides slightly, friction performs negative work. If it executes a small hop, the forces perform no work. Therefore, $W' \leq 0$ for every small virtual displacement. Accordingly, by Fourier's inequality, the brick is in equilibrium when it rests on the table. The virtual displacements are required to be small, since otherwise we could admit a virtual displacement for which the brick slides to the edge of the table and falls off. Then the virtual work of all the forces (friction and gravity) might be positive.

The work W' that all the forces perform when a mechanical system experiences a virtual displacement from configuration X_0 to configuration X generally depends on the path in configuration space that connects points X_0 and X. Unless the system is provided with an unlimited source of energy, the set of all values of W' corresponding to all paths from X_0 to X is bounded above; hence it possesses a least upper bound W (see footnote, p. 4). In other words, $W' \leq W$ for any admissible path from X_0 to X, and there is no number less than W with this property. Since W depends only on the terminal configurations X_0 and X, and not on an intermediate path, it is designated as $W(X_0, X)$. This means that W is a function of X_0 and X. The function W does not necessarily exist if the system is infinite. Although all our experience is limited to finite systems, there are important problems of elasticity, plasticity, and fluid mechanics concerning media that are conceived to extend to infinity. To analyze an infinite system, we may consider finite free bodies cut from it.

If $W' \leq 0$ for every admissible path that leads from X_0 to X, $W(X_0, X) \leq 0$, and vice versa. Consequently, Fourier's inequality may be expressed as follows:

The point X_0 in configuration space represents an equilibrium state if $W(X_0, X) \leq 0$ for all points X in a neighborhood of X_0.

We may specify that virtual displacements are performed adiabatically. This means, for example, that if a brick on a table receives a horizontal virtual displacement the heat generated by friction does not pass into the brick. For an adiabatic virtual displacement, the heat flux Q is zero. Consequently, by Eq. (1-4), $W_i'' = -\Delta U$, where ΔU is the increment of internal energy. Therefore, since $W' = W_i' + W_e'$, Fourier's inequality may be expressed by the relation, $W_e' \leq \Delta U$. Accordingly, *equilibrium exists if there is no small adiabatic virtual displacement for which the virtual work of the external forces exceeds the increment of internal energy.*

1-6. THE PRINCIPLE OF VIRTUAL WORK. Fourier's inequality expresses a condition that is sufficient but not necessary for equilibrium. For example, a ball may be balanced on a dome, but gravity performs positive work if the ball rolls slightly. However, the virtual work of gravity is an infinitesimal of higher order than the displacement of the ball. Numerous examples of this type suggest that Fourier's inequality may be broadened. Seemingly, equilibrium prevails if W' is stationary, in the sense that it is an infinitesimal of higher order than the displacement in configuration space. This equilibrium criterion includes some cases that are not covered by Fourier's condition $W' \leq 0$.

To phrase the preceding statement in mathematical terms, let $\mathbf{X}$ be any point in a neighborhood of point $\mathbf{X}_0$ in configuration space. Let s be the distance between these points (Sec. 1-1). Introducing the function $W(\mathbf{X}_0, \mathbf{X})$ (see Sec. 1-5), we may express the equilibrium criterion as follows:

The point $\mathbf{X}_0$ in configuration space is an equilibrium point if $\lim W/s \leq 0$ *for every admissible path of approach of $\mathbf{X}$ to $\mathbf{X}_0$.*

This statement expresses a law that is known as the principle of virtual work or the principle of virtual displacements. We adopt it as a hypothesis without proof.

The condition $\lim W/s \leq 0$ is merely sufficient for equilibrium; one may see easily that it is not generally necessary. For example, suppose that a particle that moves on the x-axis is subjected to the force $F = F_0 \operatorname{sgn} x$, where F_0 is a positive constant and sgn x (read "signum x") is the function defined by sgn $x = 1$ if $x > 0$, sgn $0 = 0$, and sgn $x = -1$ if $x < 0$. Then, for a virtual displacement from the origin, $W = F_0 |x| = F_0 s$, and $\lim W/s = F_0 > 0$. Nevertheless, by Newton's law, the point $x = 0$ is an equilibrium configuration for the particle.

Variational Form of the Principle of Virtual Work. It is often possible to express the work function $W(\mathbf{X}_0, \mathbf{X})$ in the form

$$W = \delta W + \frac{1}{2!} \delta^2 W + O(s^3) \tag{1-5}$$

where δW and $\delta^2 W$ are, respectively, linear and quadratic homogeneous functionals in the geometric variables or functions that define the virtual displacement. The remainder $O(s^3)$ (read "order of s cubed") denotes a quantity less in absolute value than Ks^3, where K is a positive constant and s is the magnitude of the virtual displacement in configuration space (17). If the system has only a finite number of degrees of freedom, the resolution of W into the form of Eq. (1-5) is accomplished by Taylor's theorem. If the system has an infinite number degrees of freedom, δW and $\delta^2 W$ are to be interpreted as in the calculus of variations (Chap. 3). The terms δW and $\delta^2 W$ are called the first and second variations of W.

According to Eq. (1-5), the equilibrium criterion $\lim W/s \leq 0$ is represented by the inequality $\delta W \leq 0$. The relation $\delta W \leq 0$ is the usual form of expression of the principle of virtual work, but the relation $\lim W/s \leq 0$ is more general, since W cannot always be resolved into the form of Eq. (1-5). For example, if an object of weight P is displaced horizontally on a table, the virtual work is $W = -\mu P \sqrt{(\delta x)^2 + (\delta y)^2}$, where μ is the coefficient of friction, and $(\delta x, \delta y)$ are the virtual increments of rectangular coordinates that locate the object. This function cannot be expressed in the form of Eq. (1-5), such that δW is a linear expression in δx and δy.

Unchecked Systems. The peculiarity in the preceding example of a particle that is subjected to the force $F = F_0 \operatorname{sgn} x$ is that the force changes discontinuously when the virtual displacement is performed. If the forces do not change discontinuously when a virtual displacement is executed, the system is said to be *unchecked*.* For an unchecked system, the condition $\lim W/s \equiv 0$ is necessary for equilibrium. The sign $\equiv$ signifies that the equality holds for all admissible virtual displacements. The derivation of the equation $\lim W/s \equiv 0$ is immediate, for, if a system is in equilibrium, the resultant force on each particle is zero. Consequently, if the force that acts on a particle varies continuously with the displacement of the particle (i.e. if the system is unchecked), the work that it performs is an infinitesimal of higher order than the displacement of the particle, since the force acquires only infinitesimal magnitude during an infinitesimal virtual displacement. Accordingly, if the system consists of a finite number of particles, the condition $\lim W/s \equiv 0$ is necessary for equilibrium. If the system contains an infinite number of particles, this argument logically requires the hypothesis that the order of limits be interchangeable; that is, the limit of the sum of the amounts of work performed on the individual particles equals the sum of the limits.

The condition $\lim W/s \equiv 0$ is necessary and sufficient for equilibrium of an unchecked system. This relation may be expressed in the more conventional form $\delta W \equiv 0$, provided that δW exists.

1-7. GENERALIZED FORCE. For a system with a finite number of generalized coordinates $(x_1, x_2, \cdots, x_n)$, the virtual work δW corresponding to a virtual displacement from point $(x_1, x_2, \cdots, x_n)$ to a neighboring point $(x_1 + \delta x_1, x_2 + \delta x_2, \cdots, x_n + \delta x_n)$ is represented by a linear form in the increments of the coordinates; that is,

$$\delta W = Q_1 \, \delta x_1 + Q_2 \, \delta x_2 + \cdots + Q_n \, \delta x_n \tag{1-6}$$

where $Q_1, Q_2, \cdots, Q_n$ are certain functions of $x_1, x_2, \cdots, x_n$. The customary notation δx_i is used instead of dx_i to denote that the displacement is virtual. In general, the expression $Q_1 \, \delta x_1 + Q_2 \, \delta x_2 + \cdots + Q_n \, \delta x_n$ is not an exact differential. By analogy to the expression $F_x \, dx + F_y \, dy + F_z \, dz$ that represents the work performed by a force (F_x, F_y, F_z) when the particle on which it acts undergoes the displacement (dx, dy, dz),

---

* The phrase "reversible system" is used ordinarily instead of "unchecked system," but this terminology may cause confusion with the thermodynamic meaning of reversibility.

the functions $Q_1, Q_2, \cdots, Q_n$ are called "components of generalized force."* For a nonholonomic system, the components of generalized force are not determined uniquely, since there are constraints represented by nonintegrable differential relations of the type

$$\phi_1 \, \delta x_1 + \phi_2 \, \delta x_2 + \cdots + \phi_n \, \delta x_n = 0$$

These relations, or linear combinations of them, may be added to Eq. (1-6) without violation of the condition that the left side of Eq. (1-6) is δW.

The terms Q_i do not necessarily have the dimension of force; in fact, they frequently do not all have the same dimension. Their dimensions are determined by the fact that $Q_i \, \delta x_i$ has the dimension of work. Consequently, if x_i denotes a length, Q_i has the dimension of force, and if x_i denotes an angle Q_i has the dimension of the moment of a force.

The principle of virtual work ($\delta W = 0$) yields the following conclusion:

Any configuration of a mechanical system for which the components of generalized force all vanish is an equilibrium configuration.

This principle is a formal generalization of the Newtonian law that a particle is in equilibrium if the force that acts on the particle is zero. Since it merely expresses a sufficient condition for equilibrium, it is equally valid for holonomic and nonholonomic systems. The principle of virtual work also yields the conclusion that vanishing of all components of generalized force is a necessary condition for equilibrium of an unchecked holonomic system.

Generalized External Force. The virtual work corresponding to an arbitrary displacement of a mechanical system may be separated into a sum, $W = W_e + W_i$, where W_e and W_i are the least upper bounds of the virtual work of the external and internal forces, respectively (see Sec. 1-3). It may be possible to represent W_e as a sum

$$W_e = \delta W_e + \frac{1}{2!} \, \delta^2 W_e + O(s^3)$$

analogous to the decomposition of W [Eq. (1-5)]. If the system has a finite number of degrees of freedom, the first variation δW_e is a linear form in $\delta x_1, \delta x_2, \cdots, \delta x_n$; that is,

$$\delta W_e = P_1 \, \delta x_1 + P_2 \, \delta x_2 + \cdots + P_n \, \delta x_n \qquad (1\text{-}7)$$

where $P_1, P_2, \cdots, P_n$ are certain functions of $x_1, x_2, \cdots, x_n$. In general, the expression $P_1 \, \delta x_1 + P_2 \, \delta x_2 + \cdots + P_n \, \delta x_n$ is not an exact differential. By an obvious extension of the terminology introduced previously, the

* A different definition of generalized force is sometimes used in dynamics (Sec. 7-5).

functions $P_1, P_2, \cdots, P_n$ are called "components of generalized external force." If the variables x_i denote lengths and angles, the variables P_i may usually be identified as the components of prescribed external forces and couples that act on the system.

Even though the loads on a structure or mechanism are constants, the generalized external force components P_i need not be constants. For example, consider the system shown in Fig. 1-3. Let θ be the generalized coordinate, and let the force F be constant. If θ receives an increment $\delta\theta$,

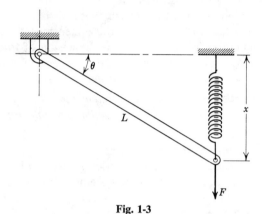

Fig. 1-3

the first variation of the work of the force F is $\delta W_e = FL \cos \theta \, \delta\theta = P \, \delta\theta$, where P is the generalized external force. Hence $P = FL \cos \theta$. Accordlingy, P depends on θ, even though F is constant. On the other hand, if the vertical displacement x is taken as the generalized coordinate, the virtual work of the load F is $F \, \delta x = P \, \delta x$, whence, $F = P$. Thus it is seen that the question of constancy of the generalized force depends not only on the nature of the loads but also on the choice of the coordinates.

1-8. POTENTIAL ENERGY. A mechanical system is said to be "conservative" if the virtual work W' vanishes for a virtual displacement that carries the system completely around any closed path. (W' is defined in Sec. 1-5). If this condition is satisfied only for virtual displacements that are executed with infinitesimal speed, the system is said to be "conservative in the statical sense;" if it is satisfied regardless of the speed, the system is said to be "conservative in the kinetic sense." A system that is conservative in the kinetic sense is necessarily conservative in the statical sense.

For a conservative system, the virtual work W' corresponding to a virtual displacement from point $\mathbf{X_0}$ to point $\mathbf{X_1}$ is annulled by the virtual work corresponding to the displacement from $\mathbf{X}_1$ to $\mathbf{X}_0$, since the two displacements are effected by a circuit about a closed path. Since this is true irrespective of the path that leads from $\mathbf{X}_0$ to $\mathbf{X}_1$, even though the return path is the same in all cases, the virtual work W' for the transition from $\mathbf{X}_0$ to $\mathbf{X}_1$ is independent of the path. In other words, the virtual work W' corresponding to any displacement of a conservative system from point $\mathbf{X}_0$ to point $\mathbf{X}_1$ depends only on the terminal configurations $\mathbf{X}_0$ and $\mathbf{X}_1$. There is no sliding friction in a conservative system, since the work of sliding friction depends on the path; for example, the work that friction performs on an object that is dragged depends on the path as well as on the end points. Viscous friction (frictional forces proportional to the speed) may exist in a system that is conservative in the statical sense but not in a system that is conservative in the kinetic sense. For a conservative system, $W' = W$, where W' and W are defined as in Sec. 1-5.

Since the virtual work W' that corresponds to a virtual displacement of a conservative system from a given configuration $\mathbf{X}_0$ to a variable configuration $\mathbf{X}$ is independent of the path, it may be denoted by $W' = -V(\mathbf{X}_0, \mathbf{X})$. If $\mathbf{X}_0$ is a fixed prescribed configuration, the function $V(\mathbf{X}_0, \mathbf{X})$ is called the "potential energy" of the system in configuration $\mathbf{X}$. This name was introduced by the Scottish scientist W. J. M. Rankine (134). The point $\mathbf{X}_0$ is called the "zero configuration" for potential energy.

The zero configuration $\mathbf{X}_0$ merely affects an additive constant in the potential energy. For proof, suppose that the zero configuration is $\mathbf{X}_1$ instead of $\mathbf{X}_0$. Since the work $-V(\mathbf{X}_1, \mathbf{X})$ that the forces perform when the system travels from $\mathbf{X}_1$ to $\mathbf{X}$ is independent of the path, we may suppose that the path passes through $\mathbf{X}_0$. Consequently, $V(\mathbf{X}_1, \mathbf{X}) = V(\mathbf{X}_1, \mathbf{X}_0) + V(\mathbf{X}_0, \mathbf{X})$. Since $\mathbf{X}_0$ and $\mathbf{X}_1$ are fixed points, this yields $V(\mathbf{X}_1, \mathbf{X}) = V(\mathbf{X}_0, \mathbf{X}) +$ constant. It will be seen later that an additive constant in the potential energy is irrelevant, since only changes of potential energy are significant. Consequently, the zero configuration $\mathbf{X}_0$ need not be designated explicitly. The potential energy of a conservative system is accordingly denoted by $V(\mathbf{X})$. This notation indicates that V is a scalar point function in configuration space.

It is essential to observe that the function V does not exist for a non-conservative system. Although the function W, defined in Sec. 1-5, exists for any mechanical system with bounded energy, it does not necessarily satisfy the condition which characterizes potential energy; namely, $V(\mathbf{X}_1, \mathbf{X}) = V(\mathbf{X}_1, \mathbf{X}_0) + V(\mathbf{X}_0, \mathbf{X})$. This fact is exemplified by an object that slides on the floor. If the object experiences a frictional force of constant magnitude F, the virtual work W' corresponding to any path is $W' = -FL$, where L is the length of the path. Consequently, for given terminal points, W' attains a maximum when the object describes a straight path; in this case, $W' = W$. Accordingly, $W(\mathbf{X}_1, \mathbf{X}) = -Fs$, where s is the distance between points $\mathbf{X}_1$ and $\mathbf{X}$. Consequently, if points $\mathbf{X}_0$, $\mathbf{X}_1$, and $\mathbf{X}$ are not collinear, $W(\mathbf{X}_1, \mathbf{X}) > W(\mathbf{X}_1, \mathbf{X}_0) + W(\mathbf{X}_0, \mathbf{X})$. This inequality conflicts with the preceding relation that was derived for potential energy; hence in the present example W exists but there is no potential energy function.

Since an additive constant is irrelevant, the potential energy V of a conservative system may often be calculated conveniently by integration of the expression for the total differential dV. For example, if the distance r between the earth and the sun increases the infinitesimal amount dr, the virtual work performed by the force of attraction between the bodies is $dW' = -dV = -(kmm'/r^2)\,dr$, where m and m' are the masses of the earth and the sun, and k is the gravitational constant. This relationship follows from Newton's law of gravitation. The natural motions of the earth and the sun may be considered to be stopped while the virtual displacement dr is performed. Integration of the preceding differential equation yields $V = -kmm'/r$. An inconsequential additive constant of integration is omitted.

If a mechanical system lies in the gravitational field of the earth, a part of its potential energy results from the weights of the particles. Suppose that the particles are enumerated and that their weights and elevations are, respectively, $w_1, w_2, \cdots$, and $z_1, z_2, \cdots$. If the elevations receive increments $\Delta z_1, \Delta z_2, \cdots$, the negative work that gravity performs is

$$\Delta V = w_1\,\Delta z_1 + w_2\,\Delta z_2 + \cdots$$

provided that the weights w_i are constants. Hence

$$\Delta V = \Delta(w_1 z_1 + w_2 z_2 + \cdots) = \Delta(w\bar{z})$$

where w is the weight of the entire system and $\bar{z}$ is the elevation of the center of gravity of the system. The last relation follows from the definition of the center of gravity. Consequently, $V = w\bar{z}$. This equation means that *if the acceleration of gravity is constant the part of the potential energy of a system that results from the gravitational field of the earth is equal to the weight of the system, multiplied by the elevation of its center of gravity.* Since the zero configuration of potential energy is irrelevant, the datum plane from which elevations are measured need not be specified.

Conservative External and Internal Forces. The work W' that all the forces perform when a system experiences a virtual displacement from configuration $\mathbf{X}_0$ to configuration $\mathbf{X}$ may be separated into a sum, $W' = W_e' + W_i'$, where W_e' is the work performed by the external forces and W_i' is the work performed by the internal forces. It may happen that W_e'

is independent of the path from X_0 to X; then the external forces are said to be conservative. In this case W_e' depends only on the terminal configurations X_0 and X; hence it is denoted by $-\Omega(X)$. The initial configuration X_0 is not designated, since it merely affects an additive constant in Ω. The point function Ω is called the "potential energy of the external forces."

Likewise, it may happen that W_i' is independent of the path; then the internal forces are said to be conservative, and W_i' is denoted by $-U(X)$. The point function U is called "potential energy of the internal forces." If the internal and external forces are both conservative, $V = U + \Omega$.

As in the general definition of potential energy, we may define the internal and external forces as conservative in the statical sense or the kinetic sense. A mechanical system may be said to be "elastic" if the internal forces are conservative in the kinetic sense. Then the function $U(X)$ is commonly called "strain energy."

Aside from an irrelevant additive constant, the strain energy of a system is equal to the internal energy. Consequently, if the internal forces are conservative, the internal energy is a point function in configuration space. However, the internal energy may depend on a parameter that does not change appreciably during the mechanical process under consideration. For example, in problems of thermoelasticity temperature occurs as a parameter in the strain-energy expression.

By the first law of thermodynamics, the strain energy of a system in a given configuration is equal to the work that we must perform in the absence of external forces to lead the system slowly and adiabatically from the zero configuration to the given configuration.

In problems of elastic structures the potential energy of the external forces is frequently confused with the strain energy. For example, if a linearly elastic beam carries a constant concentrated weight, P, the potential energy of the external force is $-Pu$, and not $-\frac{1}{2}Pu$ (where u is the deflection at the point of application of the load). This fact is apparent if we recognize that the potential energy of the external load results from gravity; it has nothing to do with properties of the beam.

Systems with Finite Degrees of Freedom. For a conservative system with finite degrees of freedom, the potential energy V is a function of the generalized coordinates $(x_1, x_2, \cdots, x_n)$. The total differential of V is

$$dV = \frac{\partial V}{\partial x_1}\delta x_1 + \frac{\partial V}{\partial x_2}\delta x_2 + \cdots + \frac{\partial V}{\partial x_n}\delta x_n$$

Also, $-dV$ is the work performed by all the forces when the system receives an infinitesimal displacement $(\delta x_1, \delta x_2, \cdots, \delta x_n)$. Therefore, if the system is holonomic, so that the virtual displacement is arbitrary, Eq. (1-6) yields

$$Q_i = -\frac{\partial V}{\partial x_i} \qquad (1\text{-}8)$$

Equation (1-8) determines the components of generalized force if the potential energy function V is known.

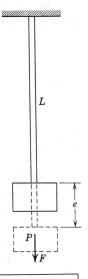

More generally, if only the internal forces are conservative, the increment of potential energy of internal forces corresponding to increments $(\delta x_1, \delta x_2, \cdots, \delta x_n)$ of the coordinates is

$$dU = \frac{\partial U}{\partial x_1}\,\delta x_1 + \frac{\partial U}{\partial x_2}\,\delta x_2 + \cdots + \frac{\partial U}{\partial x_n}\,\delta x_n$$

The total work of all the forces is

$$Q_1\,\delta x_1 + Q_2\,\delta x_2 + \cdots + Q_n\,\delta x_n$$
$$= -dU + P_1\,\delta x_1 + P_2\,\delta x_2 + \cdots + P_n\,\delta x_n$$

where $P_1, P_2, \cdots, P_n$ are the components of generalized external force. Therefore, if the system is holonomic,

$$Q_i = P_i - \frac{\partial U}{\partial x_i} \qquad (1\text{-}9)$$

If the external forces are also conservative, $V = U + \Omega$, and Eqs. (1-8) and (1-9) yield $P_i = -\partial\Omega/\partial x_i$.

Example. Suppose that a uniform vertical elastic rod with cross-sectional area A and modulus of elasticity E is fastened at its upper end. The rod carries a weight P at its lower end, and a magnet pulls on the weight with a force F given by the formula $F = C(a - e)^{-2}$, where C and a are constants and e is the extension of the rod (Fig. 1-4).

Fig. 1-4

Since the virtual work corresponding to any cycle of displacement is zero, the system is conservative. If we stretch the rod slowly, the tension N that we apply is given by the formula of elasticity, $N = EAe/L$, where L is the free length of the rod. Therefore, the work that we perform against internal forces is

$$U = \int N\,de = \int \frac{EAe}{L}\,de$$

Accordingly,

$$U = \frac{EAe^2}{2L} \qquad (1\text{-}10)$$

This equation determines the strain energy of the rod.

The potential energy of the external forces is the negative work of these forces. Hence

$$d\Omega = -\left[P + \frac{C}{(a-e)^2}\right] de$$

Integration yields

$$\Omega(e) = -Pe - \frac{C}{a-e}$$

The total potential energy of the system is $V = U + \Omega$.

1-9. PROPERTIES OF CONSERVATIVE SYSTEMS.

Consider a real or natural motion of a mechanical system rather than the virtual displacements that have been discussed. If the system travels from configuration X_1 to configuration X, the work W' of all the forces that act on the system during the transition equals the increment ΔT of kinetic energy that the system acquires between X_1 and X; this is the law of kinetic energy (Sec. 1-3). If the system is conservative in the kinetic sense, W' is also equal to the loss of potential energy $-\Delta V$, where ΔV denotes $V(X) - V(X_1)$; this follows from the definition of potential energy. Consequently,

$$\Delta T + \Delta V = 0 \quad \text{or} \quad T + V = \text{constant} \qquad (1\text{-}11)$$

Equation (1-11) expresses the law of kinetic energy for a conservative system. The sum $T + V$ is called the "total mechanical energy" of the system. Equation (1-11) is accordingly known as the "law of conservation of mechanical energy." The phrase "conservative system" is derived from the fact that mechanical energy is conserved in such a system. The differential equation of motion of any conservative system with one degree of freedom is determined by the equation $T + V = $ constant. Many illustrative applications of this principle may be found in elementary books on mechanics.

Principle of Stationary Potential Energy. For a conservative system, the principle of virtual work (Sec. 1-6) is expressed by the equation $\lim \Delta V/s = 0$, since $W = -\Delta V$. It is frequently possible to express ΔV in the form

$$\Delta V = \delta V + \frac{1}{2!}\delta^2 V + O(s^3) \qquad (1\text{-}12)$$

in which δV and $\delta^2 V$ are the first and second variations discussed in Art. 1-6.

If δV exists, the principle of virtual work $(\lim \Delta V/s = 0)$ is equivalent to the condition $\delta V = 0$; this is tantamount to the equation $\delta W = 0$ discussed in Sec. 1-6. If the system has n generalized coordinates x_i, $\delta V = \sum (\partial V/\partial x_i)\, \delta x_i$. If the system is holonomic, arbitrary virtual displacements δx_i are permitted. Then the condition that δV vanishes for all admissible virtual displacements is equivalent to the equations $\partial V/\partial x_i = 0$; $i = 1, 2, \cdots, n$. If the system has an enumerably infinite set of coordinates x_i, this equilibrium criterion may still be employed, unless questions of convergence preclude it. Of course, the equations $\partial V/\partial x_i = 0$ are necessary equilibrium conditions only for unchecked systems (Sec. 1-6). For example, if a block on a frictionless inclined plane is butted against a step or ledge on the plane, the system is conservative and it is in equilibrium, but δV does not vanish identically.

For any stationary system with finite degrees of freedom, the vanishing of the components Q_i of generalized force is sufficient for equilibrium (Sec. 1-7). Therefore, by Eq. (1-9), an elastic system with finite degrees of freedom is in equilibrium if

$$P_i = \frac{\partial U}{\partial x_i} \tag{1-13}$$

This condition is also necessary for equilibrium if the system is holonomic and unchecked.

In the theory of elastic structures U is frequently a quadratic form in the variables x_i; that is,

$$U = \tfrac{1}{2} \sum_{i=1}^{n} \sum_{j=1}^{n} a_{ij} x_i x_j \tag{1-14}$$

In this case Eq. (1-13) yields

$$\sum_{j=1}^{n} a_{ij} x_j = P_i \tag{1-15}$$

If the components P_i of generalized external force are constants, as usually happens, Eq. (1-15) is linear; its solution represents the equilibrium configuration. It is to be noted that the coefficients on the left side of Eq. (1-15) are the rows in the matrix (a_{ij}).

1-10. POTENTIAL ENERGY OF A SYSTEM OF PARTICLES. The theory of particles that attract or repel each other finds applications in several fields of physics and engineering and even in pure mathematics.

If two particles attract each other with a force F that depends only on the distance r between the particles, the work that is performed on the system when r increases an infinitesimal amount dr is $F\, dr$. Consequently, the potential energy of the two particles due to their mutual attraction is

$$V = \int F \, dr \tag{1-16}$$

This indefinite integral may be augmented by an arbitrary additive constant. Equation (1-16) applies for particles that repel each other if F is then considered to be negative.

An attraction between two particles is called a "bond." The number of bonds in a system of attracting particles equals the number of distinct pairs of particles, although the bonds between particles that are comparatively far apart are frequently negligible. A system of n particles contains $n(n - 1)/2$ bonds.

Let the bonds in a system of particles be numbered $1, 2, \cdots, m$. If we give the system an infinitesimal displacement, the increment of potential energy (i.e. the work that we perform) is $dV = \sum F_i \, dr_i$, where the sum extends from 1 to m. Here, F_i is the tension in the ith bond and r_i is the distance between the corresponding particles. Consequently, the total potential energy of a system of attracting particles is the sum of the potential energies of all the bonds. This conclusion signifies that a system of attracting particles is conservative, irrespective of the law of attraction.

Force Fields. Suppose that several fixed particles attract or repel a movable particle. The force **F** that acts on the movable particle depends only on the location of that particle. In other words, **F** is a single-valued vector function of the coordinates (x, y, z) of the movable particle. Accordingly, **F** is called a "vector point function" or a "vector field;" more specifically, a "force field." Electrical fields and gravitational fields are examples of force fields.

If (x, y, z) are rectangular coordinates and if F_x, F_y, F_z are the x, y, z components of the vector **F**, an infinitesimal vector (dx, dy, dz) in the direction of vector **F** conforms to the differential relations

$$\frac{dx}{F_x} = \frac{dy}{F_y} = \frac{dz}{F_z} \tag{1-17}$$

The general integral of these differential equations represents a system of curves, called "lines of force." The lines of force are everywhere tangent to the vectors of the force field. The theory of integration of equations of the form of Eq. (1-17) is discussed in books on differential equations (35).

The potential energy V of a movable particle in a force field that is generated by fixed particles is the sum of the potential energies of the bonds between the movable particle and the fixed particles. An additive constant that represents the potential energy of the bonds among the fixed particles is irrelevant. Evidently, V is a function of x, y, z. If we give the movable particle an infinitesimal displacement (dx, dy, dz), the work that we perform is the total differential of V; that is,

$$dV = \frac{\partial V}{\partial x} dx + \frac{\partial V}{\partial y} dy + \frac{\partial V}{\partial z} dz \tag{a}$$

Also, since the force that we exert is equal to $-\mathbf{F}$,

$$dV = -F_x \, dx - F_y \, dy - F_z \, dz \tag{b}$$

Since Eqs. (a) and (b) are valid for any infinitesimal displacement,

$$F_x = -\frac{\partial V}{\partial x}, \qquad F_y = -\frac{\partial V}{\partial y}, \qquad F_z = -\frac{\partial V}{\partial z} \tag{1-18}$$

or

$$\mathbf{F} = -\text{grad } V \tag{1-18a}$$

A force field that possesses a potential energy function is said to be "conservative." Equation (1-18) shows that a conservative force field is determined completely by its potential energy function. The lines of force are evidently normal to the equipotential surfaces, $V = \text{const.}$ Equation (1-18) is a special case of Eq. (1-8).

Newtonian Potential. According to Newton's law of gravitation, the force of attraction between any two point masses, m and m', is

$$F = k \frac{mm'}{r^2} \tag{1-19}$$

in which k is a constant that depends on the units of force, length, and mass. The same relationship applies for the force between electric charges (Coulomb's law).

Let the units be such that $k = 1$. Then, by Eq. (1-16), the potential energy of two particles that attract each other in conformity with the Newtonian law is

$$V = -\frac{mm'}{r} \tag{1-20}$$

Accordingly, the potential energy of a unit mass in a force field that is generated by fixed point masses $m_1, m_2, \cdots, m_n$ is

$$V = -\sum_{i=1}^{n} \frac{m_i}{r_i} \tag{1-21}$$

where r_i is the distance between the ith fixed mass and the movable mass. Instead of considering isolated fixed masses, we may consider a continuous fixed distribution of mass with variable density ρ. Then Eq. (1-21) is replaced by

$$V = -\int \frac{\rho \, dS}{r} \tag{1-22}$$

in which r is the distance from a particle of the distributed mass to the movable particle. The distributed mass may be extended along a line or curve; it may be spread over a surface or it may be distributed throughout a region of space. Accordingly, the integral in Eq. (1-22) may be a line integral, a surface integral, or a volume integral, and, correspondingly, dS denotes an element of length, area, or volume. Surface integrals have particular significance in electrostatics, since electric charges accumulate on the boundaries between contiguous materials.

The functions defined by Eqs. (1-21) and (1-22) are called "Newtonian potential functions." Since they are scalars, they are more convenient for analytical purposes than the corresponding forces. For example, to compute the gravitational attraction of a three-dimensional body on an isolated particle, we compute the Newtonian potential of the body by means of Eq. (1-22). Then the force on the particle is determined by Eq. (1-18).

Aside from their physical significance, Newtonian potential functions are important in mathematics because they are solutions of Laplace's differential equation

$$\frac{\partial^2 V}{\partial x^2} + \frac{\partial^2 V}{\partial y^2} + \frac{\partial^2 V}{\partial z^2} = 0 \tag{1-23}$$

This statement is verified by the observation that if (a, b, c) is any fixed point, the following function satisfies Eq. (1-23):

$$\frac{1}{r} = [(x - a)^2 + (y - b)^2 + (z - c)^2]^{-\frac{1}{2}} \tag{1-24}$$

Any function that satisfies Eq. (1-23) is said to be "harmonic." Any linear combination of terms of the type of Eq. (1-24) is harmonic; therefore, the Newtonian potential function generated by fixed point masses [Eq. (1-21)] is harmonic. Likewise, since ρ is a function of the parameters (a, b, c), Newtonian potential functions generated by continuous distributions of mass are harmonic outside of the masses. Innumerable solutions of the three-dimensional Laplace equation* are obtained by Eq. (1-22), since the density function ρ and the regions occupied by the

---

* For Laplace's equation in two variables x and y, the function $\log r$ replaces the function $1/r$.

attracting masses are arbitrary. In fact, a complete theory of the important partial differential equations of Laplace and Poisson may be based on the Newtonian potential function. This surprising ramification of Newton's gravitational theory constitutes a branch of mathematics known as "potential theory" (41, 78).

Example. Let a spherical surface of radius a carry a constant positive density ρ of electric charge. The Newtonian potential at a distance x from the center of the sphere is to be determined $(x > a)$ (Fig. 1-5). By Eq. (1-22),

$$dV = -\frac{\rho \, dS}{r} = -\frac{2\pi\rho a^2 \sin\theta \, d\theta}{r} \tag{c}$$

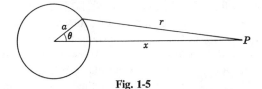

Fig. 1-5

Also, by the law of cosines,

$$r^2 = a^2 + x^2 - 2ax \cos\theta$$

Differentiation of this relation yields

$$2r \, dr = 2ax \sin\theta \, d\theta$$

Hence, by Eq. (c),

$$dV = -\frac{2\pi\rho a \, dr}{x}$$

Therefore,

$$V = -\frac{2\pi\rho a}{x} \int_{x-a}^{x+a} dr = -\frac{4\pi\rho a^2}{x}$$

The total charge on the sphere is $e = 4\pi\rho a^2$. Hence $V = -e/x$. The force acting on a negative unit charge at point P is $F = -dV/dx = -e/x^2$. The negative sign indicates that the force is opposite in sense to the x-axis.

This result shows that the force exerted by the charged sphere on an external point charge is the same as though all the charge on the sphere were concentrated at the center of the sphere. Likewise, the gravitational attraction of a homogeneous or stratified spherical body such as the earth is the same as though all the mass of the body were concentrated at the center of the body, since the solid sphere may be decomposed into homogeneous spherical shells of infinitesimal thickness, and the preceding rule applies for each of these shells.

1-11. STABILITY. In a practical sense, an equilibrium configuration of a mechanical system is said to be stable if accidental forces, shocks, vibrations, eccentricities, imperfections, inhomogeneities, residual stresses, or other probable irregularities do not cause the system to depart excessively or disastrously from that configuration. In a mathematical sense, stability is usually interpreted to mean that infinitesimal disturbances will cause only infinitesimal departures from the given equilibrium configuration. The mere fact that an assemblage is stable in this refined sense does not necessarily signify that it is safe from an engineering viewpoint. For example, a ship may be stable in the sense that small waves will not capsize it, yet it may not be stable enough for an ocean voyage. Similarly, a convex shell-like structure, such as a shallow-domed roof, can be in a state of stable equilibrium, yet a jolt may cause it to "snap through" into a badly deformed shape. Such illustrations indicate that the infinitesimal theory of stability must be used with discretion. Nevertheless, the infinitesimal theory has had innumerable practical applications.

In the following the word "stability" is used with the classical meaning; that is, infinitesimal disturbances of a stable system cause only infinitesimal displacements in configuration space. We might suppose that a system is in a state of unstable equilibrium if an arbitrarily small amount of energy, supplied from external sources, can cause the system to experience a large displacement. However, this is not true. For example, a table that rests on the floor is certainly in a state of stable equilibrium. Nevertheless, theoretically we can move the table as far as we please without any expenditure of work. To do this, we lift the table slightly so that the reaction of the floor ceases to act, and we then displace it horizontally. Certainly, if external forces must perform positive work to produce any small displacement of a given system, equilibrium is stable, since infinitesimal disturbances then cannot produce large effects. However, according to the preceding example, this condition is merely sufficient for stability; it is not necessary.

The theory of stability of conservative systems is inherently simpler than the theory for nonconservative systems. In view of the law of conservation of mechanical energy, $T + V = $ constant, an infinitesimal increment of T is accompanied by only an infinitesimal increment of V. Consequently, if a motionless conservative system is in a configuration of minimum potential energy, an infinitesimal initial velocity causes only an infinitesimal displacement in configuration space. Accordingly, stable equilibrium exists. Conversely, if the value of the potential energy is not a minimum and if the system is holonomic, an impulse that directs the system along a path for which V decreases causes the kinetic energy to increase continually (since $T + V$ = constant); therefore, equilibrium is not stable. The principle of stability of conservative holonomic systems consequently may be expressed as follows:

> *A conservative holonomic system is in a configuration of stable equilibrium if, and only if, the value of the potential energy is a relative minimum.*

This conclusion is known as the "law of minimum potential energy." In a qualitative way it indicates the meaning of weak stability. The stability of a

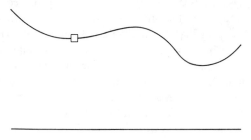

Fig. 1-6

system is said to be "weak" if the system can depart a long distance from the given equilibrium state with only a small increase of potential energy. A bead at a minimum point of a frictionless rigid wire (Fig. 1-6) is in a state of weak stability if only a small hill separates it from a lower minimum.

An illustration of the principle of minimum potential energy is provided by weights that are connected by any frictionless mechanism with rigid members. The potential energy of the system is the weight of the system multiplied by the height of the center of gravity (Sec. 1-8). Consequently, since the weight is constant, a motionless conservative holonomic system with constant internal energy is in a state of stable equilibrium under the action of a constant gravitational field if, and only if, the height of the center of gravity of the system is a relative minimum. A marble in a cup illustrates this principle. Another example is a frictionless chain of rigid bars with hinged ends. The bars assume positions such that the elevation of the center of gravity of the system is a relative minimum. The catenary (curve of a heavy uniform flexible inextensional sagging cord) is a limiting case of this type of linkage. Among all curves of given length and given end points, the catenary has the lowest centroid. With the aid of the calculus of variations, the equation of the catenary may be derived from this principle (Sec. 3-7). A stationary cup of water also illustrates the foregoing principle. The surface of the water is plane and horizontal, since this configuration provides the lowest elevation to the center of gravity of the water. Actually, the surface of the water curves up slightly where it intersects the wall. This phenomenon is explained by the fact that there is a small contribution to the potential energy from a peculiar molecular action at a free surface of a liquid. Surface tension is a manifestation of this action.

For systems with finite degrees of freedom, the mathematical implications of the principle of minimum potential energy are comparatively simple. The potential energy V is a single-valued function of the generalized coordinates $x_1, x_2, \cdots, x_n$. Also, V depends on the external load on the system. When the type of loading is prescribed, the load is determined by a single scalar parameter p. For the present p is regarded as a given constant, and it need not be considered explicitly. The coordinates x_i are considered as regular (Sec 1-2) in a region R of configuration space that includes the configurations of interest. Only holonomic systems are considered. For values of the x's in R, the function V and its partial derivatives to the third order with respect to the x's are postulated to be continuous functions of the x's. Then, by Taylor's theorem, the increment of V corresponding to increments $h_1, h_2, \cdots, h_n$ of the x's is

$$\Delta V = \sum h_i V_i(x) + \frac{1}{2!} \sum \sum h_i h_j V_{ij}(x)$$
$$+ \frac{1}{3!} \sum \sum \sum h_i h_j h_k V_{ijk}(x + \theta h) \tag{1-25}$$

where $0 < \theta < 1$. Subscripts on V denote partial derivatives, and x stands collectively for all the x's; for example, $V_{ij}(x)$ denotes $\partial^2 V / \partial x_i \, \partial x_j$ at the point $(x_1, x_2, \cdots, x_n)$. For brevity, Eq. (1-25) is written

$$\Delta V = \delta V + \tfrac{1}{2} \delta^2 V + R_3 \tag{1-26}$$

Since the derivatives of V with respect to the x's are continuous in R, the principle of virtual work signifies that a necessary and sufficient condition for equilibrium is that δV vanish for all h_i; that is, $\partial V / \partial x_i = 0$, $i = 1, 2, \cdots, n$. We consider a point x_i' that is a solution of these equations.

The second variation of the potential energy may be written

$$\delta^2 V = \sum \sum a_{ij} h_i h_j, \qquad a_{ij} = V_{ij}(x') \tag{1-27}$$

Accordingly, $\delta^2 V$ is a quadratic form in the variables $h_1, h_2, \cdots, h_n$. Although R_3 is not exactly a cubic form in the h's, it differs from a cubic form only by infinitesimals of the fourth order, since the coefficients $V_{ijk}(x + \theta h)$ approach constant values as the h's approach zero. Consequently, if $\delta^2 V$ is positive definite (see Appendix A-1), there exists a deleted neighborhood of point x_i' throughout which $|\delta^2 V| > 2 |R_3|$. Accordingly,

$\dfrac{d^2 V}{\partial x_i^2} > 0$ stability, < 0 instability, $= 0$ poi.

$\dfrac{\partial V}{\partial x_i} = 0$ For equalibrium.

in this neighborhood, ΔV is positive; hence the value of V at point x_i' is a relative minimum. Conversely, if $\delta^2 V$ is negative definite, negative semi-definite, or indefinite, ΔV takes negative values in any neighborhood of x_i', and therefore the value of V is not a minimum; hence equilibrium is not stable. If $\delta^2 V$ is positive semidefinite, or if $\delta^2 V$ vanishes identically, the value of V may or may not be a relative minimum; the question cannot be decided without further investigations of the function V.

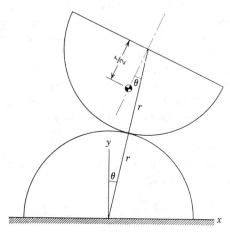

Fig. 1-7

Example. A thin hemispherical shell of constant thickness is balanced on a hemispherical dome of the same radius as the shell. The stability of the configuration is to be investigated. If the shell rolls slightly, the configuration is as illustrated in Fig. 1-7. The center of gravity of the shell lies at the midpoint of the radius. Consequently, by Fig. 1-7, the elevation of the center of gravity is

$$y = 2r \cos \theta - \tfrac{1}{2}r \cos 2\theta$$

Equilibrium is stable if, and only if, the configuration $\theta = 0$ provides a relative minimum to y, since the potential energy is mgy. Expanding the cosines in power series, we obtain

$$y = r\left(\frac{3}{2} - \frac{\theta^4}{4} + \cdots\right)$$

where the dots indicate terms of higher degree in θ. Comparing this equation with the series

$$\Delta y = \delta y + \frac{1}{2!}\delta^2 y + \frac{1}{3!}\delta^3 y + \frac{1}{4!}\delta^4 y + \cdots$$

we obtain

$$\delta y = \delta^2 y = \delta^3 y = 0, \qquad \delta^4 y = -6r\theta^4$$

$$r\frac{\theta^4}{4} = \frac{1}{(4)(3)(2)(1)} \delta^4 y \quad \therefore \quad \delta^4 y = -6r\theta^4$$

For sufficiently small values of θ, the sign of Δy is evidently the same as the sign of $\delta^4 y$; namely, negative. Therefore, for $\theta = 0$, the value of y is not a relative minimum, and accordingly equilibrium is not stable. This example is somewhat unusual, since ordinarily the character of $\delta^2 V$ determines whether or not equilibrium is stable. The peculiarity in the present case is that $\delta^2 V$ vanishes identically.

PROBLEMS

1. Prove the triangle law for displacements in configuration space.

2. A wheel of diameter D rolls on a straight track. Calculate the magnitude of the displacement in configuration space corresponding to an angular displacement θ of the wheel.

3. A gas flows steadily through a tube of circular cross section. The velocity is distributed parabolically on a cross section, and the velocity at the center is three times the velocity at the wall. Express the rate of displacement in configuration space ds/dt in terms of the mean velocity V.

4. A frictionless incompressible fluid flows at constant mass rate through a conduit of variable cross section. Consider the fluid between two cross sections, 1 and 2. By applying the law of kinetic energy to this fluid, derive Bernoulli's equation for liquids.

5. A body that slides on a straight horizontal bar oscillates under the action of a restoring force that is proportional to the displacement. Assuming that the coefficient of friction is a constant, show by the law of kinetic energy that the difference between the amplitudes of successive oscillations is a constant.

6. An artificial satellite travels about the earth in a circular orbit at elevation h above the surface of the earth. Denoting the radius of the earth by r, determine the ratio of the kinetic energy of the satellite to the work required to lift it to altitude h.

7. A particle describes the path $x = t$, $y = t^2$, $z = t^3$, in which t denotes time. The motion is resisted by a force that is opposite to the velocity and that is proportional to the speed. Calculate the work that the resisting force performs during the interval $t = 1$ to $t = 3$.

8. A particle that moves in space is subjected to a force of constant magnitude that is always directed toward the origin. Prove that the work the force performs on the particle during any displacement is the product of the magnitude of the force and the reduction of the distance from the origin.

9. Two mass particles are connected by a rigid massless rod. Prove by direct application of Newton's laws that the tension in the rod performs no net work on the particles when the system moves freely in any way.

10. A particle in the (x, y) plane moves in the force field $F_x = -ky$, $F_y = kx$, (k = constant). Prove that when the particle describes any closed path in

the counterclockwise sense, the work performed on the particle is $2kA$, where A is the area enclosed by the path.

11. Neglecting friction, determine the mechanical advantage of the differential pulley by the principle of virtual work (Fig. P1-11).

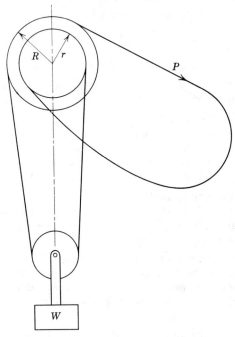

Fig. P1-11

12. The thrust force F has constant magnitude, and it is always directed along the axis of the lower bar. The hinges are frictionless. Prove that the system is nonconservative (Fig. P1-12).

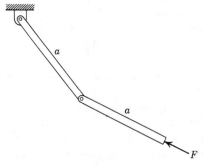

Fig. P1-12

13. The deflection of a simple beam is represented by the sine series, $y = \Sigma a_n \sin n\pi x/L$, where L is the length of the beam. The coefficients a_n are regarded as generalized coordinates. The beam carries a nonuniform distributed load $p(x)$. Show that the components P_i of generalized external force are proportional to the coefficients in the sine series for $p(x)$. Determine the proportionality constant.

14. A flexible string has its ends fixed at the points $O:(0, 0)$ and $Q:(L, 0)$ in the (x, y) plane. The string is deflected so that it passes through the point $P:(x, y)$. Segments OP and PQ are straight. The tension F in the string is independent of the location of point P. Is there a strain-energy function $U(x, y)$? If so, derive it.

15. The force of attraction between two identical atoms is $F = ar^{-2} - br^{-10}$, in which a and b are positive constants and r is the distance between the atoms. (a) Supposing that the two atoms oscillate on a fixed straight line, express dr/dt as a function of r by means of the principle of kinetic energy. (b) Determine the equilibrium configuration by the principle of stationary potential energy.

16. Let (x, y) be fixed rectangular coordinates with the y-axis directed downward. A body of mass m is hung from a support at the origin by a nonlinear spring with free length L_0. The tension in the spring is $k(L - L_0)^3$, where L is the length of the stretched spring. The body swings in the (x, y) plane. Adopting the coordinates (x, y) as generalized coordinates of the body and neglecting the mass of the spring, write the equation of conservation of mechanical energy.

17. A thin hoop of radius R hangs over a fixed horizontal shaft of radius r. The hoop swings in its plane without slipping on the shaft. By means of the principle of conservation of mechanical energy, derive the differential equation of motion of the hoop.

18. Free vibrations of a beam of length L are represented by the equation, $y = c \sin(n\pi x/L) \sin \omega t$, where ω is a constant. The mass of the beam per unit length is a constant ρ. The potential energy of the beam is

$$U = \tfrac{1}{2}EI \int_0^L y_{xx}^2 \, dx,$$

where subscripts x denote partial derivatives. Write the equation of conservation of mechanical energy. By equating the mechanical energies for times $t = 0$ and $t = \pi/(2\omega)$, derive an equation for the angular frequency ω.

19. A uniform stick of length 5 in. rests across the rim of a frictionless fixed hemispherical bowl that is 4 in. in diameter. Calculate the angle of inclination of the stick to the horizontal by the principle of stationary potential energy (Fig. P1-19).

20. A straight beam of length L is bent into an arc of a circle by a constant bending moment M. The potential energy of the internal forces is $EIL/2R^2$, where R is the radius of curvature of the axis. Derive a formula for the total

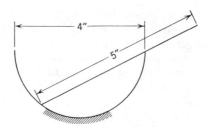

Fig. P1-19

potential energy of the beam. Derive the relation between the bending moment and the curvature by the principle of stationary potential energy.

21. A homogeneous hemispherical solid rests on a plane inclined at angle α to the horizontal. The body may roll, but it cannot slide. Determine the equilibrium configuration (angle of roll θ relative to the plane) by the principle of stationary potential energy.

22. A spherical soap bubble contains a constant mass of air. The pressure-volume relation of the air is $pv^{1.4}$ = constant. The potential energy of the air is $-\int p \, dv$. The potential energy of the soap film is $2kS$, where k is the surface tension and S is the area of the soap film. Express the potential energy of the system in terms of the radius r of the bubble. Hence derive a formula for the radius of the bubble by means of the principle of stationary potential energy.

23. A soap film spans a circular hole of radius a in a plate. Internal pressure p causes the film to adopt the form of a segment of a sphere. The strain energy of the soap film is $2kS$, where k is the surface tension (k = constant) and S is the area of the soap film. By geometry, $S = \pi a^2(1 + x^2)$ where $x = \tan \frac{1}{2}\alpha$ (Fig. P1-23). Also, by geometry, the volume of the bubble above plane O–O is $Q = \frac{1}{6}\pi a^3(x^3 + 3x)$. Adopt x as a generalized coordinate. Regarding p as a given constant, apply the principle of stationary potential energy to

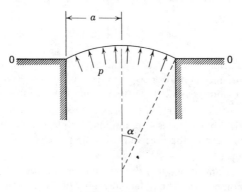

Fig. P1-23

express x as a function of u, where $u = 4k/ap$. For what range of u is x real? What is the physical interpretation of complex values of x? Show that when two real solutions exist, they determine bubbles with equal radii r.

24. A spherical shell carries a constant density of electric charge on its inner surface. Prove that the potential energy of a point charge within the shell is independent of its location. What conclusion can be drawn concerning the force on the point charge?

25. Prove that the Newtonian potential of a thin uniform disk of radius a at a point of a straight line perpendicular to the plane of the disk through its center and at a distance x from the center is $-2\pi\rho(\sqrt{a^2 + x^2} - x)$, where ρ is the mass per unit area. If $a \to \infty$, what is the force on a unit mass at point x?

26. A wire that is bent into a circle carries a constant density of electric charge. Derive the potential at any point on the axis of symmetry that is perpendicular to the plane of the circle. Hence, compute the force that acts on a unit charge that lies on the axis.

27. Calculate the Newtonian potential of a thin straight uniform rod of length $2L$ at any point outside of the rod. Calculate the attraction of the rod on a point mass that lies in the plane perpendicular to the rod at its center. Calculate the limiting value of this attraction as the length of the rod becomes infinite. Take origin at center.

28. A homogeneous body consists of a cone of height h attached to a hemispherical base with radius r (Fig. P1-28). The body rests on a table in an upright position. By the principle of minimum potential energy, determine the range of h/r for which this configuration is stable.

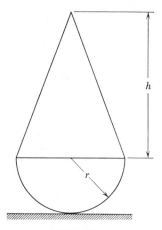

Fig. P1-28

29. Prove by the principle of minimum potential energy that among all attitudes of a floating body for which the weight of displaced liquid equals the weight

of the body the stable equilibrium configurations are those for which $Z_{cg} - Z_{cb}$ is a relative minimum, where Z_{cg} is the height of the center of gravity of the body, and Z_{cb} is the height of the centroid of the submerged part of the body (called the "center of buoyancy"). Let $Z_{cg} > Z_{cb}$. *Hint.* Suppose that the liquid is contained in a tank. Note that the attitude of the body affects the potential energy of the liquid.

30. A long homogeneous timber with a rectangular cross section of width b and depth a floats in water. The specific gravity of the timber is $\frac{1}{2}$. Show that if $b/a > \sqrt{3/2}$ the timber floats with its wide face horizontal and that if $1 < b/a < \sqrt{3/2}$ it does not float with its face horizontal. Determine the angle of the wide face to the horizontal if $b/a = \sqrt{4/3}$. How does it float if $b/a = 1$? *Hint.* Use the theorem stated in Prob. 29.

31. A long homogeneous timber of rectangular cross section floats in water with its top face horizontal. Prove that a necessary and sufficient condition for the timber to be in stable equilibrium is $s^2 - s + b^2/6a^2 > 0$, where s is the specific gravity of the wood, b is the width of the cross section, and a is the depth of the cross section. *Hint.* Use the theorem stated in Prob. 29.

2 Elastic beams and frames

The methods that I expound require neither constructions nor geometrical nor mechanical reasoning, but only algebraic operations, subject to an exact and invariable procedure.

J. L. LAGRANGE.

The principle of virtual work finds its most important engineering applications in the statical problems of deformable bodies. Some illustrative applications to deformable systems with enumerable degrees of freedom are presented in this chapter.

2-1. STRAIN ENERGY OF BEAMS, COLUMNS, AND SHAFTS.
The engineering theory of straight beams is based on the assumption that the strains of the longitudinal fibers of a beam are distributed linearly on any cross section. This assumption is implied by the common statement that plane cross sections remain plane.

In Fig. 2-1 the ξ- and η-axes represent the principal axes of inertia through the centroid of the cross section of a straight beam. Consequently,

$$\int_A \xi \, dA = 0, \qquad \int_A \eta \, dA = 0, \qquad \int_A \xi\eta \, dA = 0 \qquad \text{(a)}$$

where A denotes the cross-sectional area. The bending moment **M**, being represented by a vector that lies in the plane of the cross section, may be resolved into components M_ξ and M_η on the ξ- and η-axes. It is assumed that the fiber stress at the point (ξ, η) is determined by a linear equation, $\sigma = a\xi + b\eta + c$. Then the statical relations $M_\xi = \int \sigma\eta \, dA$, $M_\eta = -\int \sigma\xi \, dA$, $N = \int \sigma \, dA$, in conjunction with Eq. (a), determine the constants a, b, c. Thus we obtain

$$\sigma = \frac{-M_\eta\xi}{I_\eta} + \frac{M_\xi\eta}{I_\xi} + \frac{N}{A} \qquad \text{(2-1)}$$

Here, N denotes the net tension, and (I_ξ, I_η) are the moments of inertia of the cross section about the ξ- and η-axes, respectively.

By Eq. (1-10) and Hooke's law, $\sigma = E\epsilon$, the strain energy of an element of the beam of length ds and cross-sectional area dA is $\sigma^2\, dA\, ds/2E$. Consequently, the strain energy of a beam due to bending and direct tension is

$$U = \frac{1}{2E} \int_0^L ds \int \sigma^2\, dA \qquad (2\text{-}2)$$

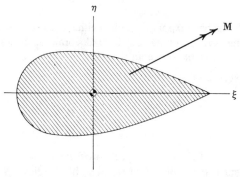

Fig. 2-1

where L is the length of the beam. Since $\int \xi\, dA = \int \eta\, dA = 0$ and $\int \xi\eta\, dA = 0$, Eqs. (2-1) and (2-2) yield

$$U = \frac{1}{2E} \int_0^L \left(\frac{M_\xi^2}{I_\xi} + \frac{M_\eta^2}{I_\eta} + \frac{N^2}{A} \right) ds \qquad (2\text{-}3)$$

Equation (2-3) shows that if a beam is bent simultaneously in both of its principal planes the contributions to the strain energy from these two components of bending are additive. Also, Eq. (2-3) shows that the strain energy due to direct tension may be added to the strain energy of bending. In view of these conclusions, we consider only pure bending in one principal plane; that is, we set $N = 0$ and $M_\eta = 0$. Also, for simplicity, we write $M_\xi = M$ and $I_\xi = I$. Then Eq. (2-3) yields

$$U = \int_0^L \frac{M^2\, ds}{2EI} \qquad (2\text{-}4)$$

It is to be noted that the superposition of strain-energy components implied by Eq. (2-3) is a special circumstance; in general, the effects of several loads on the strain energy of a structure are not simply additive. For example, if an elastic bar of length L and cross-sectional area A is subjected to a tensile force N_1, the strain energy [by Eq. (1-10)] is $N_1^2 L/2EA$. If a tensile force N_2 is superposed on the force N_1, the strain energy becomes $(N_1 + N_2)^2 L/2EA$. The latter expression is not the sum of the strain energies due to the forces N_1 and N_2 acting separately. This circumstance is illustrated graphically if we plot a straight line representing the load-extension graph of the member and note that the area under the line represents the strain energy.

It is shown in elementary beam theory that when curvature due to shear is neglected the bending moment M and the curvature $1/R$ of the deformed centroidal axis of a straight elastic beam are related by

$$M = \frac{EI}{R} \tag{2-5}$$

Equations (2-4) and (2-5) yield

$$U = \int_0^L \frac{EI \, ds}{2R^2} \tag{2-6}$$

If primes denote derivatives with respect to arc length s, Eq. (2-6) may be written

$$U = \tfrac{1}{2} \int_0^L EI[(x'')^2 + (y'')^2] \, ds \tag{2-7}$$

Equation (2-7) is valid for arbitrarily large deflections, provided that yielding does not occur.

If the x-axis coincides with the undeformed centroidal axis of the beam, y denotes the lateral deflection. If the deflection is small, s is approximately equal to x. Then x'' is approximately zero, and Eq. (2-7) yields the well-known approximation

$$U = \tfrac{1}{2} \int_0^L EI(y'')^2 \, dx \tag{2-8}$$

Strain Energy Due to Shear. Besides the strain energy of bending, there is strain energy due to shearing stresses on the cross-sectional planes. The shearing stress distribution on a cross-sectional plane is approximated by the elementary formula of beam theory

$$\tau = \frac{SQ}{Ib} \tag{2-9}$$

where S is the shearing force on the cross section, b is the width of the cross section at ordinate η, and Q is the moment of the area above the line with ordinate η about the ξ-axis. Also, by elasticity theory, the strain energy per unit volume due to the shearing stress τ is $\tau^2/2G$, where G is the shear modulus. Introducing τ from Eq. (2-9) and integrating throughout the volume of the beam, we obtain the following formula for the strain energy due to shearing forces:

$$U_S = \int_0^L \frac{\kappa S^2 \, dx}{2GA} \tag{2-10}$$

Here, κ is a dimensionless constant, defined by

$$\kappa = \frac{A}{I^2} \int \frac{Q^2 \, d\eta}{b} \tag{2-11}$$

For a rectangular cross section, $\kappa = 1.20$. Except for short beams or beams with thin webs, the energy of shear is usually small compared to the energy of bending.

If shear is significant, the strain energy of bending is not proportional to the square of the curvature, since the curvature is augmented by the shear deformation. Then Eq. (2-8) is inapplicable. The deflection separates into the sum of the displacements caused by shear and by bending. The slope y', being the derivative of the deflection, likewise separates into a sum of two terms—slope due to shear and slope due to bending. The slope due to shear will be denoted by β. Since the stress-strain relation is linear, the strain energy of shear is

$$U_S = \tfrac{1}{2} \int_0^L \beta S \, dx$$

With Eq. (2-10), this yields

$$\beta = \frac{\kappa S}{AG} \tag{2-12}$$

Shear deformation is frequently important when beam theory is applied to fabricated structures. For example, a long Vierendeel truss or a latticed strut may be regarded as a beam. Evidently, Eq. (2-12) does not apply for such structures. Likewise, a hull of a ship or a wing of an airplane may be analyzed as a beam. In such cases the shear is carried primarily by the plates or the skin, and very little is transmitted to the stringers. Also, a slight buckling of the plates of a ship or the spar webs of an airplane wing may contribute materially to shear deformation. Although Eq. (2-12) does not apply to these complex structures, a simple proportion between S and β may still be admissible; that is,

$$S = \lambda\beta \cdot \tag{2-13}$$

where λ is a constant (or a function of x) that must be determined by detailed considerations of the structure. The factor λ has the dimension of force. For a monocoque beam, Eqs. (2-12) and (2-13) yield $\lambda = AG/\kappa$. Since the shear energy per unit length is $\frac{1}{2}S\beta$, Eq. (2-13) yields

$$U_S = \frac{1}{2}\int_0^L \lambda\beta^2\, dx \qquad (2\text{-}14)$$

The slope caused by bending is $y' - \beta$, where y is the total deflection. Consequently, the curvature due to bending is $\pm M/EI = y'' - \beta'$. Therefore, the strain energy of bending is

$$U_b = \int_0^L \frac{M^2}{2EI}\, dx = \frac{1}{2}\int_0^L EI(y'' - \beta')^2\, dx$$

With Eq. (2-14), this yields the following formula for the total strain energy:

$$U = \frac{1}{2}\int_0^L \left[EI(y'' - \beta')^2 + \lambda\beta^2\right] dx \qquad (2\text{-}15)$$

In the statical problems of beams the functions y and β must be determined to minimize the total potential energy among functions that satisfy the end conditions and the continuity requirements. Since the slope at a clamped end of a beam results entirely from shear deformation, the boundary conditions at a clamped end are $y = 0$ and $y' = \beta$. Since the bending moment is $M = \pm EI(y'' - \beta')$, the boundary conditions at a simply supported end are $y = 0$ and $y'' = \beta'$. Since the shear is $S = dM/dx$, the boundary conditions at a free end are $y'' = \beta'$ and $y''' = \beta''$. At a point where a concentrated load is applied, y' and β are generally discontinuous, but $y' - \beta$ and $y'' - \beta'$ are continuous, since the bending moment is continuous. However, $y''' - \beta''$, being proportional to the shear, is discontinuous.

If shear deformation is negligible, $\beta = 0$, and Eq. (2-15) reduces to Eq. (2-8).

Torsion. If a linearly elastic bar of length L is subjected to torque T, the resulting angular displacement θ of the ends (total twist) is expressed by the formula

$$T = \frac{GJ\theta}{L} \qquad (2\text{-}16)$$

where J is a section constant and G is the shear modulus. The product GJ is called "torsional stiffness." For a solid shaft of circular cross section or for a hollow shaft of annular cross section, J is the polar moment of inertia of the cross section about its center. For any other shape of cross section, J is less than the polar moment of inertia. Formulas for J are developed in the literature on the torsion problem (85).

The strain energy due to twisting is $U_T = \int T\,d\theta$. Consequently, by Eq. (2-16),

$$U_T = \frac{GJ\theta^2}{2L} = \frac{LT^2}{2GJ} \tag{2-17}$$

Strain energy due to twisting may be added to the strain-energy components of bending, tension, and transverse shear.

2-2. BEAM COLUMNS ANALYZED BY FOURIER SERIES.

Functions that represent statical deflections of beams can be represented by Fourier series. The coefficients in the series may be regarded as generalized

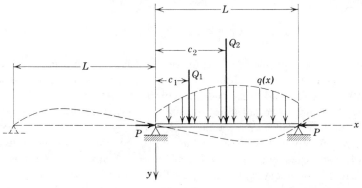

Fig. 2-2

coordinates. Ordinarily, the external forces that act on a structure are conservative. Accordingly, the coefficients in the Fourier series for the deflection of an elastic beam may usually be determined by the principle of stationary potential energy. The main precaution to be observed in this technique is that the condition for termwise differentiation of a Fourier series shall not be violated.

Figure 2-2 represents a simply supported uniform beam column that carries an axial load P, a few concentrated lateral loads $Q_1, Q_2, \cdots$, and an arbitrary distributed load $q(x)$. The deflection function has physical significance only in the range $0 \leq x \leq L$, but, for the sake of representation by a sine series, it is extended as an odd function into the range $-L \leq x \leq 0$. This extension is indicated by the dashed curve in Fig. 2-2. The deflection is represented by

$$y = \sum_{n=1}^{\infty} b_n \sin \frac{n\pi x}{L} \tag{a}$$

This series automatically satisfies the end conditions $y = y'' = 0$. The conditions for termwise differentiation are satisfied (9), and therefore

$$y' = \frac{\pi}{L} \sum n b_n \cos \frac{n\pi x}{L}, \qquad y'' = -\frac{\pi^2}{L^2} \sum n^2 b_n \sin \frac{n\pi x}{L} \qquad (b)$$

The strain energy of bending is given by Eq. (2-8). Squaring the series for y'', we obtain a double series. However, the integrals of the cross products cancel because of the relations

$$\int_0^L \sin \frac{m\pi x}{L} \sin \frac{n\pi x}{L} \, dx = \int_0^L \cos \frac{m\pi x}{L} \cos \frac{n\pi x}{L} \, dx = \begin{cases} 0 & \text{if} \quad m \neq n \\ \dfrac{L}{2} & \text{if} \quad m = n \end{cases} \qquad (c)$$

where m and n are any positive integers. Consequently, Eqs. (b) and (2-8) yield

$$U = \frac{\pi^4 EI}{4L^3} \sum_{n=1}^{\infty} n^4 b_n^2 \qquad (2\text{-}18)$$

It is assumed that the beam bends without any change of the length of the center line. In the theory of flexure this condition characterizes the centroidal axis. Since $dx^2 + dy^2 = ds^2$, $dx/ds = \sqrt{1 - (dy/ds)^2}$. Expanding the square root by the binomial series and retaining only the first two terms, we obtain $dx/ds = 1 - \frac{1}{2}(dy/ds)^2$. The distance between the ends of the bent beam is

$$L_1 = \int_0^L \frac{dx}{ds} \, ds$$

The potential energy of the axial load is $\Omega_P = -P(L - L_1)$. Hence, by the preceding relation,

$$\Omega_P = -\frac{P}{2} \int_0^L \left(\frac{dy}{ds}\right)^2 ds$$

This is further approximated by

$$\Omega_P = -\frac{P}{2} \int_0^L (y')^2 \, dx \qquad (2\text{-}19)$$

where y' denotes dy/dx. Substituting Eq. (b) into Eq. (2-19) and utilizing Eq. (c), we obtain

$$\Omega_P = -\frac{\pi^2 P}{4L} \sum_{n=1}^{\infty} n^2 b_n^2 \qquad (d)$$

The potential energy of the lateral loads $Q_1, Q_2, \cdots$ is

$$\Omega_Q = -Q_1 y_1 - Q_2 y_2 - \cdots$$

where y_i is the deflection at the point of application of load Q_i. Hence, by Eq. (a),

$$\Omega_Q = -Q_1 \sum_{n=1}^{\infty} b_n \sin \frac{n\pi c_1}{L} - Q_2 \sum_{n=1}^{\infty} b_n \sin \frac{n\pi c_2}{L} - \cdots \tag{e}$$

The potential energy of the distributed load is

$$\Omega_q = -\int_0^L qy \, dx$$

Hence, by Eq. (a),

$$\Omega_q = -\sum_{n=1}^{\infty} b_n \int_0^L q \sin \frac{n\pi x}{L} \, dx \tag{f}$$

The potential energy of the system is $V = U + \Omega_P + \Omega_q + \Omega_Q$. Since the coefficients b_i are generalized coordinates, the condition of stationary potential energy yields $\partial V/\partial b_i = 0$. By evaluating this derivative and then replacing i by n, we obtain

$$b_n = \frac{\displaystyle\int_0^L q \sin (n\pi x/L) \, dx + Q_1 \sin (n\pi c_1/L) + Q_2 \sin (n\pi c_2/L) + \cdots}{(\pi^4 EI/2L^3)n^4 - (\pi^2 P/2L)n^2} \tag{g}$$

If any coefficient b_n becomes infinite, the column experiences unlimited deflections when an arbitrarily small lateral load acts. Infinite deflections are obviously impossible. The infinity results from quadratic approximations in the preceding theory; for example, from Eq. (2-8) and from the binomial expansion of the square root. Nevertheless, the Euler critical load for a column without lateral loads is determined by the condition that some coefficient b_n will become infinite. Evidently, b_1 is the first coefficient to become infinite. The infinity occurs when the denominator in Eq. (g) becomes zero with $n = 1$. Hence the buckling load is

$$P_e = \frac{\pi^2 EI}{L^2} \tag{2-20}$$

This is the Euler column formula. Accordingly, Eq. (g) may be written as follows:

$$b_n = \frac{2L^3}{\pi^4 EI} \frac{\displaystyle\int_0^L q \sin (n\pi x/L) \, dx + Q_1 \sin (n\pi c_1/L) + Q_2 \sin (n\pi c_2/L) + \cdots}{n^4 - (Pn^2/P_e)} \tag{2-21}$$

Substituting Eq. (2-21) into Eq. (a), we obtain the sine series for the deflection curve.

Reactions of Intermediate Supports. The foregoing theory serves also for the analysis of a beam column with intermediate supports. The reactions of intermediate supports are to be reckoned among the concentrated lateral loads Q_i. Expressing the deflections at the supports by means of Eqs. (a) and (2-21), introducing the condition $y = 0$ for rigid supports (or reaction $= ky$ for a spring support, where k is the spring constant), and utilizing the over-all equations of statical equilibrium, we obtain enough equations to solve for the reactions.

Other End Conditions. A cantilever beam may be regarded as half a symmetrically loaded simple beam (Fig. 2-3). Consequently, the preceding

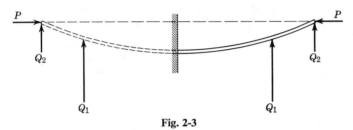

Fig. 2-3

theory applies to cantilever beams and cantilever beam columns. The center load Q on the simple beam represents twice the shear applied by the support of the cantilever. It may be determined by elementary statics. If there is no lateral load on the end of the cantilever, the reactions at the ends of the equivalent simple beam are zero.

A beam with one end clamped and the other end simply supported may be regarded as a cantilever beam with an undetermined lateral load R at its outer end. The load R represents the reaction of the simple support. It must be determined to make the deflection at the tip equal to zero.

Beams with clamped ends or elastically restrained ends may be analyzed by Fourier series, although special methods must be used to ensure that the end conditions are satisfied. Some special techniques and tables for facilitating the use of Fourier series in problems of deflections of beams have been presented by Ruffner (143).

The theory of twisting of beams by torques applied at intermediate sections may be treated by Fourier series in the same manner as flexure problems. This approach to problems of twisting of I-beams has been investigated by Goldberg (108).

Applications of Fourier series to various problems of buckling of columns are presented in the book by Timoshenko (84).

Example. A uniformly loaded beam column has simple supports at the ends and at the center. Hence $q = $ constant and $Q_1 = -R$, where R

is the reaction of the center support. The other Q's are zero. Equation (2-21) accordingly yields

$$b_n = \frac{2L^3}{\pi^4 EI} \frac{(qL/n\pi)(1 - \cos n\pi) - R \sin n\pi/2}{n^4 - rn^2}, \qquad r = \frac{P}{P_e} \qquad \text{(h)}$$

The reaction R is determined by the condition that $y = 0$ for $x = L/2$. Hence, by Eqs. (a) and (h),

$$R \sum_{n=1}^{\infty} \frac{\sin^2 n\pi/2}{n^4 - rn^2} = \frac{2qL}{\pi} \sum_{n=1}^{\infty} \frac{\sin n\pi/2}{n^5 - rn^3} \qquad \text{(i)}$$

If $P = 0$, Eq. (i) yields $R = (\frac{5}{8})qL$. If r approaches 1, Eq. (i) yields $R = (2/\pi)qL$. Equation (i) applies also for negative values of P; that is, for tension of the beam. For example, if $r = -10$, Eq. (i) yields $R = 0.576qL$. These results indicate that the axial load has only a small effect on the center reaction.

2-3. CURVED BEAMS. A circular ring or a uniformly curved beam that is symmetrical about the plane of its centroidal axis is considered deformed in its plane. The following notations are used (see Fig. 2-4):

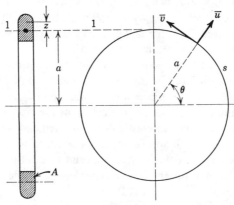

Fig. 2-4

    a the radius of the undeformed centroidal axis
    A the area of the cross section of the ring
    θ an angular coordinate that locates cross sections of the ring (Fig. 2-4)
    I the moment of inertia of a cross section of the ring about an axis through its centroid (axis 1 in Fig. 2-4)

z an ordinate in a cross section of the ring measured from the centroidal axis (positive outward, Fig. 2-4); since the cross sections are assumed to be inextensional, z is invariant under the deformation

$\bar{u}, \bar{v}$ radial and circumferential components of displacement of a particle on the centroidal axis; these variables are functions of θ alone, the positive sense of $\bar{u}$ is outward, and the positive sense of $\bar{v}$ is the sense of increasing θ (Fig. 2-4)

s arc length measured on the centroidal axis (Fig, 2-4); primes denote derivatives with respect to s

A dimensionless constant Z of the Winkler theory of curved beams is defined by

$$\int \frac{dA}{a+z} = \frac{A}{a}(1+Z), \qquad \int \frac{z\, dA}{a+z} = -ZA, \qquad \int \frac{z^2\, dA}{a+z} = ZAa \qquad (2\text{-}22)$$

The equivalence of these three definitions may be verified by expanding the integrands by division and noting that $\int z\, dA = 0$, since z is measured from the centroid. If $a \gg z$, the following approximation may be obtained by discarding z from the denominator of the third of the equations (2-22):

$$Z \approx \frac{I}{Aa^2} \qquad (2\text{-}23)$$

The last of the equations in (2-22) is most convenient for evaluation of Z by numerical integration.

The radial and circumferential components of displacement at a distance z from the centroidal axis are denoted by u, v. It is assumed that u does not vary on a cross section; hence $\bar{u} = u$. Following the Winkler theory (74), we assume that the cross sections remain plane and normal to the centroidal axis under the deformation. Then, if $\bar{v} = 0$, the displacement v results entirely from rotation of the cross section; hence, to first-degree terms, $v = -zu'$, where u' denotes du/ds. If $\bar{v} \neq 0$, an additional displacement results from $\bar{v}$; this is equal to $\bar{v}(1 + z/a)$. Accordingly, the linearized displacement equations are

$$\bar{u} = u, \qquad v = -zu' + \left(1 + \frac{z}{a}\right)\bar{v} \qquad (a)$$

The strain of a fiber at distance z from the centroidal axis is denoted by ϵ. The part of the strain that results from the circumferential displacement v is dv/dS, where dS is a line element of a fiber at distance z from the centroidal axis. Since $dS = (1 + z/a)\, ds$, $dv/dS = av'/(a+z)$, where v' denotes dv/ds. Also, the radial displacement u causes the strain $u/(a+z)$.

Hence the net strain is

$$\epsilon = \frac{u + av'}{a + z} \tag{b}$$

Equations (a) and (b) yield

$$\epsilon = \bar{v}' + \frac{u - azu''}{a + z} \tag{c}$$

The net tension in the beam is $N = E\int \epsilon \, dA$. With Eqs. (2-22) and (c), this yields

$$N = \frac{EA}{a}[(u + a\bar{v}') + Z(u + a^2 u'')] \tag{2-24}$$

The bending moment M about the centroidal axis of the cross section is defined as positive if it tends to reduce the radius of curvature. Then, $M = E\int z\epsilon \, dA$. With Eqs. (2-22) and (c), this yields

$$M = -EAZ(u + a^2 u'') \tag{2-25}$$

If hoop stress predominates, we obtain the strain energy of the beam by integrating the strain energies of the fibers. A volume element of the beam is $(a + z) \, dA \, d\theta$. Consequently, if the material obeys Hooke's law, the strain energy is

$$U = \tfrac{1}{2}E\int d\theta \int (a + z)\epsilon^2 \, dA \tag{2-26}$$

Equations (c), (2-22), and (2-26) yield

$$U = \int \frac{EA}{2a^2}[(u + a\bar{v}')^2 + Z(u + a^2 u'')^2] \, ds \tag{2-27}$$

Equations (2-24), (2-25), and (2-27) remain approximately valid for a curved beam with a noncircular centroidal axis, if a denotes the radius of curvature of the centroidal axis at section s. For the treatment of buckling problems of rings and arches by the principle of minimum potential energy, a refinement of Eq. (2-27) that includes cubic terms in u and $\bar{v}$ is required (125).

Equations (2-24), (2-25), and (2-27) yield

$$U = \int \frac{1}{2a^2 EA}\left[(M + aN)^2 + \frac{M^2}{Z}\right] ds \tag{2-28}$$

Ring Subjected to Concentrated Radial Loads. Because the stresses, the moments, and the displacements are periodic functions, ring problems are well adapted to treatment by Fourier series.

Figure 2-5 represents a uniform ring that is subjected to ν equal and equally spaced radial loads of magnitude P. The displacement components u and v evidently have the fundamental period $2\pi/\nu$. If θ is measured from a load, u is an even function and $\bar{v}$ is an odd function. Hence

$$u = \frac{a_0}{2} + \sum_{n=1}^{\infty} a_n \cos \nu n\theta,$$

$$\bar{v} = \sum_{n=1}^{\infty} b_n \sin \nu n\theta \qquad \text{(d)}$$

Therefore,

$$u + a^2 u'' = \frac{a_0}{2} + \sum_{n=1}^{\infty} (1 - \nu^2 n^2) a_n$$
$$\cos \nu n\theta$$
$$\qquad \text{(e)}$$
$$u + a\bar{v}' = \frac{a_0}{2} + \sum_{n=1}^{\infty} (a_n + \nu n b_n)$$
$$\cos \nu n\theta$$

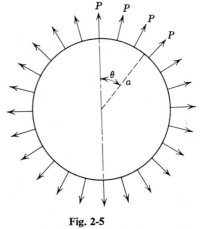

Fig. 2-5

where primes denote derivatives with respect to arc length. Accordingly, by Eq. (2-27),

$$U = \frac{\pi EA}{4a}(1 + Z)a_0^2 + \frac{\pi EA}{2a} \sum_{n=1}^{\infty} [(a_n + \nu n b_n)^2 + Z(\nu^2 n^2 - 1)^2 a_n^2] \qquad \text{(f)}$$

The potential energy of the external loads is $\Omega = -\nu P\, u(0)$. Hence, by Eq. (d),

$$\Omega = -\tfrac{1}{2}\nu P a_0 - \nu P \sum_{n=1}^{\infty} a_n \qquad \text{(g)}$$

The total potential energy is $V = U + \Omega$. The principle of stationary potential energy yields $\partial V/\partial a_i = \partial V/\partial b_i = 0$. With Eqs. (f) and (g), these conditions yield

$$a_n = \frac{\nu P a}{\pi EAZ(\nu^2 n^2 - 1)^2} \quad \text{for} \quad n > 0 \qquad \text{(h)}$$

$$a_0 = \frac{\nu P a}{\pi EA(1 + Z)}, \qquad b_n = -\frac{a_n}{\nu n}$$

Equation (e) accordingly yields $u + a\bar{v}' = a_0/2$. Hence Eqs. (e), (h), (2-24), and (2-25) yield the following expressions for the net tension and the bending moment at section θ:

$$N = \frac{\nu P}{2\pi} - \frac{\nu P}{\pi} \sum_{n=1}^{\infty} \frac{\cos \nu n\theta}{\nu^2 n^2 - 1} \tag{i}$$

$$M = -\frac{\nu P a Z}{2\pi(1+Z)} + \frac{\nu P a}{\pi} \sum_{n=1}^{\infty} \frac{\cos \nu n\theta}{\nu^2 n^2 - 1} \tag{j}$$

The tension N_0 and the bending moment M_0 at a point where a load is applied are obtained from Eqs. (i) and (j) if θ is set equal to zero. The resulting series may be evaluated by means of the following series for the cotangent (45):

$$\pi \cot \pi z = \frac{1}{z} + \sum_{n=1}^{\infty} \frac{2z}{z^2 - n^2} \tag{k}$$

Setting $z = 1/\nu$ in Eq. (k), we obtain

$$\sum_{n=1}^{\infty} \frac{1}{\nu^2 n^2 - 1} = \frac{1}{2} - \frac{\pi}{2\nu} \cot \frac{\pi}{\nu} \tag{l}$$

Hence Eqs. (i) and (j) yield

$$N_0 = \frac{P}{2} \cot \frac{\pi}{\nu}, \qquad M_0 = \frac{Pa}{2}\left[\frac{\nu}{\pi(1+Z)} - \cot \frac{\pi}{\nu}\right] \tag{m}$$

With Eq. (m) the tension and the bending moment at any section may be obtained by statics.

2-4. PIN-JOINTED TRUSSES.
The bending moments in the members of a truss are often secondary effects. Analyses based on the concept of hinged joints or ball-and-socket joints are then justified. Such joints are designated as pin joints. Since the members of a pin-jointed truss are subjected to direct tensions or compressions, the strain energy of the system is determined by the displacements of the joints. Accordingly, a pin-jointed truss is essentially a system with finite degrees of freedom. From a theoretical viewpoint, problems of equilibrium of elastic pin-jointed trusses are easily solvable by the principle of stationary potential energy. The method applies to plane trusses and to space trusses, whether or not they are statically determinate. Also, the load-extension relation of the members need not be linear, provided that the proper strain-energy function is introduced. However, if the elasticity is nonlinear, the resulting algebraic equations are also nonlinear, and they are usually difficult to

Solving static indeterminate problems
- pin connected
- straight
- negligible mass
- loads applied at connections

solve. Even in cases of linear elasticity, the algebraic equations often are tedious to solve, but they may be handled readily by network or digital computers. Direct or iterative numerical methods may also be used with desk calculators (15, 23, 47).

If the displacements of the joints are so small that the members are not rotated appreciably, the extension of a member is approximately the difference of the axial components of the displacements at its ends. If the displacements of the joints are large, they still determine the extensions of the members, but the geometrical relations become nonlinear. Large deflections of trusses rarely occur in practice.

The strain energy of a uniform tensile member with modulus of elasticity E is determined by Eq. (1-10). A generalization of this equation is desirable, so that problems of self-straining (or residual stresses) may be treated by the principle of virtual work. Consider a bar of length L and constant cross-sectional area A. Let the extension of the bar in the designated initial state be e'. Let the final extension of the bar be $e' + e$. The supplemental extension e results from external loads. The tension corresponding to the extension $e' + e$ is

e' = prestress elongation.

$$N = \frac{EA}{L}(e' + e) \qquad (2\text{-}29)$$

e = extension associated with external load

By Eq. (1-10), the strain energy of the bar is

For truss sum over all bars

$$U = \frac{EA}{2L}(e' + e)^2 \qquad (2\text{-}30)$$

$e_T = e + e'$ = "total"

Pentagonal Truss. As an example, a regular pentagon-shaped truss with no residual stresses is considered to be loaded symmetrically by a vertical force F, as shown by Fig. 2-6. All members have the same cross-sectional area A. The length of an outer member is L. By geometry, the length of any cross brace is $1.61804L$. The stress-strain relation of the material is linear. The displacement components of the joints are x, y, z, w (Fig. 2-6).

Resolving the displacements into axial components, we obtain the following formulas for the extensions of the members:

$$e_1 = 0.80902y - 0.58779z + 0.80902w$$

$$e_2 = -0.95106x + 0.30902w$$

$$e_3 = -0.58779x + 0.80902y + 0.58779z$$

$$e_4 = 0.30902y - 0.95106z - 0.30902w \qquad (a)$$

$$e_5 = 2w$$

$$e_6 = 2y$$

By Eq. (2-30), the strain energy of the structure is

$$U = \frac{AE}{2L}\left(\frac{2e_1{}^2}{1.61804} + \frac{2e_2{}^2}{1.61804} + 2e_3{}^2 + 2e_4{}^2 + e_5{}^2 + \frac{e_6{}^2}{1.61804}\right)$$

The potential energy of the external forces is $-Fx$. Consequently, the total potential energy of the system is $V = U - Fx$. The condition of

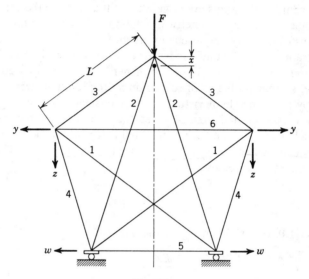

Fig. 2-6

stationary potential energy is expressed by the equations $\partial V/\partial x = \partial V/\partial y = \partial V/\partial z = \partial V/\partial w = 0$. Consequently, since

$$\frac{\partial U}{\partial x} = \frac{\partial U}{\partial e_1}\frac{\partial e_1}{\partial x} + \frac{\partial U}{\partial e_2}\frac{\partial e_2}{\partial x} + \cdots + \frac{\partial U}{\partial e_6}\frac{\partial e_6}{\partial x}$$

with similar equations for $\partial U/\partial y$, $\partial U/\partial z$, and $\partial U/\partial w$,

$$0.58779e_2 + 0.58779e_3 = -\frac{FL}{2EA}$$

$$0.50000e_1 + 0.80902e_3 + 0.30902e_4 + 0.61804e_6 = 0$$

$$-0.36327e_1 + 0.58779e_3 - 0.95106e_4 = 0$$

$$0.50000e_1 + 0.19098e_2 - 0.30902e_4 + e_5 = 0$$

(b)

Substitution of Eq. (a) into Eq. (b) yields

$$-0.9045x + 0.4755y + 0.3455z + 0.1816w = -\frac{FL}{2EA}$$

$$0.4755x - 2.3906y + 0.1123z - 0.3090w = 0$$

$$0.3455x + 0.1123y - 1.4635z = 0 \tag{c}$$

$$0.1816x - 0.3090y - 2.5590w = 0$$

The symmetry of these equations follows from the fact that the strain energy U is a quadratic form in the generalized coordinates x, y, z, w. The solution of Eq. (c) is

$$x = 0.7025\frac{FL}{EA}, \qquad y = 0.1438\frac{FL}{EA}$$

$$\tag{d}$$

$$z = 0.1769\frac{FL}{EA}, \qquad w = 0.0325\frac{FL}{EA}$$

Consequently, by Eq. (a),

$$e_1 = 0.0387\frac{FL}{EA}, \qquad e_2 = -0.6581\frac{FL}{EA}, \qquad e_3 = -0.1926\frac{FL}{EA}$$

$$e_4 = -0.1338\frac{FL}{EA}, \qquad e_5 = 0.0650\frac{FL}{EA}, \qquad e_6 = 0.2877\frac{FL}{EA}$$

Accordingly, the tensions in the members are

$$N_1 = 0.024F, \qquad N_2 = -0.407F, \qquad N_3 = -0.193F$$

$$N_4 = -0.134F, \qquad N_5 = 0.065F, \qquad N_6 = 0.178F$$

These equations may be checked by the condition of equilibrium of forces at the joints.

Simple Space Truss with Residual Tensions. The lower ends of the bars of the four-bar, pin-jointed truss shown in Fig. 2-7 are fixed and the upper joint is free. When $P = 0$, the residual tension in member 1 is N_1'. When the load P is applied, the tensions in the bars are N_1, N_2, $N_3 = N_4$. Bars 3 and 4 are identical.

Positive senses are assigned to the bars, as indicated by the arrowheads.

The following table gives the cross-sectional areas, the lengths, and the direction cosines of the members. The lengths and the direction cosines have been computed by means of the designated coordinates of the ends of the bars.

The displacement vector of the upper joint due to the load P is $\mathbf{q} = i u + j v + k w$. By symmetry, $u = 0$. The extension e_i of any member that

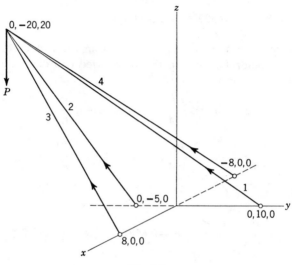

Fig. 2-7

results from the displacement $\mathbf{q}$ is the component of $\mathbf{q}$ in the direction of the member; hence

$$e_i = u \cos \alpha_i + v \cos \beta_i + w \cos \gamma_i$$

Therefore,

$$e_1 = \frac{1}{\sqrt{13}} (-3v + 2w), \qquad e_2 = \tfrac{1}{5}(-3v + 4w), \tag{e}$$

$$e_3 = e_4 = \frac{5\sqrt{6}}{18}(-v + w)$$

The initial extensions of the members due to the residual tensions are $e_1', e_2', e_3' = e_4'$. By Eq. (2-30), the total potential energy of the truss is

$$V = Pw + \tfrac{1}{2}E \sum_{i=1}^{4} \frac{A_i}{L_i}(e_i' + e_i)^2$$

By the principle of stationary potential energy, the equilibrium configuration is determined by $\partial V/\partial v = \partial V/\partial w = 0$. Hence

$$\sum_{i=1}^{4} \frac{A_i}{L_i}(e_i' + e_i)\frac{\partial e_i}{\partial v} = 0$$

$$\sum_{i=1}^{4} \frac{A_i}{L_i}(e_i' + e_i)\frac{\partial e_i}{\partial w} + \frac{P}{E} = 0$$

(f)

The derivatives $\partial e_i/\partial v$ and $\partial e_i/\partial w$ are constants; they are determined by Eq. (e).

TABLE 2-1

Member	A	L	$\cos\alpha$	$\cos\beta$	$\cos\gamma$
1	1/2	$10\sqrt{13}$	0	$-3/\sqrt{13}$	$2/\sqrt{13}$
2	1	25	0	$-3/5$	$4/5$
3	1/4	$12\sqrt{6}$	$-2\sqrt{6}/18$	$-5\sqrt{6}/18$	$5\sqrt{6}/18$
4	1/4	$12\sqrt{6}$	$2\sqrt{6}/18$	$-5\sqrt{6}/18$	$5\sqrt{6}/18$

The residual extension e_1' is expressed in terms of the residual tension N_1' by Eq. (2-29); that is,

$$e_1' = \frac{N_1'L_1}{EA_1} = \frac{20\sqrt{13}N_1'}{E}$$

(g)

The residual extensions e_2' and $e_3' = e_4'$ are determined by the condition that $v = w = 0$ if $P = 0$. Then, by Eq. (e), $e_1 = e_2 = e_3 = e_4 = 0$. Hence, by Eq. (f),

$$\sum_{i=1}^{4} \frac{A_i e_i'}{L_i}\frac{\partial e_i}{\partial v} = 0, \qquad \sum_{i=1}^{4} \frac{A_i e_i'}{L_i}\frac{\partial e_i}{\partial w} = 0$$

(h)

Since e_1' is known and $e_3' = e_4'$, Eq. (h) determines e_2' and e_3'. Introducing the derivatives $\partial e_i/\partial v$ and $\partial e_i/\partial w$ from Eq. (e), and numerical values of A_i and L_i from Table 2-1, we obtain

$$e_2' = \frac{25}{52}e_1', \qquad e_3' = e_4' = -\frac{648}{325}e_1'$$

With Eqs. (2-29) and (g), these relations yield

$$N_2' = \frac{5}{\sqrt{13}}N_1' = 1.387N_1',$$

$$N_3' = N_4' = -\frac{9\sqrt{78}}{65}N_1' = -1.223N_1'$$

(i)

In view of Eq. (h), the quantities e_i' cancel from Eq. (f). In other words, the displacement (v, w) is not affected by the residual tensions. This conclusion exemplifies the principle of superposition in the linear theory of structures. Dropping the terms e_i' from Eq. (f), eliminating e_i by means of Eq. (e), and introducing numerical values from Table 2-1, we obtain

$$0.031876v - 0.033476w = 0$$

$$-0.033476v + 0.037742w = -\frac{P}{E}$$

The solution of these equations is

$$v = -406\frac{P}{E}, \qquad w = -387\frac{P}{E} \tag{j}$$

Equations (e) and (j) yield

$$e_1 = 123\frac{P}{E}, \qquad e_2 = -65.6\frac{P}{E}, \qquad e_3 = e_4 = 13.3\frac{P}{E} \tag{k}$$

Equations (2-29), (i), and (k) yield

$$N_1 = N_1' + 1.71P, \qquad N_2 = 1.39N_1' - 2.62P$$
$$N_3 = N_4 = -1.22N_1' + 0.113P \tag{l}$$

These equations may be checked by the condition that the upper joint must be in equilibrium under the action of the load P and the tensions in the members.

2-5. FRAMES WITH TORSIONAL AND FLEXURAL MEMBERS.
To express the potential energy of an elastic frame in terms of the displacements and rotations of the joints, we first consider a single beam (Fig. 2-8). The deflected beam has the general form indicated by the dashed curve. The equation of the deflection curve is $y = y(x)$. Hence

$$y(0) = y_1, \qquad y(L) = y_2, \qquad y'(0) = \theta_1, \qquad y'(L) = \theta_2 \tag{a}$$

where (y_1, y_2) are the transverse displacements of the ends, and (θ_1, θ_2) are the angular displacements of the ends. Primes denote derivatives with respect to x. The rotation of a straight line through the ends is

$$\phi = \frac{y_2 - y_1}{L} \tag{2-31}$$

Supposing that a single lateral load F acts on the beam at the point $x = a$, we introduce (94) a unit step function $H(x)$, defined by $H = 0$ if

$x < a$ and $H = 1$ if $x > a$. Then the bending moment at any section of the beam is

$$M = -M_1 + S_1 x - HF(x - a) = -EIy'' \qquad \text{(b)}$$

Integration yields

$$y' = \theta_1 + \frac{2M_1 x}{KL} - \frac{S_1 x^2}{KL} + \frac{FH(x - a)^2}{KL} \qquad \text{(c)}$$

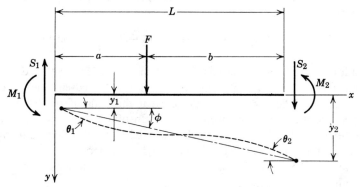

Fig. 2-8

where $K = 2EI/L$. The constant K is called the "stiffness factor" of the beam. Integration of Eq. (c) yields

$$y = y_1 + \theta_1 x + \frac{M_1 x^2}{KL} - \frac{S_1 x^3}{3KL} + \frac{FH(x - a)^3}{3KL} \qquad \text{(d)}$$

Equations (a), (c), and (d) yield two linear equations that determine M_1 and S_1. Then M_2 and S_2 may be determined by statics. Thus there results

$$M_1 = -K(2\theta_1 + \theta_2 - 3\phi) + \frac{Fab^2}{L^2}$$

$$M_2 = -K(\theta_1 + 2\theta_2 - 3\phi) - \frac{Fa^2 b}{L^2} \qquad \text{(2-32)}$$

$$S_1 = -\frac{3K}{L}(\theta_1 + \theta_2 - 2\phi) + \frac{Fb^2(L + 2a)}{L^3}$$

$$S_2 = -\frac{3K}{L}(\theta_1 + \theta_2 - 2\phi) - \frac{Fa^2(L + 2b)}{L^3} \qquad \text{(2-33)}$$

The lengths a and b that appear in these equations are designated in Fig. 2-8. Equations (2-32) are fundamental in the slope-deflection theory of elastic frames (63). These equations may be generalized by the addition of other terms analogous to those containing F if there are several lateral loads. The terms $+Fab^2/L^2$ and $-Fa^2b/L^2$ in Eqs. (2-32) are known as the "fixed-end moments", since they represent the moments that would exist at the ends if the ends were clamped (that is, if $\theta_1 = \theta_2 = y_1 = y_2 = 0$). For any type of loading, these terms are replaced by the appropriate fixed-end moments.

Eliminating M_1 and S_1 from Eq. (d) by means of Eqs. (2-32) and (2-33), we obtain

$$y = y_1 + \theta_1 x - (2\theta_1 + \theta_2 - 3\phi)\frac{x^2}{L} + (\theta_1 + \theta_2 - 2\phi)\frac{x^3}{L^2} + \text{constant} \quad (2\text{-}34)$$

The additive constant represents the part of the deflection that does not depend on $y_1, y_2, \theta_1, \theta_2$; it may be called the "fixed-end deflection," since it is the deflection that would exist if the ends of the beam were clamped and the lateral load were applied. Insofar as the potential energy of the external forces is concerned, the fixed-end deflection contributes only a constant amount; hence it may be disregarded.

Equations (2-8), (b), (2-32), and (2-33) yield the following formula for the strain energy of bending:

$$U = K[\theta_1{}^2 + \theta_1\theta_2 + \theta_2{}^2 - 3\phi(\theta_1 + \theta_2) + 3\phi^2] + \text{constant} \quad (2\text{-}35)$$

where the additive constant is independent of y_1, y_2, θ_1, and θ_2. Naturally, this constant depends on the lateral load F, If there are several lateral loads or distributed loads, only the constant is altered. This constant is irrelevant, since our purpose is to determine $y_1, y_2, \theta_1, \theta_2$. Then the shears and moments at the ends are determined by Eqs. (2-32) and (2-33).

If the beam (Fig. 2-8) is a part of a structure, the variables $y_1, y_2, \theta_1, \theta_2$ are generalized coordinates. Note that ϕ is expressed in terms of y_1 and y_2 by Eq. (2-31). The potential energy of the structure, as determined by Eq. (2-35), is a quadratic form in the generalized coordinates; that is,

$$U = \tfrac{1}{2}\sum_{i=1}^{n}\sum_{j=1}^{n}a_{ij}x_i x_j \quad (2\text{-}36)$$

The coordinates $(x_1, x_2, \cdots, x_n)$ here represent rotations or displacements of the joints or angular displacements ϕ of the members. Equations (1-15) are the equations of equilibrium. For plane frames, these equations are the same as those provided by the slope-deflection theory (63). After the x's are determined, the bending moments and the shears may be computed by the slope-deflection equations [Eqs. (2-32) and (2-33)].

To systematize the procedure, it is convenient to set up the matrix (a_{ij}) and to augment this matrix by the column $(P_1, P_2, \cdots, P_n)$, the components of generalized external force that appear in Eq. (1-15). If the x's receive increments dx_i, the incremental work of the external forces is $\Sigma P_i\, dx_i$; this relation defines the coefficients P_i. Alternatively, we have $P_i = -\partial \Omega/\partial x_i$, where Ω is the potential energy of the external forces.

Simple Bent. Figure 2-9 represents a simple plane frame, called a "bent." The upper joints are gusseted and the lower joints are hinged. The frame deforms in its plane. For each member, $EI = 10^8$ lb-in^2. Hence, for a vertical member, $K = 10^6$ lb-in., and, for the horizontal member, $K = 2 \times 10^6$ lb-in. The numbers in Fig. 2-9 are labels for the joints. The rotations of the joints (positive clockwise) are x_1, x_2, x_3, x_4. The rotation of the straight-line segment 12 or 34 (positive clockwise) is x_5. Accordingly, the horizontal displacement of the top beam is $200x_5$. Although the horizontal displacements of joints 2 and 3 are not exactly equal, they differ only by second-order effects. These effects will be neglected; they are unimportant unless buckling is imminent.

The strain energy of bending of member 12 is denoted by U_{12}; similar notations are used for the other members. By Eq. (2-35),

$$10^{-6}U_{12} = x_1{}^2 + x_1x_2 + x_2{}^2 - 3x_5(x_1 + x_2) + 3x_5{}^2$$

$$10^{-6}U_{23} = 2(x_2{}^2 + x_2x_3 + x_3{}^2) \qquad\qquad (e)$$

$$10^{-6}U_{34} = x_3{}^2 + x_3x_4 + x_4{}^2 - 3x_5(x_3 + x_4) + 3x_5{}^2$$

The potential energy of the load F is $\Omega = -Fy_a$, where y_a is the deflection at the point of application of the load. Referring to Eq. (2-34), we have for the horizontal member $y_1 = 0, \phi = 0, L = 100, x = 25, \theta_1 = x_2,$ $\theta_2 = x_3$. Hence, aside from the additive constant, $y_a = (75/16)(3x_2 - x_3)$. Consequently, since $P_i = -\partial\Omega/\partial x_i$, the components of generalized external force are $P_1 = 0, P_4 = 0, P_5 = 0, P_2 = (225/16)F, P_3 = -(75/16)F$.

In forming the matrix (a_{ij}), we must observe the factor $\frac{1}{2}$ in Eq. (2-36). Accordingly, the terms on the principal diagonal of the matrix are twice the sums of the corresponding coefficients from Eq. (e). However, the other terms are the sums of the corresponding coefficients from Eq. (e), since in Eq. (2-36) each nonsquared product of the x's occurs twice; for example, the terms containing x_1 and x_2 are $\frac{1}{2}a_{12}x_1x_2 + \frac{1}{2}a_{21}x_2x_1$.

The accompanying symmetric matrix (Table 2-2) of coefficients is obtained from Eq. (e). For convenience in numerical computations, the coefficients a_{ij} and the components P_i are reduced by the factor 10^{-6}; this procedure does not alter the solution of the equations. Also, since the rotations, moments, and shears are directly proportional to F, we may assign any convenient numerical value to F. To cancel the factor $75/16$ in the equations for P_2 and P_3, we set $F = (16/75) \times 10^6$ lb. The equations

TABLE 2-2

	x_1	x_2	x_3	x_4	x_5	P
x_1	2	1	0	0	-3	0
x_2	1	6	2	0	-3	3
x_3	0	2	6	1	-3	-1
x_4	0	0	1	2	-3	0
x_5	-3	-3	-3	-3	12	0

corresponding to this matrix are $2x_1 + x_2 - 3x_5 = 0$, $x_1 + 6x_2 + 2x_3 - 3x_5 = 3$, etc. The solution of these equations is

$$x_1 = -5/42, \qquad x_2 = 31/42, \qquad x_3 = -17/42, \qquad x_4 = 19/42,$$
$$x_5 = 1/6$$

These are the angular displacements in radians, corresponding to $F = (16/75) \times 10^6$ lb.

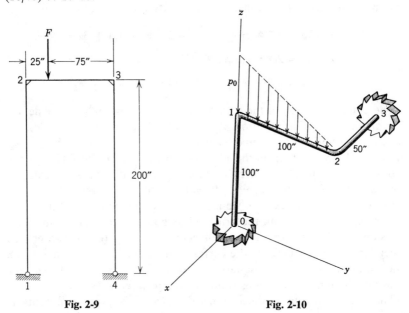

Fig. 2-9   Fig. 2-10

By Eq. (2-32), the moments applied to the joints by the members (positive clockwise) are $M_{12} = 0$, $M_{21} = -(6/7) \times 10^6$ lb-in., $M_{23} = (6/7) \times 10^6$ lb-in., $M_{32} = -(6/7) \times 10^6$ lb-in., $M_{34} = (6/7) \times 10^6$ lb-in., $M_{43} = 0$. By Eq. (2-33), the horizontal forces applied to the vertical members by the hinges are $S_1 = -(3/7) \times 10^4$ lb, $S_4 = (3/7) \times 10^4$ lb.

The five equations obtained from the respective rows of the preceding matrix may be interpreted as follows: first equation—no bending moment at joint 1. Second equation—net moment applied to joint 2 by the members is zero. Third equation—net moment applied to joint 3 by the members is zero. Fourth equation—no bending moment at joint 4. Fifth equation—sum of shears in the vertical members is zero. These interpretations follow immediately from Eqs. (2-32) and (2-33). In the slope-deflection method the preceding equilibrium conditions are used to set up the equations that determine the x's . Although the foregoing results apply only for $F = (16/75) \times 10^6$ lb, the values corresponding to any value of F may be obtained by direct proportioning.

Space Frame. The space frame shown in Fig. 2-10 consists of three members respectively parallel to the coordinate axes. The ends 0 and 3 are clamped. At joints 1 and 2 the members are welded together. A triangularly distributed load is applied to member 12. The cross sections are symmetrical, so that EI has a constant value for all axes through the centroid of a cross section. Also, EI has the same value for all members. The torsional stiffness of a member is $GJ = (3/4)EI$. This relation is typical for solid or hollow prismatic bars. Setting $EI/50 = C$, we obtain $K_{01} = K_{12} = C$, $K_{23} = 2C$, $(GJ/L)_{01} = (GJ/L)_{12} = 0.375C$, $(GJ/L)_{23} = 0.75C$.

The displacement vectors of joints 1 and 2 are denoted by $\mathbf{q}_1$ and $\mathbf{q}_2$. The (x, y, z) projections of these vectors are (q_{1x}, q_{1y}, q_{1z}) and (q_{2x}, q_{2y}, q_{2z}). Since the members are practically inextensional, $q_{1z} = 0$, $q_{2x} = 0$, and $q_{1y} = q_{2y}$. Because the rotations of joints 1 and 2 are small, they may be represented by vectors $\boldsymbol{\theta}_1$ and $\boldsymbol{\theta}_2$, in accordance with the right-hand-screw rule. The following generalized coordinates are introduced:

$$100x_1 = q_{1x}, \qquad 100x_2 = q_{1y} = q_{2y}, \qquad 100x_3 = q_{2z}$$

$$x_4 = \theta_{1x}, \qquad x_5 = \theta_{1y} \qquad x_6 = \theta_{1z}, \qquad x_7 = \theta_{2x}$$

$$x_8 = \theta_{2y}, \qquad x_9 = \theta_{2z}$$

The deflection of member 12 in the positive z-direction is denoted by $w(y)$. By Eq. (2-34),

$$w = x_4 y - 0.01(2x_4 + x_7 - 3x_3)y^2 + 0.0001(x_4 + x_7 - 2x_3)y^3 + \text{constant.}$$

The distributed load is $p = p_0(1 - 0.01y)$. The potential energy of the infinitesimal load $p\,dy$ is $pw\,dy$. Hence the potential energy of the entire external load is

$$\Omega = \int_0^{100} pw \, dy$$

With the preceding equations for p and w, this yields

$$\Omega = \frac{500 p_0}{3} (9x_3 + 3x_4 - 2x_7)$$

Since the components of generalized external force are determined by $P_i = -\partial\Omega/\partial x_i$, we obtain

$$P_3 = -1500 p_0, \qquad P_4 = -500 p_0, \qquad P_7 = \frac{1000}{3} p_0$$

The other P's are zero.

The strain energies of the three members are

$$U_{01} = C(x_5{}^2 - 3x_1 x_5 + 3x_1{}^2) + C(x_4{}^2 + 3x_2 x_4 + 3x_2{}^2) + 0.1875 C x_6{}^2$$

$$U_{12} = C[x_4{}^2 + x_4 x_7 + x_7{}^2 - 3x_3(x_4 + x_7) + 3x_3{}^2]$$
$$\quad + C[x_6{}^2 + x_6 x_9 + x_9{}^2 - 3x_1(x_6 + x_9) + 3x_1{}^2] + 0.1875 C(x_8 - x_5)^2$$

$$U_{23} = 2C(x_8{}^2 + 6x_3 x_8 + 12x_3{}^2) + 2C(x_9{}^2 - 6x_2 x_9 + 12x_2{}^2) + 0.375 C x_7{}^2$$

These expressions represent strain energies of bending in two different planes and strain energy of twisting [see Eqs. (2-17) and (2-35)].

By transferring C to the right side of the equations and representing U in the form of Eq. (2-36), we obtain the following matrix (a_{ij}), (Table 2-3) augmented by the column P_i/C:

TABLE 2-3

	x_1	x_2	x_3	x_4	x_5	x_6	x_7	x_8	x_9	P_i/C
x_1	12	0	0	0	−3	−3	0	0	−3	0
x_2	0	54	0	3	0	0	0	0	−12	0
x_3	0	0	54	−3	0	0	−3	12	0	$-1500p_0/C$
x_4	0	3	−3	4	0	0	1	0	0	$-500p_0/C$
x_5	−3	0	0	0	2.375	0	0	−0.375	0	0
x_6	−3	0	0	0	0	2.375	0	0	1	0
x_7	0	0	−3	1	0	0	2.75	0	0	$333.33p_0/C$
x_8	0	0	12	0	−0.375	0	0	4.375	0	0
x_9	−3	−12	0	0	0	1	0	0	6	0

The algebraic equations corresponding to this matrix are

$$12x_1 - 3x_5 - 3x_6 - 3x_9 = 0$$
$$54x_2 + 3x_4 - 12x_9 = 0$$
$$54x_3 - 3x_4 - 3x_7 + 12x_8 = -1500 p_0/C, \text{ etc.}$$

The solution of these equations may be obtained easily with a desk computer, since there are numerous zeros in the matrix. Using an elimination scheme (23), we obtain

$$x_1 = 68.97p_0/C, \qquad x_2 = 35.98p_0/C, \qquad x_3 = -98.42p_0/C$$
$$x_4 = -252.2p_0/C, \qquad x_5 = 131.5p_0/C, \qquad x_6 = 45.49p_0/C$$
$$x_7 = 105.6p_0/C, \qquad x_8 = 281.2p_0/C, \qquad x_9 = 98.96p_0/C$$

Having determined the displacements and rotations of the joints, we may obtain the bending and twisting moments in the members by means of Eq. (2-16) and the slope-deflection equations [Eq. (2-32)]. For member 12, the terms containing F in Eqs. (2-32) must be replaced by the fixed-end moments for a beam with a triangularly distributed load; namely, by $+ p_0L^2/20$ and $-p_0L^2/30$. Since the fixed-end moments are known for the concentrated load F, they may be determined for any distributed load by the superposition principle.

The examples in this section and in Sec. 2-4 are representative of all problems of small deflections of linearly elastic frames. The strain energy is always represented by a quadratic form,

$$U = \tfrac{1}{2}\sum\sum a_{ij}x_i x_j$$

where the x's denote displacements or rotations of the joints. Usually, the external forces are conservative. Then the components of generalized external force are $P_i = -\partial\Omega/\partial x_i$, where Ω is the potential energy of the external forces. The principle of stationary potential energy yields

$$\sum_{j=1}^{n} a_{ij}x_j = P_i$$

The foregoing theory supplies a practical method for constructing the matrices (a_{ij}) and (P_i). After the x's are determined by the last equations, the torques, bending moments, and tensions in the members may be computed directly.

PROBLEMS

1. For analytical purposes, a uniform beam of length L and stiffness EI is approximated by a chain of n equal rigid links with linear springs in the hinges. The spring constant for a hinge is k. The hinges are frictionless. Express k in terms of EI, L, and n, so that the strain energies due to constant bending moments M are equal for the beam and the linkage. How do the relative angular displacements of the ends compare for the two systems?

2. An airplane wing is regarded as a cantilever beam of length L. The stiffness of the wing is assumed to be given by $EI = a_0 - a_1 x^2$, where x is the distance of a cross section of the wing from the root and (a_0, a_1) are constants. The

air load per unit length is approximated by $p = b_0 - b_1 x$, where (b_0, b_1) are constants. The deflection is approximated by $y = Ax^2 + Bx^3$, where (A, B) are constants. Derive the expression for the total potential energy of the wing. Hence derive linear algebraic equations that determine the coefficients (A, B) by the principle of stationary potential energy.

3. Set $\sigma = a\xi + b\eta + c$, where ($a$, b, c) are constants, and derive Eq. (2-1) by the conditions of statics:

$$M_\eta = -\int \sigma\xi \, dA, \qquad M_\xi = \int \sigma\eta \, dA, \qquad N = \int \sigma \, dA$$

Generalize Eq. (2-1) for the case in which the origin is at the centroid of the cross section but $I_{\xi\eta} \neq 0$. Generalize Eq. (2-3) for this case.

4. Calculate the factor κ in Eq. (2-11) for an I-beam without fillets, having the following dimensions: flange width, 5 in., flange thickness, 1 in., over-all depth, 10 in., web thickness, 0.25 in.

5. A simple beam carries a linearly distributed load that varies from zero at one end to p_0 at the other end. Calculate the deflection function in the form of a sine series.

6. A simple beam carries a concentrated load at the center. If the beam is subjected to a tensile force equal to three times the Euler critical load, by what percentage is the deflection at the center reduced?

7. Set $P/P_e = r$, and derive a formula for R/F for the beam shown in Fig. P2-7.

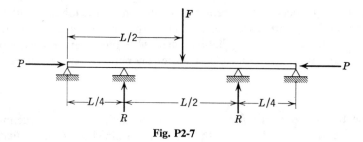

Fig. P2-7

8. A uniform elastic beam with clamped ends carries concentrated loads F at the points $x = L/4$, $x = L/2$, and $x = 3L/4$. Show that the boundary conditions and symmetry conditions are satisfied by $y = \Sigma a_n$ vers $2n\pi x/L$, where vers $\theta = 1 - \cos\theta$. Derive formulas for the coefficients a_n by the principle of stationary potential energy.

9. Show that if ν is an even number the formula for N_0 in Eq. (m), Sec. 2-3, is in agreement with the result obtained by statics from a free-body diagram of half the ring.

10. A uniform elastic ring is subjected to three equal radial concentrated loads P, 120° apart. By the theory of Sec. 2-3, determine the net tension and the bending moment at a point diametrically opposite to one of the loads.

11. Assuming that the shear q is proportional to $|\sin\theta|$ and that the ring is in

equilibrium, derive the coefficients in the series for the displacement components $(\bar{u}, \bar{v})$. Suppose that the reacting shear q acts at the centroidal axis of the ring (Fig. P2-11).

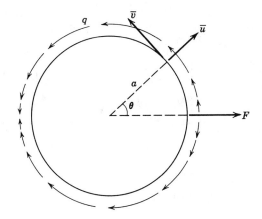

Fig. P2-11

12. Derive the formula for the section constant Z of a curved beam with an elliptical cross section. Show that Z is independent of the width of the cross section.

13. Let $\sigma = E\epsilon$, and write Eq. (c) of Sec. 2-3 in the form $\sigma = (\alpha + \beta z)/(a + z)$. Express the constants (α, β) in terms of the bending moment M and the net tension N (applied at the centroid) by means of the equations $N = \int \sigma \, dA$, $M = \int \sigma z \, dA$. Thus derive the Winkler formula which expresses σ in terms of M, N, z, a, A, Z.

14. The tension N in any bar of the semicircular frame is $N = k \sqrt{e}$, where k is a constant and e is the extension of the bar. The joints are hinged. By the principle of stationary potential energy, determine the deflection at the center and the tensions in the bars (Fig. P2-14).

15. The members of the regular hexagonal pin-jointed frame have equal cross-sectional areas. Select and define suitable generalized coordinates, making use of symmetry conditions. Determine the deformation (Fig. P2-15).

16. The rigid horizontal beam is hinged at the left end and is supported by three wires, as shown in Fig. P2-16. For each wire, $E = 10^7$ lb/in.^{2} and $A = 0.04$ in.^{2} By the principle of stationary potential energy, determine the angular deflection of the beam.

In Probs. 17 to 21 the numbers adjacent to the members denote the relative values of EA. In each case calculate the unknown tensions in the members by the principle of stationary potential energy. Check the results by the condition of statical equilibrium of forces and moments. Assume small deflections and linear elasticity.

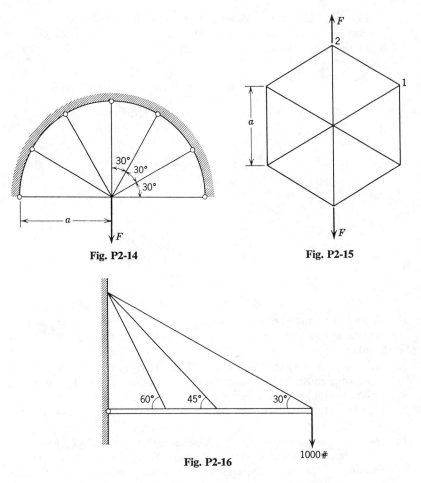

Fig. P2-14   Fig. P2-15

Fig. P2-16

17. Beam is rigid and uniform. It weighs 300 lb. When it is lowered into position in a horizontal attitude, the wires become taut simultaneously (Fig. P2-17).

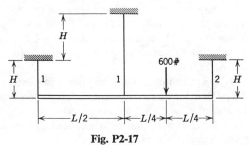

Fig. P2-17

18. Fig. P2-18.

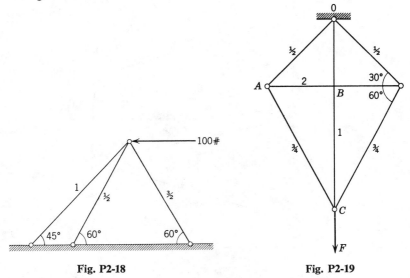

Fig. P2-18   Fig. P2-19

19. Fig. P2-19.

20. The three wires have the same *EA*. Other members of the frame are rigid (Fig. P2-20).

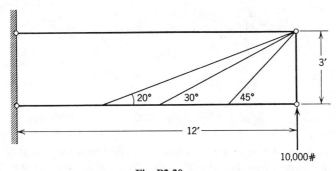

Fig. P2-20

21. Fig. P2-21.

22. The four wires have the same *EA*. If the slab is lowered into place in a horizontal attitude, the wires become taut simultaneously. The slab weighs 4000 lb, and a 2000-lb concentrated load is applied on the diagonal, 5 ft from the center, as indicated by Fig. P2-22. Determine the tensions in the wires. *Hint.* Let the deflection of the slab be $w = ax + by + c$ and adopt (a, b, c) as generalized coordinates.

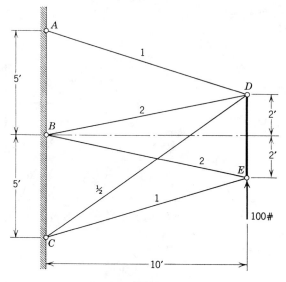

Fig. P2-21

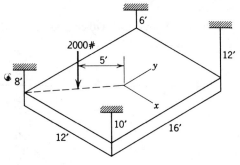

Fig. P2-22

23. The frame is a regular octahedron with cross braces on the center square. All members have the same EA. Compute the tensions in the members by the principle of stationary potential energy (Fig. P2-23).

24. Solve Prob. 20, assuming that the three wires have the same cross-sectional area and that the stress-strain relation of a wire is $\sigma = k\sqrt{\epsilon}$.

25. By means of the principle of stationary potential energy, derive the tensions in the members of the wall bracket, assuming that the displacement of the joint is small and that the stress-strain relation of the material is unknown. Show that the results agree with those obtained by balancing forces at the right-hand joint (Fig. P2-25).

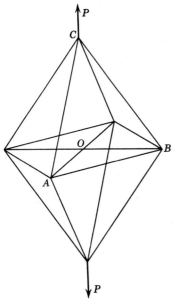

Fig. P2-23

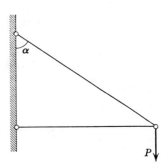

Fig. P2-25

26. Derive Eqs. (2-32) and (2-33).

27. Derive Eq. (2-34).

28. Derive Eq. (2-35).

29. The figure represents a plane elastic frame with rigid joints. The numbers adjacent to the members are the stiffness factors (lb-in.) multiplied by 10^{-7}. Set up the equations that determine the rotations of the joints and the displacements of the horizontal members by the method of stationary potential energy (Fig. P2-29).

30. Derive formulas for the rotations of the joints and the displacement of the top member of the frame shown in Fig. P2-30.

31. The three members of the frame are identical. By the principle of stationary potential energy, determine the horizontal displacement of the center member in terms of P, a, and EI (Fig. P2-31).

32. For the frame shown in Fig. P2-32, $E = 30 \times 10^6$ lb/in.^{2} and $I = 5$ in.^{4} (for the cross section of any member). Compute the rotations of joints 1 and 2 in degrees.

33. Each member of the lambda frame has length L, cross-sectional area A, modulus of elasticity E, and moment of inertia I. As a result of the external moment M applied at the vertex, the joint experiences displacement components (u, v) and a rotation θ. Taking the extensions of the members into account, derive formulas for (u, v) by the principle of stationary potential energy (Fig. P2-33).

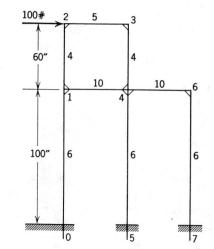

Fig. P2-29

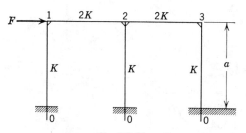

Fig. P2-30

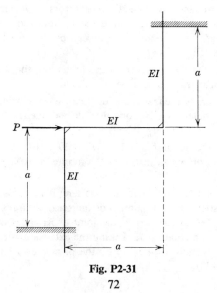

Fig. P2-31

72

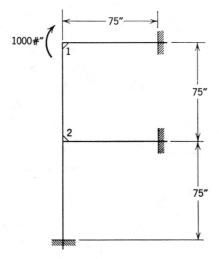

Fig. P2-32

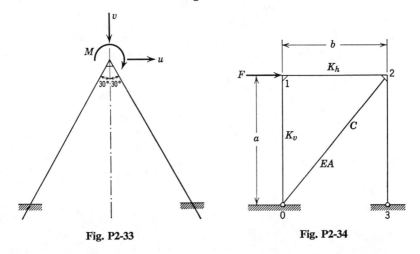

Fig. P2-33   Fig. P2-34

34. The bent is hinged at its feet. The diagonal tie has no flexural stiffness. Derive formulas for the rotation of each joint and for the displacement of the top member by the principle of stationary potential energy (Fig. P2-34).

35. All members of the frame have the same EI. The members are clamped at the feet. Calculate the horizontal reactions at the supports. (Fig. P2-35).

36. The stiffness factors of the vertical and horizontal members of the two-story bent are respectively K_v and K_h. The diagonal tie has cross-sectional area A and modulus of elasticity E; it has no flexural stiffness. Derive the equations that determine the translations and rotations of the joints by the principle of stationary potential energy (Fig. P2-36).

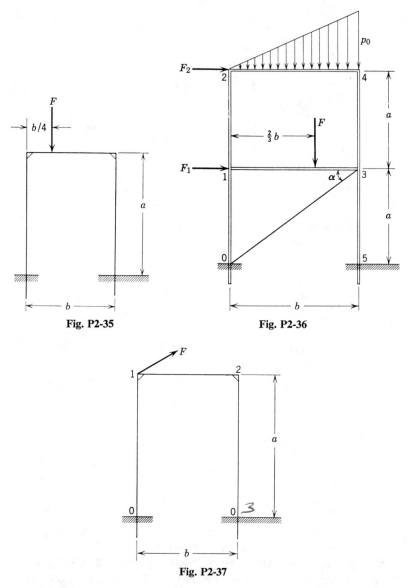

Fig. P2-35   Fig. P2-36

Fig. P2-37

37. The bent is loaded normal to its plane by a force F at one corner. Represent the rotations of the joints by vectors with components parallel to the members. The torsional stiffness of any member is GJ. The stiffness factors of the vertical and horizontal members are, respectively, K_v and K_h. By the principle of stationary potential energy, derive the equations that determine the rotations and translations of the joints (Fig. P2-37).

3 Methods of the calculus of variations

The order of historical events clearly shows the true position of the variational principle: It stands at the end of a long chain of reasoning as a satisfactory and beautiful condensation of the results.

MAX BORN

3-1. CANTILEVER BEAM. The calculus of variations is a mathematical tool by which the principles of stationary or minimum potential energy may be applied to systems with infinitely many degrees of freedom. Consider, for example, a uniform elastic cantilever beam with a load P at its end (Fig. 3-1). By Eq. (2-8) an approximate expression for the potential energy of the system is

$$V = -Py_1 + \tfrac{1}{2}EI \int_0^L (y'')^2 \, dx \qquad \text{(a)}$$

where y_1 is the deflection at the free end.

Certain restrictions are imposed on the deflection $y(x)$. In the first place, y and y' must be continuous functions in the interval $(0, L)$, since the beam may not be fractured. Furthermore, the second derivative y'' must exist, and it must be an integrable function, for otherwise Eq. (a) has no meaning. By elementary beam theory we see that all derivatives of y are continuous in the interval $(0, L)$. However, in more subtle problems of the calculus of variations the continuity properties of the solution are not known in advance, and this circumstance accounts for peculiar difficulties in the subject.

Besides the continuity conditions, the function $y(x)$ must satisfy the boundary conditions at the clamped end, $y(0) = y'(0) = 0$. These are called "forced boundary conditions," since they are mathematical expressions of the constraints. On the other hand, certain boundary conditions for the free end that will be deduced as necessary conditions for minimum potential energy are called "natural boundary conditions."

Let us designate as class Γ all functions $y(x)$ that satisfy the preceding continuity conditions and forced boundary conditions. Any function of

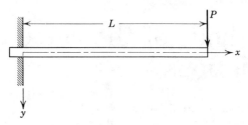

Fig. 3-1

class Γ is called an admissible deflection function. The principle of minimum potential energy signifies that if the beam is given any sufficiently small virtual displacement that belongs to class Γ the corresponding increment ΔV of potential energy is positive. In other words, the stable equilibrium configuration $y(x)$ provides a relative minimum to V.

For convenience, the virtual displacement is represented by $\epsilon \eta(x)$, where ϵ is an arbitrary infinitesimal and $\eta(x)$ is an arbitrary function of class Γ. Then $y(x) + \epsilon \eta(x)$ is also a function of class Γ. The potential energy of the beam, corresponding to the deflection $y + \epsilon \eta$, is

$$V + \Delta V = -P(y_1 + \epsilon \eta_1) + \tfrac{1}{2} EI \int_0^L (y'' + \epsilon \eta'')^2 \, dx$$

With Eq. (a), this yields

$$\Delta V = -\epsilon P \eta_1 + \epsilon EI \int_0^L y'' \eta'' \, dx + \tfrac{1}{2} EI \epsilon^2 \int_0^L (\eta'')^2 \, dx \qquad (b)$$

We write Eq. (b) in the form

$$\Delta V = \delta V + \tfrac{1}{2} \delta^2 V \qquad (c)$$

in which δV is the part of the expression for ΔV that is linear in ϵ and $\tfrac{1}{2} \delta^2 V$ is the part that is quadratic in ϵ. Accordingly,

$$\delta V = -\epsilon P \eta_1 + \epsilon EI \int_0^L y'' \eta'' \, dx \qquad (d)$$

$$\delta^2 V = \epsilon^2 EI \int_0^L (\eta'')^2 \, dx \qquad (e)$$

Now δV is proportional to ϵ and $\delta^2 V$ is proportional to ϵ^2. Accordingly, if δV is not zero, $2|\delta V| > |\delta^2 V|$ for sufficiently small values of ϵ. In view of Eq. (c), the term δV then controls the sign of ΔV. Since δV changes sign when ϵ changes sign, ΔV cannot be positive for all small values of ϵ unless δV vanishes for all admissible functions $\eta(x)$. Accordingly, identical vanishing of δV is necessary for V to have a relative minimum. In the present case this condition is also sufficient, since $\delta^2 V$ is obviously positive for any nonzero function η''.

By two integrations by parts, Eq. (d) is transformed to the following form:

$$\delta V = -\epsilon P \eta(L) + \epsilon EI y'' \eta' \Big|_0^L - \epsilon EI y''' \eta \Big|_0^L$$

$$+ \epsilon EI \int_0^L y'''' \eta \, dx$$

Since $\eta(0) = \eta'(0) = 0$, this equation reduces to

$$\delta V = -\epsilon [P + EI\, y'''(L)] \eta(L) + \epsilon EI\, y''(L)\, \eta'(L)$$

$$+ \epsilon EI \int_0^L y'''' \eta \, dx \tag{f}$$

In view of Eq. (f), the following conditions are necessary* and sufficient for δV to vanish for all admissible functions $\eta(x)$:

$$y''''(x) = 0 \tag{g}$$

$$y''(L) = 0, \qquad P + EI y'''(L) = 0 \tag{h}$$

The equations in (h) are the natural boundary conditions of the problem. They may be identified as the moment condition and the shear condition for the free end. Equation (g) is known as the Euler equation of the problem.

The general solution of Eq. (g) is $y = A_0 + A_1 x + A_2 x^2 + A_3 x^3$, in which the A's are constants. The forced boundary conditions yield $A_0 = A_1 = 0$. The natural boundary conditions yield

$$A_2 = \frac{PL}{2EI}, \qquad A_3 = \frac{-P}{6EI}$$

* The necessity for these conditions is proved in Sec. 3-2.

Hence

$$y = \frac{Px^2(3L - x)}{6EI} \qquad \text{(i)}$$

The equilibrium configuration represented by Eq. (i) is stable, since $\delta^2 V$ is always positive in this example.

3-2. EULER'S EQUATION. The preceding example suggests the following general problem of the calculus of variations:* To determine the function $y(x)$ that minimizes a definite integral,

$$V = \int_{x_0}^{x_1} F(x, y, y', y'', \cdots, y^{(n)}) \, dx \qquad \text{(3-1)}$$

in which F is a given function of $n + 2$ variables, and $y(x)$ is a function that satisfies certain continuity conditions and forced boundary conditions.

To refine this problem (6), we require that the function F be continuous with all its partial derivatives to order $n + 1$ for all real values of y, y', y'', $\cdots$, $y^{(n)}$ and for all values of x in the interval $x_0 \leq x \leq x_1$. The admissible functions $y(x)$ constitute a class Γ. Any function of class Γ is defined as single-valued and continuous with its derivatives to order n in the interval $x_0 \leq x \leq x_1$. In some cases these limitations are too restrictive. For example, the admissible values of y might be restricted to a certain range; then the properties of the function F for values outside of that range are irrelevant. Also, the curve $y = y(x)$ that minimizes V might have cusps (discontinuities in y'), or it might have a vertical tangent at either end ($y' = \infty$). Furthermore, V might be an improper integral; for example, F might become infinite for $x = x_0$. It is difficult to take all conditions into account. Although the discussion is limited to regular types of variational problems, the results, in some cases, can be extended to problems with certain irregularities.

Besides the continuity conditions, a function of class Γ may be required to satisfy certain linear forced boundary conditions. More precisely, the values of y, y', y'', $\cdots$ at the end points (x_0, x_1) may be required to satisfy several relations of the type

$$a_0 + a_1 y + a_2 y' + a_3 y'' + \cdots + a_n y^{(n-1)} = 0$$

in which $a_0, a_1, \cdots$ are constants. By requiring the forced boundary conditions to be linear, we ensure that $\epsilon \eta(x)$ is an admissible variation

---

* An admirably brief, comprehensive, and rigorous treatment of the topics in the calculus of variations that are important for physical applications is presented in the treatise by Courant and Hilbert (13).

whenever $\eta(x)$ is an admissible variation. The statement that $\epsilon\, \eta(x)$ is an admissible variation signifies that $y + \epsilon\eta$ is a function of class Γ, provided that y is a function of class Γ.

Any admissible function $y = f(x)$ defines a curve C in the range (x_0, x_1). Accordingly, Γ may be called the "class of admissible curves."

The values of the integral V for all admissible curves have a least upper bound A and a greatest lower bound B, provided that the values of V are bounded above and below (see footnote on p. 4). The problem is to find the curve (or curves) of class Γ, if any exists, for which $V = B$. This curve is said to provide an "absolute minimum" to V in class Γ. The case of a maximum ($V = A$) requires no special treatment, since the substitution of $-F$ for F converts it into the case of a minimum.

It is a striking fact that many problems of the calculus of variations that seemingly are formulated properly do not possess solutions. For example, consider the curve of class Γ that connects the point (0, 0) with the point (1, 1) and that has the least moment of inertia about the x-axis. Analytically, the problem is to find the curve of class Γ that minimizes the integral

$$\int_0^1 y^2 \sqrt{1 + (y')^2} \, dx$$

Without undertaking an analysis of the problem, we can perceive that it has no solution. Among continuous curves that which consists of segments of the two straight lines $y = 0$ and $x = 1$ provides an absolute minimum to the moment of inertia. However, this curve does not belong to class Γ, since it has no continuous tangent. Nevertheless, the value $\frac{1}{3}$, which is the moment of inertia of this curve about the x-axis, is evidently the greatest lower bound of the values of the moment of inertia for curves of class Γ. There is no curve belonging to class Γ that attains this lower bound. The problem has a solution in class Γ that is expressible by elliptic functions if other end points are chosen for the curve; for example, if the end points are $(\pm 1, 5)$.

Without verifying the existence of the solution of a variational problem, we may investigate necessary conditions for a minimum. Before considering the absolute minimum B, we must consider relative minima. The curve $C\!:\! y = y(x)$ is said to provide a "relative minimum" to V if there is a strip of width 2ρ (bounded by curves with ordinates $y + \rho$ and $y - \rho$), in which every other curve $\bar{C}$ of class Γ conforms to the inequality $\bar{V} \geq V$ (Fig. 3-2). In other words, if curve C furnishes a relative minimum to V among curves of class Γ, there is a positive number ρ such that $\bar{V} \geq V$ for all curves $\bar{C}$ of class Γ in the ρ-strip, $|\bar{y} - y| < \rho$. More specifically, these conditions are said to define an "improper relative minimum." If $\bar{V} > V$ (that is, if the case $\bar{V} = V$ is excluded), the minimum is said to be "proper."

Suppose that the curve $C:y = y(x)$ provides a relative minimum to V, and suppose that y'' is the derivative of highest order occurring in the integrand of Eq. (3-1). Let $\bar{C}$ be a comparison curve with ordinate $y + \epsilon\,\eta(x)$, in which $\eta(x)$ is any admissible variation and ϵ is an arbitrary

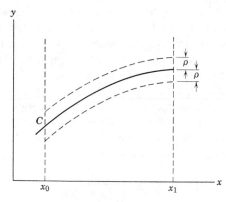

Fig. 3-2

constant. Then, if the minimum is proper, $V(\bar{C}) > V(C)$ for sufficiently small ϵ. Denoting $V(\bar{C})$ by $V + \Delta V$, we obtain by Eq. (3-1)

$$\Delta V = \int_{x_0}^{x_1} [F(x, y + \epsilon\eta, y' + \epsilon\eta', y'' + \epsilon\eta'') - F(x, y, y', y'')]\, dx$$

Expanding the integrand by Taylor's theorem, we obtain

$$\Delta V = \int_{x_0}^{x_1} \epsilon\left(\eta\frac{\partial}{\partial y} + \eta'\frac{\partial}{\partial y'} + \eta''\frac{\partial}{\partial y''}\right)F\, dx$$

$$+ \frac{1}{2!}\int_{x_0}^{x_1} \epsilon^2\left(\eta\frac{\partial}{\partial y} + \eta'\frac{\partial}{\partial y'} + \eta''\frac{\partial}{\partial y''}\right)^2 F\, dx + 0(\epsilon^3) \qquad \text{(a)}$$

Here the symbol $0(\epsilon^3)$ denotes a quantity that is less in absolute value than $K\epsilon^3$, where K is a positive constant that depends on η. Equation (a) is conventionally represented in the following form:

$$\Delta V = \delta V + \tfrac{1}{2}\delta^2 V + R_3 \qquad \text{(b)}$$

The terms δV and $\delta^2 V$ are called the first and second variations of V; they represent, respectively, the first and second integrals in Eq. (a).

If δV is not zero, it controls the sign of ΔV when ϵ is sufficiently small. Consequently, since the sign of δV is reversed by a reversal of the sign of ϵ, a necessary condition for ΔV to be positive for all small real values of ϵ is that δV vanish identically. Accordingly, this condition is necessary for V to possess a relative minimum. When δV vanishes, a sufficient condition for a relative minimum is that $\delta^2 V$ be positive definite; that is, $\delta^2 V$ shall be positive for all admissible nonzero variations $\eta(x)$.

Integration by parts transforms the first variation to the following form:

$$\delta V = \epsilon \left[\eta \left(\frac{\partial F}{\partial y'} - \frac{d}{dx} \frac{\partial F}{\partial y''} \right) \Big|_{x_0}^{x_1} + \eta' \frac{\partial F}{\partial y''} \Big|_{x_0}^{x_1} \right.$$
$$\left. + \int_{x_0}^{x_1} \left(\frac{\partial F}{\partial y} - \frac{d}{dx} \frac{\partial F}{\partial y'} + \frac{d^2}{dx^2} \frac{\partial F}{\partial y''} \right) \eta \, dx \right] \tag{3-2}$$

A necessary condition that δV vanish for all admissible variations $\eta(x)$ is that the terms outside the integral in Eq. (3-2) vanish independently; that is,

$$\eta \left(\frac{\partial F}{\partial y'} - \frac{d}{dx} \frac{\partial F}{\partial y''} \right) \Big|_{x_0}^{x_1} + \eta' \frac{\partial F}{\partial y''} \Big|_{x_0}^{x_1} = 0 \tag{3-3}$$

This condition follows from the fact that η may be chosen so that the integral vanishes irrespective of the values of η and η' at the end points (x_0, x_1). Equation (3-3) must be satisfied for all values of the constants $\eta(x_0)$, $\eta(x_1)$, $\eta'(x_0)$, $\eta'(x_1)$ that satisfy the forced boundary conditions. In general, these conditions impose certain restrictions on the function y at the end points. These restrictions are known as the "natural boundary conditions;" they are necessary conditions in order that V shall take a minimum value.

Equations (3-2) and (3-3) yield

$$\int_{x_0}^{x_1} \left(\frac{\partial F}{\partial y} - \frac{d}{dx} \frac{\partial F}{\partial y'} + \frac{d^2}{dx^2} \frac{\partial F}{\partial y''} \right) \eta \, dx = 0 \tag{3-4}$$

Since Eq. (3-4) is satisfied by all variations η that conform to class Γ, we may infer that the integrand is zero. To verify this conjecture, we formalize its statement as follows:

If $M(x)$ is a continuous function in the interval (x_0, x_1), and if

$$\int_{x_0}^{x_1} \eta M \, dx = 0$$

for all functions η that lie in class Γ, then M is identically zero. To prove this lemma, assume tentatively that M is not identically zero. For definiteness, say that M is positive at a point x' in the interval (x_0, x_1). Then, since M is continuous, there exists a subinterval (ξ_0, ξ_1) of the interval (x_0, x_1), containing the point x', in which $M > 0$. Choose η to be a function of the type shown in Fig. 3-3; that is, η is a continuous function with continuous first and second derivatives such that $\eta = 0$ outside of the interval (ξ_0, ξ_1) and η is positive in the interval (ξ_0, ξ_1). Then η conforms to class Γ, and $\int_{x_0}^{x_1} \eta M \, dx > 0$. Since this inequality is contrary to the hypothesis that the integral is zero, the lemma is proved.

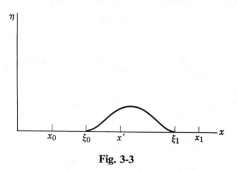

Fig. 3-3

Identifying M as the parenthetical expression in the integrand in Eq. (3-4), we obtain

$$\frac{\partial F}{\partial y} - \frac{d}{dx}\frac{\partial F}{\partial y'} + \frac{d^2}{dx^2}\frac{\partial F}{\partial y''} = 0 \qquad (3\text{-}5)$$

Equation (3-5) is known as the "Euler equation" of the variational problem.

An integral is said to be stationary if its first variation vanishes. Equations (3-3) and (3-5) are necessary and sufficient conditions for V to be stationary, but they are merely necessary for V to have a minimum value. Any solution of the Euler equation is called an "extremal." The curve or curves that provide relative minima to the integral V in class Γ are always included in the system of extremals of the integrand.

Equation (3-5) may be generalized without difficulty. If the function F contains derivatives of y to the third order, the Euler equation (13) is

$$\frac{\partial F}{\partial y} - \frac{d}{dx}\frac{\partial F}{\partial y'} + \frac{d^2}{dx^2}\frac{\partial F}{\partial y''} - \frac{d^3}{dx^3}\frac{\partial F}{\partial y'''} = 0 \qquad (3\text{-}6)$$

The extension to cases in which higher derivatives of y occur is obvious. The signs in the Euler equation alternate.

If the integrand F contains several unknown functions $y(x)$, $z(x)$, etc. a necessary condition for a stationary value of V is that the separate Euler equations for $y, z, \cdots$ be satisfied.

Weierstrass (6) has shown that if a function $y(x)$ with isolated discontinuities in its first or second derivatives provides a minimum value to V it must satisfy the Euler equation in any interval between successive discontinuities. He also derived supplementary conditions that must be satisfied at the points of discontinuity of the derivatives.

3-3. VARIATIONAL NOTATION. In the derivation of Euler's equation (Sec. 3-2), the function $y = f(x)$ was augmented by an infinitesimal function $\epsilon\,\eta(x)$. The curve $y = f(x) + \epsilon\,\eta(x)$ lies infinitesimally close to the curve $y = f(x)$. In practical applications of the calculus of variations the incremental function $\epsilon\,\eta(x)$ is often denoted by δy. Accordingly, the function $y = f(x)$ may receive two kinds of increments—the differential dy due to an increment dx and a variation δy that represents an infinitesimal function added to the function $y = f(x)$.

By definition, $\delta y' = \epsilon\,\eta'(x)$, $\delta y'' = \epsilon\,\eta''(x)$, etc. where primes denote derivatives. Hence $\delta y' = d(\delta y)/dx$, $\delta y'' = d^2(\delta y)/dx^2$, etc. In other words, the operators δ and d/dx are interchangeable.

If $F(x, y, y', y'', \cdots, y^{(n)})$ is a given function and if y is a function of x, the first variation of F is defined by

$$\delta F = \frac{\partial F}{\partial y}\,\delta y + \frac{\partial F}{\partial y'}\,\delta y' + \cdots + \frac{\partial F}{\partial y^{(n)}}\,\delta y^{(n)} \qquad (3\text{-}7)$$

The second variation of F is defined by $\delta^2 F = \delta(\delta F)$. More generally, the kth variation of F is defined by $\delta^k F = \delta(\delta^{k-1}F)$. Hence

$$\delta^k F = \left(\delta y\,\frac{\partial}{\partial y} + \delta y'\,\frac{\partial}{\partial y'} + \cdots + \delta y^{(n)}\,\frac{\partial}{\partial y^{(n)}}\right)^k F \qquad (3\text{-}8)$$

Equation (3-8) signifies that the differential operator is to be raised to the kth power as an algebraic quantity and then to be applied to the function F. By Taylor's theorem the increment of F corresponding to the variation δy is

$$\Delta F = \delta F + \frac{1}{2!}\,\delta^2 F + \frac{1}{3!}\,\delta^3 F + \cdots + \frac{1}{k!}\,\delta^k F + O(\epsilon^{k+1}) \qquad (3\text{-}9)$$

If $V = \displaystyle\int_a^b F\,dx$, where a and b are constants, the kth variation of V is defined by

$$\delta^k V = \int_a^b \delta^k F\,dx \qquad (3\text{-}10)$$

Hence by Eq. (3-9)

$$\Delta V = \delta V + \frac{1}{2!}\,\delta^2 V + \frac{1}{3!}\,\delta^3 V + \cdots + \frac{1}{k!}\,\delta^k V + O(\epsilon^{k+1})$$

3-4. SPECIAL FORMS OF THE EULER EQUATION. In the simplest variational problems derivatives higher than the first order do not occur in the integrand. Then Eq. (3-1) reduces to

$$V = \int_{x_0}^{x_1} F(x, y, y') \, dx \tag{3-11}$$

Certain simplifications result if any of the variables in Eq. (3-11) are missing.

Case 1: y' absent. If y' does not occur in the integrand F of Eq. (3-11). the Euler equation (3-6) reduces to

$$\frac{\partial F}{\partial y} = 0 \tag{3-12}$$

Since F is now a function of x and y only, Eq. (3-12) ordinarily yields only one extremal, $y = f(x)$. Equation (3-12) means that the value of y corresponding to any value of x provides a stationary value to $F(x, y)$. This condition is merely necessary for a minimum of V; a sufficient condition is that the value of y corresponding to each value of x provide a minimum to $F(x, y)$. An extremal with this property does not necessarily exist. Even if it does exist, it can satisfy the forced boundary conditions only by accident. Consequently, if y' is absent, the variational problem possesses a solution only under exceptional circumstances.

Case 2: y absent. If y does not occur in the integrand F, the Euler equation for Eq. (3-11) reduces to

$$\frac{d}{dx}\left(\frac{\partial F}{\partial y'}\right) = 0$$

Consequently,

$$\frac{\partial F}{\partial y'} = \text{constant} \tag{3-13}$$

Equation (3-13) is an ordinary differential equation of the first order. Its general solution contains an arbitrary constant in addition to the constant on the right side of Eq. (3-13). Consequently, there is a two-parameter family of extremals. If the variational problem is suitably formulated, this system contains at least one extremal that satisfies the boundary conditions. This extremal provides a stationary value to V, but, without further investigations, we have no assurance that the value of V thus obtained is a minimum.

A sufficient condition that the value of V be a minimum is that the value of y' corresponding to each value of x provide a minimum value to F. In turn, this condition requires that the constant on the right side of Eq. (3-13) be zero.

Case 3: x absent. If x does not occur explicitly in the integrand of Eq. (3-11), the Euler equation may be transformed as follows: differentiate the expression $F - y'(\partial F/\partial y')$ with respect to x, noting that F is a function of y and y' and that y is a function of x alone. Then,

$$\frac{d}{dx}\left(F - y'\frac{\partial F}{\partial y'}\right) = y'\left(\frac{\partial F}{\partial y} - \frac{d}{dx}\frac{\partial F}{\partial y'}\right) \tag{a}$$

In view of the Euler equation, the right side of Eq. (a) is zero. Consequently, Eq. (a) yields

$$F - y'\frac{\partial F}{\partial y'} = \text{constant} \tag{3-14}$$

Equation (3-14) is a special form of the Euler equation that applies when x does not occur explicitly in the integrand of Eq. (3-11).

3-5. THE DIFFERENTIAL EQUATION OF BEAMS. As in the case of finite degrees of freedom, we postulate that a stationary value of potential energy is necessary and sufficient for equilibrium of an unchecked holonomic conservative system with an infinite number of degrees of freedom. However, a necessary and sufficient condition for stable equilibrium of a holonomic conservative system is that the value of the potential energy be a proper relative minimum.

We defer consideration of the stability problem and concentrate on equilibrium criteria. Consider an elastic beam with any type of support and with a variable distributed load $p(x)$, in addition to a few concentrated loads $F_1, F_2, \cdots$. By Eq. (2-8), the total potential energy of the system is approximately

$$V = \int_0^L \left[\tfrac{1}{2}EI(y'')^2 - py\right] dx - F_1 y_1 - F_2 y_2 - \cdots$$

The desired deflection function has discontinuities in its third derivatives at the points of load concentration. Nevertheless, the integrand must satisfy the Euler equation in any interval between consecutive loads. If δV is evaluated by the method of Sec. 3-2, the concentrated loads F_i do not enter the integrand, and accordingly they do not affect the Euler equation, although they do enter into the natural boundary conditions for a segment of the beam.

Denoting the integrand by F, we obtain

$$\frac{\partial F}{\partial y} = -p, \qquad \frac{\partial F}{\partial y'} = 0, \qquad \frac{\partial F}{\partial y''} = EIy''$$

Accordingly, the Euler equation yields

$$(EIy'')'' = p \tag{3-15}$$

Equation (3-15) is the fundamental differential equation in the small-deflection theory of elastic beams. It applies for slightly tapered beams, if EI is a suitable function of x. It yields immediately the differential equation for freely vibrating beams. If a beam vibrates freely, its only load is the inertial force; that is, $p = -\rho\, \partial^2 y/\partial t^2$, where ρ is the mass per unit length. Consequently, the differential equation of a freely vibrating uniform beam is

$$EI\frac{\partial^4 y}{\partial x^4} + \rho\frac{\partial^2 y}{\partial t^2} = 0 \tag{3-16}$$

Equation (3-16) is modified for forced vibrations caused by a time-dependent distributed load $p(x, t)$, if the right side is set equal to p instead of 0.

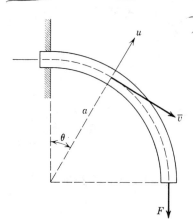

Fig. 3-4

3-6. CURVED CANTILEVER BEAM.

Figure 3-4 represents a curved cantilever beam that is loaded by a force F at the centroid of its end section. By Eq. (2-27), the total potential energy is

$$V = -F\bar{v}(\pi/2) + \frac{EA}{2a}\int_0^{\pi/2}[(u + \bar{v}')^2 + Z(u + u'')^2]\, d\theta \tag{a}$$

where primes denote derivatives with respect to θ.

If $(u, \bar{v})$ are given variations $(\delta u, \delta\bar{v})$, V takes an increment ΔV. By Eqs. (3-7) and (3-10), the part that is linear in $(\delta u, \delta\bar{v})$ is

$$\delta V = -F\,\delta\bar{v}\Big|^{\pi/2} + \frac{EA}{a}\int_0^{\pi/2}[(u + \bar{v}')(\delta u + \delta\bar{v}')$$

$$+ Z(u + u'')(\delta u + \delta u'')]\, d\theta \tag{b}$$

Integration by parts yields

$$\delta V = -F\, \delta\bar{v}\Big|^{\pi/2} - \frac{EAZ}{a}(u' + u''')\, \delta u\Big|_0^{\pi/2}$$

$$+ \frac{EAZ}{a}(u + u'')\, \delta u'\Big|_0^{\pi/2} + \frac{EA}{a}(u + \bar{v}')\, \delta\bar{v}\Big|_0^{\pi/2} + \frac{EA}{a} \tag{c}$$

$$\times \int_0^{\pi/2} \{[(u + \bar{v}') + Z(u'''' + 2u'' + u)]\, \delta u - (u' + \bar{v}')\, \delta\bar{v}\}\, d\theta = 0$$

The forced boundary conditions are

$$u = \bar{v} = u' = 0 \quad \text{for} \quad \theta = 0 \tag{d}$$

Hence, by Eq. (c), the natural boundary conditions are

$$u' + u''' = 0 \quad \text{for} \quad \theta = \frac{\pi}{2}$$

$$u + u'' = 0 \quad \text{for} \quad \theta = \frac{\pi}{2} \tag{e}$$

$$-F + \frac{EA}{a}(u + \bar{v}') = 0 \quad \text{for} \quad \theta = \frac{\pi}{2}$$

From the integral terms of Eq. (c), we obtain the Euler equations,

$$u + \bar{v}' + Z(u'''' + 2u'' + u) = 0 \tag{f}$$

$$u' + \bar{v}'' = 0 \tag{g}$$

Equation (g) yields

$$u + \bar{v}' = -KZ \tag{h}$$

where K is a constant. Hence Eq. (f) yields

$$u'''' + 2u'' + u = K \tag{i}$$

The general solutions of Eqs. (h) and (i) are

$$u = A_1 \sin\theta + A_2 \cos\theta + A_3\, \theta \sin\theta + A_4\, \theta \cos\theta + K \tag{j}$$

$$\begin{aligned} \bar{v} = -K\theta(1 + Z) + A_1 \cos\theta - A_2 \sin\theta - A_3(\sin\theta - \theta\cos\theta) \\ - A_4(\cos\theta + \theta\sin\theta) + A_5 \end{aligned} \tag{k}$$

The forced boundary conditions yield

$$A_2 + K = 0, \qquad A_1 - A_4 + A_5 = 0, \qquad A_1 + A_4 = 0 \tag{l}$$

The natural boundary conditions yield

$$K = \frac{-Fa}{EAZ}, \qquad -2A_4 + K = 0, \qquad A_3 = 0 \tag{m}$$

Hence

$$A_1 = -\tfrac{1}{2}K, \qquad A_2 = -K, \qquad A_3 = 0, \qquad A_4 = \tfrac{1}{2}K$$

$$A_5 = K, \qquad K = \frac{-Fa}{EAZ} \tag{n}$$

Equations (j) and (k) now yield

$$u = \frac{Fa}{2EAZ}(\sin\theta + 2\cos\theta - \theta\cos\theta - 2) \tag{o}$$

$$\bar{v} = \frac{Fa}{2EAZ}[2(1 + Z)\theta + 2\cos\theta - 2\sin\theta + \theta\sin\theta - 2] \tag{p}$$

Since $\sigma = E\epsilon$, Eq. (c) of Sec. 2-3 now yields

$$\sigma = \frac{F}{A}\left[1 + \frac{z(1 - \sin\theta)}{Z(a + z)}\right] \tag{q}$$

The bending moment is

$$M = Fa(1 - \sin\theta) \tag{r}$$

Consequently, Eq. (q) yields

$$\sigma = \frac{F}{A} + \frac{Mz}{AZa(a + z)} \tag{s}$$

The net tension and the bending moment are given by the equations,

$$N = \int \sigma\, dA, \qquad M = \int z\sigma\, dA \tag{t}$$

Hence, by Eqs. (q) and (2-22),

$$N = F\sin\theta, \qquad M = Fa(1 - \sin\theta) \tag{u}$$

These equations agree with elementary statics.

If we approximate Z by Eq. (2-23), we obtain by Eq. (s)

$$\sigma = \frac{F}{A} + \frac{a}{a + z}\frac{Mz}{I} \tag{v}$$

If the factor $a/(a + z)$ is approximated by 1, Eq. (v) reduces to the elementary formula for straight beams:

$$\sigma = \frac{F}{A} + \frac{Mz}{I} \tag{w}$$

Equations (t) and (w) yield $N = F$. This relation disagrees grossly with statics. Accordingly, it is not permissible to discard z from the denominator in Eq. (v) if the strains are used to calculate the net tension in the beam. This is true because the strains caused by the net tension and the bending moment often have quite different orders of magnitude. Relatively small inaccuracies in one of these strain components may consequently obliterate the other one entirely. However, the linearizing of the expressions in z ordinarily causes only small percentages of error in the stresses if the depth of the cross section is small compared to the radius a. Consequently, complete linearization with respect to z may be acceptable if the stresses are desired only for the prediction of failure of the material. The foregoing considerations apply also to curved shells, and they suggest why the theory of shells has proved to be very sensitive to approximations.

3-7. ISOPERIMETRIC PROBLEMS. Consider two integrals,

$$V = \int_{x_0}^{x_1} F(x, y, y', y'') \, dx, \qquad W = \int_{x_0}^{x_1} G(x, y, y', y'') \, dx \qquad \text{(a)}$$

in which F and G are given functions with continuous partial derivatives to the third order and y is an unknown function of class Γ (Sec. 3-2). A subset Γ' of class Γ is defined by the supplementary condition $W = C$, in which C is a given constant. Conceivably, there may exist a function y that provides a stationary value to V among functions of class Γ'. The determination of a function y with this property is the isoperimetric problem of the calculus of variations. Literally, the word "isoperimetric" means "constant perimeter." Its use in the calculus of variations stems from the fact that the earliest isoperimetric problem dealt with the form of a curve of given length that encloses maximum area.

Isoperimetric problems may be treated conveniently by the Lagrange-multiplier method. We multiply the integral W by an unspecified constant λ and add the result to V. Thus we obtain a modified integral

$$\bar{V} = \int_{x_0}^{x_1} (F + \lambda G) \, dx \qquad \text{(b)}$$

Suppose that a function $y = f(x, \lambda)$ provides a stationary value to $\bar{V}$ in class Γ. By proper choice of λ, we may be able to place the function $f(x, \lambda)$ in class Γ'. Then, since $\bar{V}$ is stationary in class Γ, it is also stationary in class Γ', but, in class Γ', W is constant. Therefore, V is stationary in class Γ'.

The Catenary. The catenary provides a simple illustration of an isoperimetric problem. Consider an inextensional flexible uniform string of length L that has given end points (x_0, y_0) and (x_1, y_1). Since the differential arc length ds is equal to $\sqrt{1 + (y')^2}\, dx$, the potential energy of the string in the gravitational field of the earth is proportional to the integral

$$V = \int_{x_0}^{x_1} y\sqrt{1 + (y')^2}\, dx \tag{c}$$

The forced boundary conditions are that the end points of the required curve shall be the points (x_0, y_0) and (x_1, y_1). Furthermore, any admissible curve must conform to the relation

$$L = \int_{x_0}^{x_1} \sqrt{1 + (y')^2}\, dx \tag{d}$$

Equation (d) is an isoperimetric condition.

Multiplying the integral in Eq. (d) by a Lagrange multiplier λ and adding the result to Eq. (c), we obtain a modified potential energy integral,

$$\bar{V} = \int_{x_0}^{x_1} (y + \lambda)\sqrt{1 + (y')^2}\, dx \tag{e}$$

In the present case the integrand F does not contain x explicitly. Accordingly, the Euler equation takes the form of Eq. (3-14). Consequently,

$$y + \lambda = K\sqrt{1 + (y')^2} \tag{f}$$

where K is a constant. This equation may be written as follows:

$$\frac{K\, dy}{\sqrt{(y + \lambda)^2 - K^2}} = dx$$

Integration yields

$$y = K \cosh\left(\frac{x}{K} + C\right) - \lambda \tag{g}$$

where C is a constant of integration. Equation (g) is the general equation of the catenary.

The three constants (K, C, λ) must be chosen to satisfy the end conditions $y(x_0) = y_0$, $y(x_1) = y_1$, and the isoperimetric condition [Eq. (d)]. The last condition takes a comparatively simple form, since

$$L = \int_{x_0}^{x_1} \sqrt{1 + (y')^2}\, dx = K \sinh\left(\frac{x}{K} + C\right)\Big|_{x_0}^{x_1} \tag{h}$$

3-8. AUXILIARY DIFFERENTIAL EQUATIONS. Let a class Γ of functions and an integral V be defined as in Sec. 3-2. Let Γ' be the subset of class Γ that consists of solutions of the differential equation

$$G(x, y, y', y'') = 0 \tag{a}$$

where G is a given function. Suppose that we wish to find a function $y(x)$ (if any exists) that renders the integral V stationary among functions of class Γ'.

There are two ways to approach this problem. First, we may be able to find the general solution of Eq. (a), either in a finite form or in an infinite form, such as a series. Since Eq. (a) is of second order, its general solution ordinarily contains two arbitrary constants, C_1 and C_2. These constants may be subjected to certain restrictions to ensure that y lies in class Γ; for example, the forced boundary conditions may restrict C_1 and C_2. By definition, any solution of Eq. (a) that conforms to class Γ also conforms to class Γ'.

Having determined the general solution of Eq. (a) and having adapted the constants (C_1, C_2) to class Γ, we may evaluate the integral V by Eq. (3-1). Thus V becomes a function of C_1 and C_2. The condition that V be stationary in class Γ' is then satisfied if $\delta V = 0$ for all variations of C_1 and C_2 that are consistent with class Γ.

It may happen that the forced boundary conditions fix C_1 and C_2 so that variations of V are impossible. The foregoing case is consequently too specialized to have much importance. More significance attaches to the case of two unknown functions, $y(x)$ and $z(x)$. Then

$$V = \int_{x_0}^{x_1} F(x, y, z, y', z', y'', z'') \, dx \tag{b}$$

Class Γ now consists of pairs of functions $y(x)$, $z(x)$ that satisfy continuity requirements and linear forced boundary conditions. Let Γ' be the subset of class Γ that consists of solutions of the differential equation

$$G(x, y, z, y', z', y'', z'') = 0 \tag{c}$$

where G is a given function. Suppose that we wish to find functions y, z (if any exist) that render the integral V stationary among functions of class Γ'.

We may follow the same scheme as before if we can find the general solution of Eq. (c). Ordinarily, this solution is represented by expressions for y and z in terms of an arbitrary function $\phi(x)$ and its derivatives. Any regular function ϕ that conforms to the forced boundary conditions provides functions y, z that belong to class Γ'. The integral V may be expressed in terms of ϕ. By giving ϕ variations that conform to the forced boundary conditions, we may derive the natural boundary conditions and the Euler equation for ϕ. Any solution of the Euler equation for ϕ that satisfies the forced boundary conditions and the natural boundary conditions provides a stationary value to V in class Γ'.

The Lagrange multiplier provides a way of treating auxiliary differential equations in variational problems without the use of the general solutions of the differential equations at the outset. Referring to Eqs. (b) and (c), consider the modified integral

$$\bar{V} = \int_{x_0}^{x_1} (F + \lambda G)\, dx \qquad\qquad (d)$$

where the Lagrange multiplier λ is an unspecified function of x. Without designating the function λ, we may give the functions y, z arbitrary variations that conform to the forced boundary conditions. Thus we derive the Euler equations and the natural boundary conditions for the integral $\bar{V}$. We may then seek functions $y(x)$, $z(x)$, $\lambda(x)$ that satisfy the Euler equations, the auxiliary differential equation (c), the forced boundary conditions, and the natural boundary conditions. These functions provide a stationary value to $\bar{V}$ among functions of class Γ', but, in class Γ', $\bar{V}$ reduces to V, since $G = 0$. Therefore, the functions y, z provide a stationary value to V in class Γ'. Since, in the end, we do not require the function λ, we may sometimes dispense with its derivation if we can supply a proof that it exists. The Lagrange multiplier method does not really evade difficulties that are inherent in the auxiliary differential equation but it sometimes leads to relations that have interesting or useful interpretations.

Both of the preceding methods may be extended to problems of multiple integrals and auxiliary partial differential equations. Care must be used to include the natural boundary conditions, since one may encounter puzzling situations if he fails to heed these conditions.

3-9. FIRST VARIATION OF A DOUBLE INTEGRAL. In general, the potential energy of a shell, plate, or membrane is represented by a double integral. The principle of minimum potential energy is particularly useful for treating these systems.

Accordingly, variational problems with two independent variables (x, y) are important in mechanics. Irrespective of their physical significance, these variables may be regarded as rectangular coordinates in a plane. Then any function $w(x, y)$ is represented by a surface in a space with rectangular coordinates (x, y, w).

Consider a given finite region R in the (x, y) plane. The region R may be multiply connected; for example, it may be a ring-shaped region, or, more generally, it may be any irregular area with holes in it. The region R will be closed; that is, it shall contain all its boundary points.

A class Γ of admissible functions $w(x, y)$ is defined as follows:

(a) Any function of class Γ is continuous with its partial derivatives to the fourth order in the region R.

(b) Any function of class Γ satisfies given linear forced boundary conditions of the type $aw + b\,\partial w/\partial n = c$, in which a, b, c are given point functions on the boundary of the region R and $\partial w/\partial n$ denotes the normal derivative of w at the boundary.

Consider the integral,

$$V = \iint_R F(x, y, w, w_x, w_y, w_{xx}, w_{xy}, w_{yy}) \, dx \, dy \qquad (a)$$

in which F is a given function of eight variables, with continuous partial derivatives to the third order for all real values of w, w_x, w_y, w_{xx}, w_{xy}, w_{yy}, and for all values of (x, y) in the region R. Subscripts x and y denote partial derivatives. Let the function w receive a variation $\epsilon\eta(x, y)$, where ϵ is an arbitrary constant and η is an arbitrary function that conforms to class Γ. Then V takes an increment ΔV, which is determined as follows:

$$\Delta V = \iint_R [F(x, y, w + \epsilon\eta, w_x + \epsilon\eta_x, w_y + \epsilon\eta_y, w_{xx} + \epsilon\eta_{xx},$$

$$w_{xy} + \epsilon\eta_{xy}, w_{yy} + \epsilon\eta_{yy}) - F(x, y, w, w_x, w_y, w_{xx}, w_{xy}, w_{yy})] \, dx \, dy \qquad (b)$$

Expanding the function F to first powers of ϵ by means of Taylor's theorem, we obtain

$$\delta V = \epsilon \iint_R \left(\eta \frac{\partial F}{\partial w} + \eta_x \frac{\partial F}{\partial w_x} + \eta_y \frac{\partial F}{\partial w_y} + \eta_{xx} \frac{\partial F}{\partial w_{xx}} \right.$$

$$\left. + \eta_{xy} \frac{\partial F}{\partial w_{xy}} + \eta_{yy} \frac{\partial F}{\partial w_{yy}} \right) dx \, dy \qquad (c)$$

in which δV (the first variation of V) is the part of ΔV that is linear in ϵ.

To transform Eq. (c) by integration by parts, we employ Green's theorem (39). It asserts that the following relation is an identity for any two functions $u(x, y)$ and $v(x, y)$ that are continuous with their first derivatives in the region R:

$$\iint_R \left(\frac{\partial v}{\partial x} - \frac{\partial u}{\partial y} \right) dx \, dy = \oint_C (u \, dx + v \, dy) \qquad (3\text{-}17)$$

Here, C denotes the bounding curve of the region R. The integration on the outer boundary is to be performed in the counterclockwise sense if the coordinates (x, y) are right-handed. If the region R is multiply connected, the integration on the inner boundaries is to be performed in the clockwise sense. Equation (3-17) may be regarded as a special case of Stokes's theorem.

First set $v = \phi\psi$ and $u = 0$; second set, $v = 0$ and $u = \phi\psi$. Thus the following two integration-by-parts formulas for double integrals are obtained from Eq. (3-17):

$$\iint_R \phi\, \frac{\partial \psi}{\partial x}\, dx\, dy = -\iint_R \psi\, \frac{\partial \phi}{\partial x}\, dx\, dy + \oint_C \phi\psi\, dy$$

$$\tag{3-18}$$

$$\iint_R \phi\, \frac{\partial \psi}{\partial y}\, dx\, dy = -\iint_R \psi\, \frac{\partial \phi}{\partial y}\, dx\, dy - \oint_C \phi\psi\, dx$$

Equation (3-18) yields

$$\iint_R \eta_x\, \frac{\partial F}{\partial w_x}\, dx\, dy = -\iint_R \eta\, \frac{\partial}{\partial x}\left(\frac{\partial F}{\partial w_x}\right) dx\, dy + \oint_C \eta\, \frac{\partial F}{\partial w_x}\, dy$$

$$\iint_R \eta_y\, \frac{\partial F}{\partial w_y}\, dx\, dy = -\iint_R \eta\, \frac{\partial}{\partial y}\left(\frac{\partial F}{\partial w_y}\right) dx\, dy - \oint_C \eta\, \frac{\partial F}{\partial w_y}\, dx$$

Applying Eq. (3-18) twice, we obtain

$$\iint_R \eta_{xx}\, \frac{\partial F}{\partial w_{xx}}\, dx\, dy = \iint_R \eta\, \frac{\partial^2}{\partial x^2}\left(\frac{\partial F}{\partial w_{xx}}\right) dx\, dy + \oint_C \eta_x\, \frac{\partial F}{\partial w_{xx}}\, dy$$

$$- \oint_C \eta\, \frac{\partial}{\partial x}\left(\frac{\partial F}{\partial w_{xx}}\right) dy$$

$$\iint_R \eta_{yy}\, \frac{\partial F}{\partial w_{yy}}\, dx\, dy = \iint_R \eta\, \frac{\partial^2}{\partial y^2}\left(\frac{\partial F}{\partial w_{yy}}\right) dx\, dy - \oint_C \eta_y\, \frac{\partial F}{\partial w_{yy}}\, dx$$

$$+ \oint_C \eta\, \frac{\partial}{\partial y}\left(\frac{\partial F}{\partial w_{yy}}\right) dx$$

Depending on the order in which the two integrations by parts are performed, we may express the integral of the term containing η_{xy} in either of the following two forms:

$$\iint_R \eta_{xy} \frac{\partial F}{\partial w_{xy}}\, dx\, dy = \iint_R \eta \frac{\partial^2}{\partial x\, \partial y}\left(\frac{\partial F}{\partial w_{xy}}\right) dx\, dy + \oint_C \eta \frac{\partial}{\partial x}\left(\frac{\partial F}{\partial w_{xy}}\right) dx$$

$$+ \oint_C \eta_y \frac{\partial F}{\partial w_{xy}}\, dy$$

or

$$\iint_R \eta_{xy} \frac{\partial F}{\partial w_{xy}}\, dx\, dy = \iint_R \eta \frac{\partial^2}{\partial x\, \partial y}\left(\frac{\partial F}{\partial w_{xy}}\right) dx\, dy - \oint_C \eta \frac{\partial}{\partial y}\left(\frac{\partial F}{\partial w_{xy}}\right) dy$$

$$- \oint_C \eta_x \frac{\partial F}{\partial w_{xy}}\, dx$$

Hence by Eq. (c)

$$\delta V = \epsilon \iint_R \left[\frac{\partial F}{\partial w} - \frac{\partial}{\partial x}\left(\frac{\partial F}{\partial w_x}\right) - \frac{\partial}{\partial y}\left(\frac{\partial F}{\partial w_y}\right) + \frac{\partial^2}{\partial x^2}\left(\frac{\partial F}{\partial w_{xx}}\right) \right.$$

$$\left. + \frac{\partial^2}{\partial x\, \partial y}\left(\frac{\partial F}{\partial w_{xy}}\right) + \frac{\partial^2}{\partial y^2}\left(\frac{\partial F}{\partial w_{yy}}\right) \right] \eta\, dx\, dy$$

$$+ \epsilon \oint_C \left[\frac{\partial F}{\partial w_x} - \frac{\partial}{\partial x}\left(\frac{\partial F}{\partial w_{xx}}\right) \right] \eta'\, dy - \epsilon \oint_C \left[\frac{\partial F}{\partial w_y} - \frac{\partial}{\partial y}\left(\frac{\partial F}{\partial w_{yy}}\right) \right] \eta\, dx$$

$$\text{(3-19)}$$

$$+ \epsilon \oint_C \eta_x \frac{\partial F}{\partial w_{xx}}\, dy - \epsilon \oint_C \eta_y \frac{\partial F}{\partial w_{yy}}\, dx$$

$$+ \left\{ \begin{array}{l} \epsilon \oint_C \eta \frac{\partial}{\partial x}\left(\frac{\partial F}{\partial w_{xy}}\right) dx + \epsilon \oint_C \eta_y \frac{\partial F}{\partial w_{xy}}\, dy \\[2ex] -\epsilon \oint_C \eta \frac{\partial}{\partial y}\left(\frac{\partial F}{\partial w_{xy}}\right) dy - \epsilon \oint_C \eta_x \frac{\partial F}{\partial w_{xy}}\, dx \end{array} \right.$$

To obtain the natural boundary conditions, we must transform the line integrals in Eq. (3-19) to the form

$$\epsilon \oint_C \eta(\cdots)\, ds + \epsilon \oint_C \eta_n(\cdots)\, ds$$

where s is arc length on the curve C, and η_n denotes the normal derivative of η on curve C. The natural boundary conditions are obtained from the vanishing of the integrands of these two integrals. Rather than to treat the general case, we consider the case in which the curve C is formed from segments of coordinate lines. This is not a severe restriction, since, in physical problems, we usually seek coordinates such that the physical boundaries coincide with coordinate lines. When such coordinates are chosen, Eq. (3-19) is in a suitable form for the determination of the natural boundary conditions. On a part of the boundary represented by $x = $ constant, $dx = 0$, and those line integrals in Eq. (3-19) that contain dx vanish. In this case it is convenient to use the second form for the last two integrals in Eq. (3-19). Likewise, on a part of the boundary represented by $y = $ constant, $dy = 0$, and those line integrals in Eq. (3-19) that contain dy vanish. In this case it is convenient to use the first form for the last two integrals in Eq. (3-19). The double integral must vanish independently of the line integrals, since δV vanishes for all admissible functions $\eta(x, y)$. This condition yields the Euler differential equation. Hence the Euler equation is

$$\frac{\partial F}{\partial w} - \frac{\partial}{\partial x}\left(\frac{\partial F}{\partial w_x}\right) - \frac{\partial}{\partial y}\left(\frac{\partial F}{\partial w_y}\right) + \frac{\partial^2}{\partial x^2}\left(\frac{\partial F}{\partial w_{xx}}\right)$$
$$+ \frac{\partial^2}{\partial x\,\partial y}\left(\frac{\partial F}{\partial w_{xy}}\right) + \frac{\partial^2}{\partial y^2}\left(\frac{\partial F}{\partial w_{yy}}\right) = 0 \tag{3-20}$$

The natural boundary conditions follow immediately from the relations

$$\eta\left[\frac{\partial F}{\partial w_x} - \frac{\partial}{\partial x}\left(\frac{\partial F}{\partial w_{xx}}\right) - \frac{\partial}{\partial y}\left(\frac{\partial F}{\partial w_{xy}}\right)\right] = 0 \quad \text{on edge} \quad x = \text{constant}$$

$$\eta_x\frac{\partial F}{\partial w_{xx}} = 0 \quad \text{on edge} \quad x = \text{constant}$$

$$\eta\left[\frac{\partial F}{\partial w_y} - \frac{\partial}{\partial x}\left(\frac{\partial F}{\partial w_{xy}}\right) - \frac{\partial}{\partial y}\left(\frac{\partial F}{\partial w_{yy}}\right)\right] = 0 \quad \text{on edge} \quad y = \text{constant}$$

$$\eta_y\frac{\partial F}{\partial w_{yy}} = 0 \quad \text{on edge} \quad y = \text{constant}$$

$$\tag{3-21}$$

If the values of η, η_x, and η_y are unrestricted by forced boundary conditions, the bracketed expressions in Eq. (3-21) must vanish on the indicated parts of the boundary. Thus two natural boundary conditions are obtained for each segment of the boundary. However, it may happen that the forced boundary conditions require that η, η_x, or η_y vanish on certain segments of the boundary; then the corresponding natural boundary conditions are eliminated.

3-10. FIRST VARIATION OF A TRIPLE INTEGRAL. Problems of equilibrium of elastic solids and problems of dynamics of plates, shells, and membranes lead to variations of triple integrals. The independent variables (x, y, z) may be regarded as rectangular coordinates.

A class Γ of admissible functions $w(x, y, z)$ is defined as follows: (a) any function of class Γ is continuous with its partial derivatives to the fourth order in a given finite closed region R of (x, y, z) space; (b) any function of class Γ satisfies given linear forced boundary conditions of the type $aw + b\, \partial w/\partial n = c$, in which (a, b, c) are given point functions on the boundary of region R and $\partial w/\partial n$ denotes the normal derivative of w at the boundary.

Consider the integral

$$V = \iiint\limits_{R} F(x, y, z, w, w_x, w_y, w_z, w_{xx}, w_{yy}, w_{zz}, w_{yz}, w_{zx}, w_{xy})\, dx\, dy\, dz \qquad \text{(a)}$$

in which F is a given function of 13 variables with continuous partial derivatives to the third order for all real values of w, w_x, w_y, w_z, w_{xx}, w_{yy}, w_{zz}, w_{yz}, w_{zx}, w_{xy} and for all values of (x, y, z) in region R.

Let the function w receive a variation $\epsilon\eta(x, y, z)$, where ϵ is an arbitrary constant and η is an arbitrary function such that $w + \epsilon\eta$ lies in class Γ. Then, by Taylor's theorem,

$$\begin{aligned}
\delta V = \epsilon \iiint\limits_{R} \Bigg(& \eta\, \frac{\partial F}{\partial w} + \eta_x\, \frac{\partial F}{\partial w_x} + \eta_y\, \frac{\partial F}{\partial w_y} + \eta_z\, \frac{\partial F}{\partial w_z} \\
& + \eta_{xx}\, \frac{\partial F}{\partial w_{xx}} + \eta_{yy}\, \frac{\partial F}{\partial w_{yy}} + \eta_{zz}\, \frac{\partial F}{\partial w_{zz}} + \eta_{yz}\, \frac{\partial F}{\partial w_{yz}} \\
& + \eta_{zx}\, \frac{\partial F}{\partial w_{zx}} + \eta_{xy}\, \frac{\partial F}{\partial w_{xy}} \Bigg)\, dx\, dy\, dz
\end{aligned} \qquad \text{(b)}$$

in which δV is the part of ΔV that is linear in ϵ.

To transform Eq. (b) by integration by parts, we employ the divergence theorem (39). The procedure is essentially the same as for a double integral. The following Euler equation is obtained:

$$\begin{aligned}
& \frac{\partial F}{\partial w} - \frac{\partial}{\partial x}\left(\frac{\partial F}{\partial w_x}\right) - \frac{\partial}{\partial y}\left(\frac{\partial F}{\partial w_y}\right) - \frac{\partial}{\partial z}\left(\frac{\partial F}{\partial w_z}\right) + \frac{\partial^2}{\partial x^2}\left(\frac{\partial F}{\partial w_{xx}}\right) \\
& + \frac{\partial^2}{\partial y^2}\left(\frac{\partial F}{\partial w_{yy}}\right) + \frac{\partial^2}{\partial z^2}\left(\frac{\partial F}{\partial w_{zz}}\right) + \frac{\partial^2}{\partial y\, \partial z}\left(\frac{\partial F}{\partial w_{yz}}\right) + \frac{\partial^2}{\partial z\, \partial x}\left(\frac{\partial F}{\partial w_{zx}}\right) \\
& + \frac{\partial^2}{\partial x\, \partial y}\left(\frac{\partial F}{\partial w_{xy}}\right) = 0 \qquad \text{(3-22)}
\end{aligned}$$

If the integrand contains several unknown functions (u, v, w), each of these functions must satisfy Eq. (3-22). To derive natural boundary conditions, we must examine the surface integrals that are obtained when Eq. (b) is transformed by integration by parts.

3-11. THE RAYLEIGH-RITZ METHOD. A mechanical system with infinitely many degrees of freedom may be reduced to a system with finite degrees of freedom by means of assumptions about the nature of the deformation. This idea was seemingly employed first by Lord Rayleigh in studies of vibrations (71). It has been used extensively by Timoshenko (84) to treat buckling problems. Rayleigh's method has often been criticized because it provides no information about the accuracy of the approximation. However, all analyses of physical systems are based on conceptual models which are believed to reproduce approximately the essential features of the phenomena under consideration. The representation of a system with infinitely many degrees of freedom by a conceptual model with finite degrees of freedom is not a greater digression from the facts than many other ideas that are used in the establishment of such models. Analyses of deformation, buckling, and vibrations of complicated structures would be almost impossible if the systems were not reduced to finite degrees of freedom, particularly, if nonlinear relations are important.

As an illustration of Rayleigh's method, let us assume that the deflection of a simple beam with a load P at the center is represented by a sine curve, $y = a \sin \pi x/L$, where a is a constant. Although this equation is inexact, the deflection curve of a simple beam seemingly can be approximated closely by a sine curve. Also, the sine function satisfies the end conditions, $y = y'' = 0$. By Eq. (2-8), the strain energy corresponding to the sinusoidal deflection is

$$U = \tfrac{1}{4}\pi^4 a^2 EI/L^3$$

Since the amplitude of the sine wave is a, the potential energy of the load P is $-Pa$. The total potential energy is accordingly $V = U - Pa$. The assumption that has been used reduces the beam to a system with one degree of freedom. The generalized coordinate is a. By the principle of stationary potential energy, the value of a is determined by $dV/da = 0$. Consequently,

$$a = \frac{2PL^3}{\pi^4 EI} \quad \text{and} \quad y = \frac{2PL^3}{\pi^4 EI} \sin \frac{\pi x}{L}$$

The difference between the deflection given by this formula and by the more accurate formula obtained by integration of the equation, $M = -EIy''$, is everywhere less than 3 per cent. The agreement between the bending moments is less favorable. If the bending moment at section x is calculated with the sinusoidal approximation (by use of the equation, $M = -EIy''$), the deviation from the true bending moment ($M = \frac{1}{2}Px$) is as great as 25 per cent. At the center of the beam the moment calculated by the sine equation is 19 per cent too low.

This example illustrates a general difficulty that arises when stresses are derived from approximate expressions for the deflections. Although the deflections may be fairly accurate, the second derivatives of the approximate deflections may deviate widely from their proper values. Consequently, when deflections are used to calculate stresses, a high degree of accuracy may be required.

Ritz Method. In 1909, W. Ritz (142) refined and generalized Rayleigh's method. He considered the determination of a function $y(x)$ to minimize an integral

$$V = \int_a^b f(x, y, y', y'', \cdots)\, dx \tag{a}$$

where f is a given function. The function y is restricted to a class of functions that satisfies certain linear forced boundary conditions. In a problem of mechanics V may represent potential energy, but the interpretation of V is irrelevant for the present discussion. Ritz undertook to form a convergent sequence of approximations $y_0, y_1, y_2, \cdots$ of the form

$$y_n = a_0 g_0 + a_1 g_1 + a_2 g_2 + \cdots + a_n g_n \tag{b}$$

where the a's are undetermined constants and the g's are given functions that satisfy the forced boundary conditions. To permit the sequence to converge in the mean-square sense to the exact solution $y(x)$, Ritz required that the functions $g_i(x)$ form a complete set in the interval (a, b). To explain this statement, we suppose that $F(x)$ is a given function in (a, b) and that F is approximated by

$$F_n = b_0 g_0 + b_1 g_1 + b_2 g_2 + \cdots + b_n g_n$$

The mean-square error of this approximation is defined by

$$M_n = \int_a^b (F - F_n)^2\, dx$$

The function set $g_0, g_1, g_2, \cdots$ is said to be "complete" in the interval (a, b) if, when $F(x)$ is any given piecewise continuous function in (a, b), coefficients b_i exist so that

$$\lim_{n \to \infty} M_n = 0$$

This is not the same as the condition $F(x) = \lim\limits_{n \to \infty} F_n$, unless the operations "limit" and "integration" may be interchanged. Therefore, without supplementary investigations, there is no rigorous assurance that the Ritz method will converge to the correct result. It is known from the theory of Fourier series that the functions, $1, \cos x, \cos 2x, \cdots ; \sin x, \sin 2x, \cdots$, form a complete set in any interval of length 2π and, indeed, that any piecewise continuous function $F(x)$ can be represented exactly by a convergent series of these functions in any interval of length 2π. It is apparent that not every infinite sequence of functions is complete, for, if even one of the trigonometric functions is omitted, the preceding sequence is not complete.

We obtain from Eqs. (a) and (b), $V_n = V_n(a_0, a_1, a_2, \cdots, a_n)$, where V_n is the approximation of V that results if y is approximated by Eq. (b). The condition $\delta V = 0$ is approximated by $\partial V_n/\partial a_i = 0$; $i = 0, 1, 2, \cdots, n$. These equations determine the a's; accordingly, they give an approximation to y by means of Eq. (b). Mathematical complications often restrict the values of n to the first few positive integers. Comparisons between solutions for $n = 1, 2, 3$, etc. indicate the rate of convergence (38). Convergence is enhanced if the functions $g_n(x)$ satisfy the natural boundary conditions, in addition to the forced boundary conditions, but usually this property is not easily fulfilled.

PROBLEMS

1. Show by the calculus of variations that a straight-line segment is the shortest path between two points in a plane.

2. Show that the shortest path between two points on a circular cylinder is helical, provided that the given points do not lie on the same generator nor on the same cross section.

3. As surface coordinates on a circular cone, adopt the distance x from the vertex of the cone and the dihedral angle θ between a fixed plane and a variable plane that intersect on the axis of the cone. A path on the cone that does not coincide with a generator may be defined by an equation of the form, $x = x(\theta)$. The path is said to be a "geodesic" if it is an extremal for the distance on the conical surface between any two points on the path. Derive the general equation of the system of geodesics (excluding generators) for a cone whose generators intersect its axis at $30°$.

4. A uniform elastic simple beam carries a uniformly distributed load p (lb/in.). Derive the natural boundary conditions. Obtain the equation of the deflection curve by means of the Euler equation. Eliminate arbitrary constants of integration.

5. A uniform elastic beam that is clamped at the ends carries a linearly distributed load that varies from zero at one end to p_0 at the other. Obtain the

equation of the deflection curve by means of the Euler equation. Eliminate arbitrary constants of integration.

6. The strain energy of a special type of beam is

$$U = k \int_0^L \left(\frac{y}{a^2} + \frac{y'}{a} + y'' \right)^2 dx$$

where k and a are constants. The end $x = 0$ is clamped, and the end $x = L$ is free. The beam carries a distributed lateral load $p(x)$. State the forced boundary conditions. Derive the natural boundary conditions and the differential equation that determines the deflection.

7. Solve the problem of the curved cantilever beam (Sec. 3-6) for the case in which the load F acts horizontally to the right.

8. Solve the problem of the curved cantilever beam (Sec. 3-6) for the case in which a bending moment M is applied at the free end. Let positive M be a closing moment.

9. A uniform semicircular arch of radius a is clamped at its ends. Supposing that the arch is loaded only by its own weight, derive formulas for the radial and tangential displacement components $(u, \bar{v})$. Hence derive formulas for the bending moment and the net tension at any cross section.

10. Free vibrations of a cantilever beam of length L are defined by the equation $y = f(x) \sin nt$. By means of Eq. (3-16), derive the function $f(x)$. Prove that n is determined by the equations $k^4 = \rho n^2/EI$, $\cos kL \cosh kL = -1$. *Hint.* The boundary conditions for the free end are $y_{xx} = y_{xxx} = 0$.

11. Solve Prob. 10 for a simple beam.

12. Calculate the constants in the equation of the catenary for the case in which the curve passes through the points $(0, 0)$, $(1, 1)$, and $(-1, 1)$.

13. A catenary is symmetrical with respect to the y-axis. The arc length, measured from the vertex, is s. Show that the curve has the following parametric equations: $x = K \operatorname{arcsinh} s/K$, $y = \sqrt{s^2 + K^2} - \lambda$. Show that the curvature is $K/(s^2 + K^2)$.

14. Derive the Euler equation for the integral,

$$V = \iint_R F(x, y, u, u_x, u_y, u_{xx}, u_{xy}, u_{yy}, u_{xxx}, u_{xxy}, u_{xyy}, u_{yyy}) \, dx \, dy$$

15. A surface of revolution is generated by connecting two points (x_0, y_0) and (x_1, y_1) by a plane curve C and rotating the curve C about the x-axis. Show that the area of the surface is stationary if the curve C is a catenary. (The surface is then called a "catenoid").

16. The strain energy of an orthotropic plate is given by

$$U = K \iint (w_{xx}{}^2 + w_{xx}w_{yy} + w_{yy}{}^2) \, dx \, dy$$

where K is a constant, (x, y) are rectangular coordinates, and w is the deflection. The plate carries an arbitrary distributed load $p(x, y)$. (a) Derive the differential equation for w by the calculus of variations. (b) Letting p be the inertial load, write the differential equation for free vibrations of the plate.

17. The potential energy of a drop of water on a clean horizontal glass plate consists of three parts: (a) The potential energy of the weight of the drop. (b) The potential energy due to surface tension. This is a positive term that is proportional to the area of the free surface of the drop. (c) The potential energy due to molecular forces between the water and the glass. This is a negative term that is proportional to the area of contact between the water and the glass.

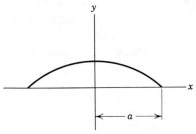

Fig. P3-17

Assuming that the ordinate of the free surface y is a single-valued function of the radial coordinate x (see Fig. P3-17), set up the expression for the total potential energy. What isoperimetric condition is imposed on admissible functions $y(x)$? Taking the isoperimetric condition into account by the Lagrange-multiplier method, write the Euler equation for the problem. Suggest a simplification of the Euler equation that is admissible if the drop is rather flat (that is, if dy/dx is small compared to 1).

18. Determine the deflection of a simple uniformly loaded beam by the Ritz method, using first one term, then two terms, and finally three terms in a sine series. Why does the coefficient of $\sin 2\pi x/L$ vanish?

19. Assuming that the deflection of a uniform cantilever beam with a load at the end is represented by the equation $y = a(1 - \cos \pi x/2L)$, where x is the distance from the fixed end, determine the tip deflection by the principle of stationary potential energy. Compare with the exact solution of the problem.

20. The strain energy of a simply supported, symmetrically loaded circular plate of radius a is

$$U = \pi D \int_0^a [r(w'')^2 + r^{-1}(w')^2 + 2\nu w' w''] \, dr$$

in which w denotes the deflection and primes denote derivatives with respect to the radial coordinate r. Calculate the deflection of a circular plate with a uniform lateral pressure p, assuming that a diametral section of the deflected middle surface is parabolic. Repeat the solution, assuming that a diametral section is one loop of a cosine curve. Compare the two results with $\nu = 0.30$.

21. Solve Prob. 20 for a plate that is loaded only by a point force at the center.

22. A uniform semicircular elastic arch with hinged ends carries a vertical concentrated load F at the center. Derive formulas for the radial and tangential displacements $(u, \bar{v})$ by the principle of stationary potential energy. *Hint.* Make use of symmetry properties, and consider only half the arch.

23. A simply supported rectangular plate of length a and width b carries a constant lateral pressure p. The deflection of the plate is approximated by the first term of a double sine series, $w = w_0 \sin \pi x/a \sin \pi y/b$, in which (x, y) are rectangular coordinates with the axes coinciding with two edges of the plate and w_0 is a constant representing the deflection at the center of the plate. The strain energy of the plate due to bending is

$$U = \tfrac{1}{2} D \iint (w_{xx} + w_{yy})^2 \, dx \, dy$$

where D is a constant called the "flexural rigidity" of the plate. By the principle of stationary potential energy, derive a formula for w_0. Without using any approximation, derive the correct differential equation for w.

24. A uniform elastic string with free length L has its ends fixed at two points at a distance L apart. A uniform lateral load q (lb/in.) is applied to the string. The strain is

$$\epsilon = u_x + \tfrac{1}{2} w_x{}^2$$

where u is the axial component of displacement and w is the lateral displacement. The cross-sectional area of the string is A and Young's modulus is E. By the principle of stationary potential energy, determine the deflection and the tension in the string.

4 Deformable bodies

As the extension, so the force.

ROBERT HOOKE

4-1. DEFORMATION OF A BODY. Some results of the general theory of strain are summarized in this section. Detailed derivations are given in the theory of elasticity (51, 61).

Let (x, y, z) be fixed rectangular coordinates. When a body is deformed, the particle that lies at the point (x, y, z) is displaced to the point $(x + u, y + v, z + w)$. The components (u, v, w) of the displacement vector are functions of (x, y, z).

If the initial length of a line element in the body is ds and the final length is ds^*, the strain of the line element is defined by

$$\epsilon = \frac{ds^* - ds}{ds} \tag{4-1}$$

In view of this definition, $\epsilon > -1$. If the initial direction cosines of the line element are (l, m, n), the strain is determined by

$$\begin{aligned}
\epsilon + \tfrac{1}{2}\epsilon^2 &= \epsilon_x l^2 + \tfrac{1}{2}\gamma_{xy} lm + \tfrac{1}{2}\gamma_{xz} ln \\
&+ \tfrac{1}{2}\gamma_{yx} ml + \epsilon_y m^2 + \tfrac{1}{2}\gamma_{yz} mn \\
&+ \tfrac{1}{2}\gamma_{zx} nl + \tfrac{1}{2}\gamma_{zy} nm + \epsilon_z n^2
\end{aligned} \tag{4-2}$$

where

$$\begin{aligned}
\epsilon_x &= u_x + \tfrac{1}{2}(u_x{}^2 + v_x{}^2 + w_x{}^2) \\
\epsilon_y &= v_y + \tfrac{1}{2}(u_y{}^2 + v_y{}^2 + w_y{}^2) \\
\epsilon_z &= w_z + \tfrac{1}{2}(u_z{}^2 + v_z{}^2 + w_z{}^2) \\
\gamma_{zy} &= \gamma_{yz} = w_y + v_z + u_y u_z + v_y v_z + w_y w_z \\
\gamma_{xz} &= \gamma_{zx} = u_z + w_x + u_z u_x + v_z v_x + w_z w_x \\
\gamma_{yx} &= \gamma_{xy} = v_x + u_y + u_x u_y + v_x v_y + w_x w_y
\end{aligned} \tag{4-3}$$

104

Subscripts (x, y, z) on (u, v, w) denote partial derivatives. The symmetric matrix,

$$\begin{bmatrix} \epsilon_x & \tfrac{1}{2}\gamma_{xy} & \tfrac{1}{2}\gamma_{xz} \\ \tfrac{1}{2}\gamma_{yx} & \epsilon_y & \tfrac{1}{2}\gamma_{yz} \\ \tfrac{1}{2}\gamma_{zx} & \tfrac{1}{2}\gamma_{zy} & \epsilon_z \end{bmatrix}$$

is known as the "strain tensor," since it obeys the tensor law of transformation when the coordinates are changed.

If two line elements have the initial direction cosines (l_1, m_1, n_1) and (l_2, m_2, n_2) and these line elements are initially perpendicular to each other (i.e. $l_1 l_2 + m_1 m_2 + n_1 n_2 = 0$), the angle θ between them after the deformation is determined by

$$(1 + \epsilon_1)(1 + \epsilon_2) \cos \theta = 2\epsilon_x l_1 l_2 + 2\epsilon_y m_1 m_2$$
$$+ 2\epsilon_z n_1 n_2 + \gamma_{yz}(m_1 n_2 + m_2 n_1) \qquad (4\text{-}4)$$
$$+ \gamma_{zx}(n_1 l_2 + n_2 l_1) + \gamma_{xy}(l_1 m_2 + l_2 m_1)$$

Here, ϵ_1 and ϵ_2 are the strains of the respective line elements; they are determined by Eq. (4-2). The angle $(\pi/2) - \theta$ is known as the "shearing strain" between the two line elements.

Under an arbitrary deformation, any infinitesimal sphere in the medium is deformed into an ellipsoid, called the "strain ellipsoid." The principal axes of the strain ellipsoid have the directions of the principal axes of strain at the center of the ellipsoid in the deformed medium. Furthermore, those radii of the infinitesimal sphere that pass into the principal axes of the strain ellipsoid are initially perpendicular to each other; they coincide with the principal axes of strain in the undeformed medium. In other words, through any point in the undeformed medium there are three mutually perpendicular line elements that remain perpendicular under the deformation. The strains of these three line elements are called the "principal strains" at the given point. They are denoted by $(\epsilon_1, \epsilon_2, \epsilon_3)$, and the corresponding values of the quantity $\phi = \epsilon + \tfrac{1}{2}\epsilon^2$ are denoted by (ϕ_1, ϕ_2, ϕ_3). The quantities (ϕ_1, ϕ_2, ϕ_3) are the three roots of the determinantal equation,*

$$\begin{vmatrix} \epsilon_x - \phi & \tfrac{1}{2}\gamma_{xy} & \tfrac{1}{2}\gamma_{xz} \\ \tfrac{1}{2}\gamma_{yx} & \epsilon_y - \phi & \tfrac{1}{2}\gamma_{yz} \\ \tfrac{1}{2}\gamma_{zx} & \tfrac{1}{2}\gamma_{zy} & \epsilon_z - \phi \end{vmatrix} = 0 \qquad (4\text{-}5)$$

Because of the symmetry of this determinant, the roots ϕ_i are always real.

* This relationship is derived in Appendix A-3

Also, since $\epsilon_i > -1$, $\phi_i > -1$. Expansion of the determinant yields

$$\phi^3 - I_1\phi^2 + I_2\phi - I_3 = 0 \tag{4-6}$$

where

$$
\begin{aligned}
I_1 &= \epsilon_x + \epsilon_y + \epsilon_z \\
I_2 &= \epsilon_y\epsilon_z + \epsilon_z\epsilon_x + \epsilon_x\epsilon_y - \tfrac{1}{4}\gamma_{yz}{}^2 - \tfrac{1}{4}\gamma_{zx}{}^2 - \tfrac{1}{4}\gamma_{xy}{}^2
\end{aligned} \tag{4-7}
$$

$$
I_3 = \begin{vmatrix}
\epsilon_x & \tfrac{1}{2}\gamma_{xy} & \tfrac{1}{2}\gamma_{xz} \\
\tfrac{1}{2}\gamma_{yx} & \epsilon_y & \tfrac{1}{2}\gamma_{yz} \\
\tfrac{1}{2}\gamma_{zx} & \tfrac{1}{2}\gamma_{zy} & \epsilon_z
\end{vmatrix}
$$

To any root ϕ_i of Eq. (4-6), there corresponds a solution of the equations,

$$
\begin{aligned}
(\epsilon_x - \phi_i)l + \tfrac{1}{2}\gamma_{xy}m + \tfrac{1}{2}\gamma_{xz}n &= 0 \\
\tfrac{1}{2}\gamma_{yx}l + (\epsilon_y - \phi_i)m + \tfrac{1}{2}\gamma_{yz}n &= 0 \\
\tfrac{1}{2}\gamma_{zx}l + \tfrac{1}{2}\gamma_{zy}m + (\epsilon_z - \phi_i)n &= 0
\end{aligned} \tag{4-8}
$$

$$l^2 + m^2 + n^2 = 1$$

The solution (l_i, m_i, n_i) of these equations is the set of direction cosines of the principal direction in the undeformed medium corresponding to the principal strain ϵ_i. The three directions corresponding to the three roots (ϕ_1, ϕ_2, ϕ_3) are mutually perpendicular.

It is shown in the theory of algebraic equations (20) that the roots of Eq. (4-6) satisfy the relations

$$
\begin{aligned}
I_1 &= \phi_1 + \phi_2 + \phi_3 \\
I_2 &= \phi_2\phi_3 + \phi_3\phi_1 + \phi_1\phi_2 \\
I_3 &= \phi_1\phi_2\phi_3
\end{aligned} \tag{4-9}
$$

Equations (4-9) are a special case of Eq. (4-7) that results if the axes (x, y, z) coincide with the principal axes. Since the principal strains are independent of the coordinate system, Eqs. (4-9) show that the quantities I_1, I_2, I_3 are independent of the coordinate system; that is, I_1, I_2, I_3 are unchanged if the coordinate axes are rotated. Consequently, the quantities I_1, I_2, I_3 are called the "first, second, and third invariants" of the strain tensor. Any other invariant of the strain tensor can be expressed in terms of I_1, I_2, and I_3.

According to the theory of rigid-body displacements (90), there exists, for any point of a deformable body, an angular displacement that carries the principal axes of strain of the undeformed body into the principal axes of strain of the deformed body. This angular displacement is called the "rotation" that the particle of the body receives. Accordingly, when a body is deformed, each particle receives a definite rotation. By the theory of rigid-body displacements, we may determine the axis of rotation and the angular displacement of any particle of the medium in terms of the displacement vector field (u, v, w).

Besides the rotation, the particle receives the translation (u, v, w). Also, it experiences the principal strains $(\epsilon_1, \epsilon_2, \epsilon_3)$ along the principal axes. This deformation is called a "dilatation." Thus the displacements in a neighborhood of a point are resolved into a translation, a rotation, and a dilatation. The translation and the rotation contribute nothing to the strains.

Volumetric Strain. Let an element of a strained medium have the initial volume V and the final volume V^*. The volumetric strain (also called cubical dilatation) is defined by

$$e = \frac{V^* - V}{V} \tag{4-10}$$

A volume element V in the form of an infinitesimal rectangular parallelepiped with its edges in the principal directions remains a rectangular parallelepiped after the deformation. The strains of the edges of the parallelepiped are the principal strains $\epsilon_1, \epsilon_2, \epsilon_3$. Consequently,

$$V^* = (1 + \epsilon_1)(1 + \epsilon_2)(1 + \epsilon_3)V \tag{4-11}$$

Therefore, by Eq. (4-10),

$$e = \epsilon_1 + \epsilon_2 + \epsilon_3 + \epsilon_2\epsilon_3 + \epsilon_3\epsilon_1 + \epsilon_1\epsilon_2 + \epsilon_1\epsilon_2\epsilon_3 \tag{4-12}$$

The expression $e + \frac{1}{2}e^2$ may be expressed in terms of $\epsilon_1, \epsilon_2, \epsilon_3$ by means of Eq. (4-12). Also, since $\phi_i = \epsilon_i + \frac{1}{2}\epsilon_i^2$, the invariants I_1, I_2, I_3 may be expressed in terms of $\epsilon_1, \epsilon_2, \epsilon_3$ by means of Eq. (4-9). Accordingly, it may be verified by routine algebra that

$$e + \tfrac{1}{2}e^2 = I_1 + 2I_2 + 4I_3 \tag{4-13}$$

Small-Displacement Theory. The foregoing equations of the theory of strain are purely geometrical and they are rigorously valid. In the small-displacement theory, Eq. (4-13) is approximated by $e = I_1$. Also, the quadratic terms in Eq. (4-3) are neglected. Hence

$$\epsilon_x = u_x, \qquad \epsilon_y = v_y, \qquad \epsilon_z = w_z$$

$$\gamma_{yz} = w_y + v_z, \qquad \gamma_{zx} = u_z + w_x, \qquad \gamma_{xy} = v_x + u_y \qquad (4\text{-}14)$$

If quadratic terms in the strains are neglected, Eqs. (4-2) show that $\epsilon_x, \epsilon_y, \epsilon_z$ are the strains of line elements that initially lie parallel to the x-, y-, and z-axes. In addition, Eq. (4-4) shows that $\gamma_{yz}, \gamma_{zx}, \gamma_{xy}$ are the shearing strains between pairs of line elements that initially lie parallel to the axes indicated by the subscripts.

It is apparent from Eqs. (4-14) that the linearized strain components satisfy the differential equations

$$\frac{\partial^2 \epsilon_z}{\partial y^2} + \frac{\partial^2 \epsilon_y}{\partial z^2} = \frac{\partial^2 \gamma_{yz}}{\partial y \, \partial z}$$

$$\frac{\partial^2 \epsilon_x}{\partial z^2} + \frac{\partial^2 \epsilon_z}{\partial x^2} = \frac{\partial^2 \gamma_{zx}}{\partial z \, \partial x}$$

$$\frac{\partial^2 \epsilon_y}{\partial x^2} + \frac{\partial^2 \epsilon_x}{\partial y^2} = \frac{\partial^2 \gamma_{xy}}{\partial x \, \partial y}$$

$$2\frac{\partial^2 \epsilon_x}{\partial y \, \partial z} + \frac{\partial^2 \gamma_{yz}}{\partial x^2} = \frac{\partial^2 \gamma_{zx}}{\partial x \, \partial y} + \frac{\partial^2 \gamma_{xy}}{\partial z \, \partial x} \qquad (4\text{-}15)$$

$$2\frac{\partial^2 \epsilon_y}{\partial z \, \partial x} + \frac{\partial^2 \gamma_{zx}}{\partial y^2} = \frac{\partial^2 \gamma_{xy}}{\partial y \, \partial z} + \frac{\partial^2 \gamma_{yz}}{\partial x \, \partial y}$$

$$2\frac{\partial^2 \epsilon_z}{\partial x \, \partial y} + \frac{\partial^2 \gamma_{xy}}{\partial z^2} = \frac{\partial^2 \gamma_{yz}}{\partial z \, \partial x} + \frac{\partial^2 \gamma_{zx}}{\partial y \, \partial z}$$

Equations (4-15) are known as the "compatibility equations" in the small-displacement theory. These equations are consequences of Eqs. (4-14). It can be shown also (57) that if given functions $(\epsilon_x, \epsilon_y, \epsilon_z, \gamma_{yz}, \gamma_{zx}, \gamma_{xy})$ satisfy Eqs. (4-15) there exist functions (u, v, w) that are solutions of Eqs. (4-14). Accordingly, in the small-displacement theory the functions $(\epsilon_x, \epsilon_y, \epsilon_z, \gamma_{yz}, \gamma_{zx}, \gamma_{xy})$ are possible strain components if, and only if, they satisfy Eqs. (4-15). The corresponding relations in the large-displacement theory for which quadratic terms in Eqs. (4-3) are retained are also known, but they are very complicated (56).

4-2. STRESS. This article summarizes some results of the theory of stress; detailed derivations are given in books on the theory of elasticity (51, 85).

The medium is considered to be in a deformed state. The stress vector **p** on any surface element S in the medium may be resolved into components p_n and τ, respectively, normal and tangent to S. These components depend not only on the location of the surface element S but also on the direction of the normal to S. Following the usual convention, we consider the normal stress p_n positive if it denotes tension and negative if it denotes compression. The tangential component τ is called "shearing stress."

If the deformed body is referred to rectangular coordinates (x, y, z), the stresses on the coordinate surfaces play an important role. The positive side of a coordinate surface $x = C$ is defined as the side for which $x > C$. Similarly, the positive sides of the y- and z-coordinate surfaces are defined. The stress vectors on the x-, y,- and z- coordinate surfaces at the point (x, y, z) are denoted by $\boldsymbol{\sigma}_x$, $\boldsymbol{\sigma}_y$, and $\boldsymbol{\sigma}_z$. These stresses are exerted by the material on the positive sides of the coordinate surfaces upon the material on the negative sides. The projections of the vectors $\boldsymbol{\sigma}_x$, $\boldsymbol{\sigma}_y$, $\boldsymbol{\sigma}_z$ on the coordinate lines are denoted by the respective rows of the following matrix, called the stress tensor:

$$\begin{bmatrix} \sigma_x & \tau_{xy} & \tau_{xz} \\ \tau_{yx} & \sigma_y & \tau_{yz} \\ \tau_{zx} & \tau_{zy} & \sigma_z \end{bmatrix}$$

For example, the first row of this matrix represents the x-, y-, and z-components of the vector $\boldsymbol{\sigma}_x$. Evidently $\sigma_x, \sigma_y, \sigma_z$ are normal stresses on the corresponding coordinate surfaces, and $\tau_{xy}, \tau_{xz}, \cdots, \tau_{zy}$ are components of the shearing stresses on the coordinate surfaces.

By balancing moments on an arbitrary free body cut from the material, we may verify that the stress tensor is symmetric; that is, $\tau_{yz} = \tau_{zy}$, $\tau_{zx} = \tau_{xz}$, $\tau_{xy} = \tau_{yx}$. These relations are always valid for an electrically and magnetically neutral medium, since effects of inertial forces and gravity are higher order infinitesimals than the tractive forces (i.e., forces due to stress) on the faces of an infinitesimal cuboid.

By virtue of the equilibrium conditions for an infinitesimal tetrahedron in the body, the stress tensor determines the stress on any surface element S in the body. If the normal to surface S has direction cosines (l, m, n) and **p** is the stress vector on S,

$$\mathbf{p} = l\boldsymbol{\sigma}_x + m\boldsymbol{\sigma}_y + n\boldsymbol{\sigma}_z \tag{4-16}$$

Denoting the components of vector **p** on the coordinate axes by (p_x, p_y, p_z), we may write Eq. (4-16) as follows:

$$
\begin{aligned}
p_x &= l\sigma_x + m\tau_{yx} + n\tau_{zx} \\
p_y &= l\tau_{xy} + m\sigma_y + n\tau_{zy} \\
p_z &= l\tau_{xz} + m\tau_{yz} + n\sigma_z
\end{aligned}
\tag{4-17}
$$

Since inertial forces and gravity forces on an infinitesimal tetrahedron are infinitesimals of higher order than the tractive forces on the faces of the element, Eqs. (4-16) and (4-17) are valid even for a vibrating or flowing body.

The normal stress p_n on a surface element S with unit normal **n** is $p_n = \mathbf{n} \cdot \mathbf{p}$. Hence, by Eqs. (4-17),

$$
\begin{aligned}
p_n = l^2\sigma_x \quad &+ lm\tau_{xy} + ln\tau_{xz} \\
&+ ml\tau_{yx} + m^2\sigma_y + mn\tau_{yz} \\
&+ nl\tau_{zx} + nm\tau_{zy} + n^2\sigma_z
\end{aligned}
\tag{4-18}
$$

The magnitude of the shearing stress τ on the surface element S is determined by the equation,

$$
\tau^2 = p^2 - p_n^{\,2} = p_x^{\,2} + p_y^{\,2} + p_z^{\,2} - p_n^{\,2}
\tag{4-19}
$$

Equation (4-18) shows that the normal stress p_n on an oblique plane element is a quadratic form in the direction cosines of the normal to the element. Consequently, by the principal-axis theory of quadratic forms (Appendix A-2), there exist three mutually perpendicular axes of principal stress through any given point, such that the normal stresses take extreme values on plane elements orthogonal to these axes. The normal stress along one principal axis is the maximum normal stress on any plane element through the given point; the normal stress along another principal axis is the minimum normal stress on any plane element through the given point. For the third principal axis, the normal stress is stationary, but it is not an absolute minimum or maximum. The theory of principal axes of stress is exactly parallel to the theory of principal axes of strain (Appendix A-3); we merely replace the quantities $\epsilon_x,\ \epsilon_y,\ \epsilon_z,\ \frac{1}{2}\gamma_{yz},\ \frac{1}{2}\gamma_{zx},\ \frac{1}{2}\gamma_{xy}$ by $\sigma_x,\ \sigma_y,\ \sigma_z,\ \tau_{yz},\ \tau_{zx},\ \tau_{xy}$. The normal stresses $\sigma_1,\ \sigma_2,\ \sigma_3$, corresponding to the principal axes of stress, are called the "principal stresses" at the given point. Planes that are perpendicular to the principal axes are called "principal planes." It is shown in the theory of stress that the shearing stresses on the principal planes are zero.

The principal stresses σ_1, σ_2, σ_3 are the roots of the equation

$$s^3 - J_1 s^2 + J_2 s - J_3 = 0 \tag{4-20}$$

where

$$
\begin{aligned}
J_1 &= \sigma_x + \sigma_y + \sigma_z \\
J_2 &= \sigma_y \sigma_z + \sigma_z \sigma_x + \sigma_x \sigma_y - \tau_{yz}{}^2 - \tau_{zx}{}^2 - \tau_{xy}{}^2
\end{aligned} \tag{4-21}
$$

$$
J_3 = \begin{vmatrix} \sigma_x & \tau_{xy} & \tau_{xz} \\ \tau_{yx} & \sigma_y & \tau_{yz} \\ \tau_{zx} & \tau_{zy} & \sigma_z \end{vmatrix}
$$

The quantities J_1, J_2, J_3 are stress invariants. They are related to the principal stresses by the equations

$$
\begin{aligned}
J_1 &= \sigma_1 + \sigma_2 + \sigma_3 \\
J_2 &= \sigma_2 \sigma_3 + \sigma_3 \sigma_1 + \sigma_1 \sigma_2 \\
J_3 &= \sigma_1 \sigma_2 \sigma_3
\end{aligned} \tag{4-22}
$$

Differential Equations of Equilibrium. Consider a region R of a deformed medium that is enclosed by a surface S and let (l, m, n) be the direction cosines of the outward normal to S. The stress vector on S is given by Eqs. (4-17). The material may be subjected to spatially distributed forces such as weight, centrifugal force, or magnetic effects. These forces are known as body forces. If the vector $\mathbf{F}$ represents the intensity of body force, the body force on a mass element $\rho \, dx \, dy \, dz$ is $\rho \mathbf{F} \, dx \, dy \, dz$, where ρ denotes mass density. Accordingly, the condition of equilibrium of forces on the material in region R is expressed by the equation

$$\iiint_R \rho(\mathbf{i} F_x + \mathbf{j} F_y + \mathbf{k} F_z) \, dx \, dy \, dz$$

$$+ \iint_S [(l\sigma_x + m\tau_{yx} + n\tau_{zx})\mathbf{i} + (l\tau_{xy} + m\sigma_y + n\tau_{zy})\mathbf{j} \tag{a}$$

$$+ (l\tau_{xz} + m\tau_{yz} + n\sigma_z)\mathbf{k}] \, dS = 0$$

Equation (a) may be transformed by the divergence theorem (39),

$$\iiint_R \operatorname{div} \mathbf{q} \, dx \, dy \, dz = \iint_S \mathbf{n} \cdot \mathbf{q} \, dS \tag{4-23}$$

Here, $\mathbf{n} = \mathbf{i}l + \mathbf{j}m + \mathbf{k}n$ and $\mathbf{q}(x, y, z)$ is an arbitrary vector point function. Letting $\mathbf{q}$ denote successively the vectors $\boldsymbol{\sigma}_x$, $\boldsymbol{\sigma}_y$, and $\boldsymbol{\sigma}_z$, we obtain from Eqs. (a) and (4-23)

$$\iiint_R \left[\mathbf{i}\left(\frac{\partial\sigma_x}{\partial x} + \frac{\partial\tau_{yx}}{\partial y} + \frac{\partial\tau_{zx}}{\partial z} + \rho F_x\right)\right.$$

$$+ \mathbf{j}\left(\frac{\partial\tau_{xy}}{\partial x} + \frac{\partial\sigma_y}{\partial y} + \frac{\partial\tau_{zy}}{\partial z} + \rho F_y\right)$$

$$\left.+ \mathbf{k}\left(\frac{\partial\tau_{xz}}{\partial x} + \frac{\partial\tau_{yz}}{\partial y} + \frac{\partial\sigma_z}{\partial z} + \rho F_z\right)\right] dx\, dy\, dz = 0$$

Since this integral vanishes for any region R within the body, the integrand is zero. Consequently,

$$\frac{\partial\sigma_x}{\partial x} + \frac{\partial\tau_{yx}}{\partial y} + \frac{\partial\tau_{zx}}{\partial z} + \rho F_x = 0$$

$$\frac{\partial\tau_{xy}}{\partial x} + \frac{\partial\sigma_y}{\partial y} + \frac{\partial\tau_{zy}}{\partial z} + \rho F_y = 0 \qquad (4\text{-}24)$$

$$\frac{\partial\tau_{xz}}{\partial x} + \frac{\partial\tau_{yz}}{\partial y} + \frac{\partial\sigma_z}{\partial z} + \rho F_z = 0$$

Equations (4-24) remain valid for a vibrating body if the body force $\mathbf{F}$ includes inertial force.

4-3. EQUATIONS OF STRESS AND STRAIN REFERRED TO ORTHOGONAL CURVILINEAR COORDINATES.

For many applications, the equations in Secs. 4-1 and 4-2 must be expressed with respect to curvilinear coordinates. To define curvilinear coordinates in general, we let (X, Y, Z) be rectangular coordinates, and we suppose that these coordinates are single-valued functions of variables (x, y, z). Then, to any set of values of (x, y, z) there corresponds a single point (X, Y, Z). Consequently, the variables (x, y, z) are themselves space coordinates. To ensure a one-to-one correspondence between the points of space and the number triplets (x, y, z), the equations for (X, Y, Z) must yield single-valued inverse functions, $x = x(X, Y, Z)$, $y = y(X, Y, Z)$, $z = z(X, Y, Z)$.

If x is held constant, X, Y, and Z become functions of the two variables (y, z). These equations define a surface, called a "coordinate surface." Accordingly, there is a family of coordinate surfaces with parameter x. Similarly, there are two other families of coordinate surfaces with the respective parameters y and z. The intersection of a surface $y = $ constant with a surface $z = $ constant is a curve on which only x varies; it is called an "x-coordinate line." Similarly, the y- and z-coordinate lines are defined. Since the coordinate surfaces and the coordinate lines are generally curved, the variables (x, y, z) are called "curvilinear coordinates."

If the coordinate surfaces are a triply orthogonal system, the coordinate lines form an orthogonal network and vice versa. Then the coordinate system is said to be "orthogonal." Rectangular, spherical, and cylindrical coordinates are examples of orthogonal coordinates.

Since the position vector of point (x, y, z) is $\mathbf{r} = \mathbf{i}X + \mathbf{j}Y + \mathbf{k}Z$, a system of curvilinear coordinates is defined by the single vector equation, $\mathbf{r} = \mathbf{r}(x, y, z)$. The total differential is $d\mathbf{r} = \mathbf{r}_x\, dx + \mathbf{r}_y\, dy + \mathbf{r}_z\, dz$. If $dy = dz = 0$, the vector $d\mathbf{r}$ is tangent to an x-coordinate line. Therefore, $\mathbf{r}_x$ is tangent to an x-coordinate line. Likewise, the vectors $\mathbf{r}_y$ and $\mathbf{r}_z$ are tangent respectively to y- and z-coordinate lines. Accordingly, the coordinates are orthogonal if, and only if,

$$\mathbf{r}_y \cdot \mathbf{r}_z = \mathbf{r}_z \cdot \mathbf{r}_x = \mathbf{r}_x \cdot \mathbf{r}_y = 0 \qquad (4\text{-}25)$$

The following discussion is restricted to orthogonal coordinates.

The distance ds between two neighboring points is determined by

$$ds^2 = d\mathbf{r} \cdot d\mathbf{r} = (\mathbf{r}_x\, dx + \mathbf{r}_y\, dy + \mathbf{r}_z\, dz)^2$$

Therefore, by Eq. (4-25),

$$ds^2 = \alpha^2\, dx^2 + \beta^2\, dy^2 + \gamma^2\, dz^2 \qquad (4\text{-}26)$$

where

$$\alpha^2 = \mathbf{r}_x \cdot \mathbf{r}_x, \qquad \beta^2 = \mathbf{r}_y \cdot \mathbf{r}_y, \qquad \gamma^2 = \mathbf{r}_z \cdot \mathbf{r}_z \qquad (4\text{-}27)$$

The factors (α, β, γ) are functions of (x, y, z); they are known as "Lamé coefficients." To a large extent, they determine the nature of the coordinate system.

In view of Eq. (4-27), the magnitudes of the vectors $\mathbf{r}_x$, $\mathbf{r}_y$, $\mathbf{r}_z$ are, respectively, α, β, γ. Consequently, the unit vectors tangent to the x-, y-, and z-coordinate lines are respectively $\mathbf{r}_x/\alpha$, $\mathbf{r}_y/\beta$, and $\mathbf{r}_z/\gamma$.

Let (u, v, w) be the projections of the displacement vector issuing from point (x, y, z) on the tangents to the respective coordinate lines at that point. Since the unit tangents to the coordinate lines are $\mathbf{r}_x/\alpha$, $\mathbf{r}_y/\beta$, and $\mathbf{r}_z/\gamma$, the displacement vector of the particle that lies initially at point (x, y, z) is $\mathbf{r}_x u/\alpha + \mathbf{r}_y v/\beta + \mathbf{r}_z w/\gamma$.

If (l, m, n) are the direction cosines of the vector $d\mathbf{r}$ relative to the local coordinate lines, $dx/ds = l/\alpha$, $dy/ds = m/\beta$, $dz/ds = n/\gamma$. The strain components are defined as for rectangular coordinates by Eq. (4-2). The derivation of the expressions for $\epsilon_x, \epsilon_y, \epsilon_z, \gamma_{yz}, \gamma_{zx}, \gamma_{xy}$ is a routine problem of tensor calculus. The following results are also derived without the use of tensors in some treatises on elasticity (61):

$$\epsilon_x = \frac{1}{\alpha}\left[u_x + \frac{\alpha_y v}{\beta} + \frac{\alpha_z w}{\gamma} + \frac{1}{2\alpha}\left(u_x + \frac{\alpha_y v}{\beta} + \frac{\alpha_z w}{\gamma}\right)^2 \right.$$

$$\left. + \frac{1}{2\alpha}\left(v_x - \frac{\alpha_y u}{\beta}\right)^2 + \frac{1}{2\alpha}\left(w_x - \frac{\alpha_z u}{\gamma}\right)^2\right]$$

$$\epsilon_y = \frac{1}{\beta}\left[v_y + \frac{\beta_z w}{\gamma} + \frac{\beta_x u}{\alpha} + \frac{1}{2\beta}\left(v_y + \frac{\beta_z w}{\gamma} + \frac{\beta_x u}{\alpha}\right)^2 \right.$$

$$\left. + \frac{1}{2\beta}\left(w_y - \frac{\beta_z v}{\gamma}\right)^2 + \frac{1}{2\beta}\left(u_y - \frac{\beta_x v}{\alpha}\right)^2\right]$$

$$\epsilon_z = \frac{1}{\gamma}\left[w_z + \frac{\gamma_x u}{\alpha} + \frac{\gamma_y v}{\beta} + \frac{1}{2\gamma}\left(w_z + \frac{\gamma_x u}{\alpha} + \frac{\gamma_y v}{\beta}\right)^2 \right.$$

$$\left. + \frac{1}{2\gamma}\left(u_z - \frac{\gamma_x w}{\alpha}\right)^2 + \frac{1}{2\gamma}\left(v_z - \frac{\gamma_y w}{\beta}\right)^2\right]$$

$$\gamma_{yz} = \frac{v_z}{\gamma} + \frac{w_y}{\beta} - \frac{\gamma_y w}{\beta\gamma} - \frac{\beta_z v}{\beta\gamma} + \frac{1}{\beta\gamma}\left(v_y + \frac{\beta_z w}{\gamma} + \frac{\beta_x u}{\alpha}\right)\left(v_z - \frac{\gamma_y w}{\beta}\right)$$

$$+ \frac{1}{\beta\gamma}\left(w_y - \frac{\beta_z v}{\gamma}\right)\left(w_z + \frac{\gamma_y v}{\beta} + \frac{\gamma_x u}{\alpha}\right) + \frac{1}{\beta\gamma}\left(u_y - \frac{\beta_x v}{\alpha}\right)\left(u_z - \frac{\gamma_x w}{\alpha}\right)$$

$$\gamma_{zx} = \frac{w_x}{\alpha} + \frac{u_z}{\gamma} - \frac{\alpha_z u}{\gamma\alpha} - \frac{\gamma_x w}{\gamma\alpha} + \frac{1}{\gamma\alpha}\left(w_z + \frac{\gamma_x u}{\alpha} + \frac{\gamma_y v}{\beta}\right)\left(w_x - \frac{\alpha_z u}{\gamma}\right)$$

$$+ \frac{1}{\gamma\alpha}\left(u_z - \frac{\gamma_x w}{\alpha}\right)\left(u_x + \frac{\alpha_z w}{\gamma} + \frac{\alpha_y v}{\beta}\right) + \frac{1}{\gamma\alpha}\left(v_z - \frac{\gamma_y w}{\beta}\right)\left(v_x - \frac{\alpha_y u}{\beta}\right)$$

$$\gamma_{xy} = \frac{u_y}{\beta} + \frac{v_x}{\alpha} - \frac{\beta_x v}{\alpha\beta} - \frac{\alpha_y u}{\alpha\beta} + \frac{1}{\alpha\beta}\left(u_x + \frac{\alpha_y v}{\beta} + \frac{\alpha_z w}{\gamma}\right)\left(u_y - \frac{\beta_x v}{\alpha}\right)$$

$$+ \frac{1}{\alpha\beta}\left(v_x - \frac{\alpha_y u}{\beta}\right)\left(v_y + \frac{\beta_x u}{\alpha} + \frac{\beta_z w}{\gamma}\right) + \frac{1}{\alpha\beta}\left(w_x - \frac{\alpha_z u}{\gamma}\right)\left(w_y - \frac{\beta_z v}{\gamma}\right)$$

$$(4\text{-}28)$$

For many applications, the quadratic terms in Eq. (4-28) are neglected. When $\epsilon_x, \epsilon_y, \epsilon_z, \gamma_{yz}, \gamma_{zx}, \gamma_{xy}$ are determined by Eq. (4-28), Eqs. (4-2), (4-4), (4-6), (4-7), (4-8), (4-9), (4-11), (4-12), and (4-13) remain valid for any orthogonal coordinates. Equations (4-28) are easily specialized for particular coordinates. For cylindrical coordinates, $x = r, y = \theta, z = z$; then, $\alpha = 1, \beta = r, \gamma = 1$. For spherical coordinates, $x = r, y = \theta = $ colatitude, $z = \phi = $ longitude; then $\alpha = 1, \beta = r, \gamma = r \sin\theta$.

With reference to any orthogonal coordinates (x, y, z), the stress notations of Sec. 4-2 are retained. For example, σ_x denotes the tensile stress on a plane element that is normal to an x-coordinate line; τ_{xy} and τ_{xz} denote the y- and z-components of the shearing stress on the same plane element. For equilibrium of moments on an infinitesimal element, the symmetry relations, $\tau_{yz} = \tau_{zy}$, $\tau_{zx} = \tau_{xz}$, $\tau_{xy} = \tau_{yx}$, must be retained for all orthogonal coordinates. Furthermore, if (l, m, n) denote the direction cosines of the normal to a plane element relative to the local coordinate lines, Eqs. (4-17), (4-18), (4-19), (4-20), (4-21), and (4-22) remain valid for any orthogonal coordinates. The differential equations of equilibrium take the following general form:

$$\frac{\partial}{\partial x} \beta\gamma\sigma_x + \frac{\partial}{\partial y} \gamma\alpha\tau_{xy} + \frac{\partial}{\partial z} \alpha\beta\tau_{xz} + \gamma\alpha_y\tau_{xy}$$

$$+ \beta\alpha_z\tau_{xz} - \gamma\beta_x\sigma_y - \beta\gamma_x\sigma_z + \rho\alpha\beta\gamma F_x = 0$$

$$\frac{\partial}{\partial x} \beta\gamma\tau_{yx} + \frac{\partial}{\partial y} \gamma\alpha\sigma_y + \frac{\partial}{\partial z} \alpha\beta\tau_{yz} + \alpha\beta_z\tau_{yz} \qquad (4\text{-}29)$$

$$+ \gamma\beta_x\tau_{yx} - \alpha\gamma_y\sigma_z - \gamma\alpha_y\sigma_x + \rho\alpha\beta\gamma F_y = 0$$

$$\frac{\partial}{\partial x} \beta\gamma\tau_{zx} + \frac{\partial}{\partial y} \gamma\alpha\tau_{zy} + \frac{\partial}{\partial z} \alpha\beta\sigma_z + \beta\gamma_x\tau_{zx}$$

$$+ \alpha\gamma_y\tau_{zy} - \beta\alpha_z\sigma_x - \alpha\beta_z\sigma_y + \rho\alpha\beta\gamma F_z = 0$$

These equations are purely statical; they must be satisfied even though creep, plastic behavior, or thermal stresses occur. If the body-force terms are generalized to include inertial force, Eq. (4-29) applies for vibrating bodies.

4-4. FIRST LAW OF THERMODYNAMICS APPLIED TO A DEFORMATION PROCESS.

The displacement vector field that carries a deformable body from a configuration $\mathbf{X}_0$ to a neighboring configuration $\mathbf{X}$ is denoted by $(\delta u, \delta v, \delta w)$. The functions $(\delta u, \delta v, \delta w)$ are to be regarded as variations of the functions (u, v, w) (see Sec. 3-3) and not as total differentials of (u, v, w). In other words, $\delta u = \epsilon\xi$, $\delta v = \epsilon\eta$, $\delta w = \epsilon\zeta$, where (ξ, η, ζ) are arbitrary functions of (x, y, z) that satisfy the forced boundary conditions and ϵ is an arbitrary infinitesimal. In the theory of plasticity only the incremental displacements $(\delta u, \delta v, \delta w)$ have significance, since the zero state of deformation has no absolute meaning. Deformations in the large may be regarded as integrated effects of infinitesimal deformations.

In the small-deformation theory the strains are determined by Eqs. (4-14). Accordingly, the variations of the strain components due to the variations (δu, δv, δw) are

$$\delta\epsilon_x = \frac{\partial}{\partial x}(\delta u), \qquad \delta\gamma_{xy} = \frac{\partial}{\partial x}(\delta v) + \frac{\partial}{\partial y}(\delta u), \cdots \tag{a}$$

If R is any region within a deformable body and S is the surface enclosing region R, the external forces acting on the material in R consist of the tractive forces due to stresses on S and the body forces acting on material in R. The work that these forces perform when the displacement (δu, δv, δw) is imposed is denoted by δW_e. It will be supposed that equilibrium conditions prevail during the displacement, and that the change of kinetic energy is zero. Then, by the first law of thermodynamics [Eq. (1-2)], $\delta W_e = \delta U - Q$, where δU is the increase of internal energy in R and Q is the heat that flows into R while the displacement (δu, δv, δw) is being performed. Also, by the law of kinetic energy (Sec. 1-3), $\delta W_e + \delta W_i = 0$, where δW_i is the work performed by internal forces in the region R.

To formulate the expression for δW_e, we let (l, m, n) be direction cosines of the outward normal to S. As in Sec. 4-2, the intensity of body force is denoted by (F_x, F_y, F_z). In view of Eqs. (4-17),

$$\delta W_e = \iint_S [(l\sigma_x + m\tau_{yx} + n\tau_{zx})\,\delta u + (l\tau_{xy} + m\sigma_y + n\tau_{zy})\,\delta v$$

$$+ (l\tau_{xz} + m\tau_{yz} + n\sigma_z)\,\delta w]\,dS$$

$$+ \iiint_R \rho(F_x\,\delta u + F_y\,\delta v + F_z\,\delta w)\,dx\,dy\,dz$$

Hence, by the divergence theorem [Eq. (4-23)],

$$\delta W_e = \iiint_R \left[\rho(F_x\,\delta u + F_y\,\delta v + F_z\,\delta w) \right.$$

$$+ \frac{\partial}{\partial x}(\sigma_x\,\delta u + \tau_{xy}\,\delta v + \tau_{xz}\,\delta w) + \frac{\partial}{\partial y}(\tau_{yx}\,\delta u + \sigma_y\,\delta v + \tau_{yz}\,\delta w)$$

$$\left. + \frac{\partial}{\partial z}(\tau_{zx}\,\delta u + \tau_{zy}\,\delta v + \sigma_z\,\delta w) \right]\,dx\,dy\,dz$$

Because of Eq. (a) and the equilibrium equations this reduces to

$$-\delta W_i = \delta W_e = \iiint\limits_R [\sigma_x \, \delta\epsilon_x + \sigma_y \, \delta\epsilon_y + \sigma_z \, \delta\epsilon_z$$

$$+ \tau_{yz} \, \delta\gamma_{yz} + \tau_{zx} \, \delta\gamma_{zx} + \tau_{xy} \, \delta\gamma_{xy}] \, dx \, dy \, dz \qquad (4\text{-}30)$$

The internal energy in region R is

$$U = \iiint\limits_R U_0 \, dx \, dy \, dz \qquad (b)$$

where U_0 is the internal energy per unit volume, called "internal-energy density." If the deformation is adiabatic ($Q = 0$), $\delta W_e = \delta U$, and since this is true for any region R we obtain from Eqs. (b) and (4-30)

$$\delta U_0 = \sigma_x \, \delta\epsilon_x + \sigma_y \, \delta\epsilon_y + \sigma_z \, \delta\epsilon_z + \tau_{yz} \, \delta\gamma_{yz} + \tau_{zx} \, \delta\gamma_{zx} + \tau_{xy} \, \delta\gamma_{xy} \quad (4\text{-}31)$$

Equation (4-31) must be augmented by a term representing heat flow if the deformation (δu, δv, δw) is not performed adiabatically.

Energy of Distortion. In the theory of plasticity it is convenient to express the components of stress in the following form:

$$\sigma_x = \sigma_x' + s, \qquad \sigma_y = \sigma_y' + s, \qquad \sigma_z = \sigma_z' + s$$

$$\tau_{yz} = \tau_{yz}', \qquad \tau_{zx} = \tau_{zx}', \qquad \tau_{xy} = \tau_{xy}' \qquad (4\text{-}32)$$

$$s = \tfrac{1}{3}(\sigma_x + \sigma_y + \sigma_z)$$

Evidently, $s = J_1/3$. The pressure is defined to be $-s$. The matrix,

$$\begin{bmatrix} \sigma_x' & \tau_{xy}' & \tau_{xz}' \\ \tau_{yx}' & \sigma_y' & \tau_{yz}' \\ \tau_{zx}' & \tau_{zy}' & \sigma_z' \end{bmatrix}$$

is called the "stress deviator." It is seen that $\sigma_x' + \sigma_y' + \sigma_z' = 0$.

When the body undergoes the infinitesimal strains ($\delta\epsilon_x$, $\delta\epsilon_y$, $\delta\epsilon_z$), the volumetric strain [Eq. (4-13)] is $\delta e = \delta\epsilon_x + \delta\epsilon_y + \delta\epsilon_z$. The increment of internal energy density due to the volumetric change is $\delta U_0^v = s \, \delta e = s(\delta\epsilon_x + \delta\epsilon_y + \delta\epsilon_z)$. The total increment of internal energy density is $\delta U_0 = \delta U_0^v + \delta U_0^d$. The term δU_0^d is called the increment of internal energy density due to distortion. Its value is

$$\delta U_0^d = \sigma_x' \, \delta\epsilon_x + \sigma_y' \, \delta\epsilon_y + \sigma_z' \, \delta\epsilon_z + \tau_{yz}' \, \delta\gamma_{yz} + \tau_{zx}' \, \delta\gamma_{zx} + \tau_{xy}' \, \delta\gamma_{xy}$$
$$(4\text{-}33)$$

4-5. STRESS-STRAIN RELATIONS OF ELASTIC BODIES. The equilibrium equations (4-24) and the symmetry requirements $\tau_{yz} = \tau_{zy}$, $\tau_{zx} = \tau_{xz}$, $\tau_{xy} = \tau_{yx}$ are the only general restrictions on the stress tensor in a stationary body. Various other restrictions arise if special properties are ascribed to the material. The materials to be considered will be elastic, in the sense that the internal forces are conservative. Then, except for an additive constant, the total internal energy in a body is equal to the potential energy of the internal forces (called "strain energy"). The strain energy U is the volume integral of a strain-energy-density function U_0. The function U_0 is obviously unaffected by a translation or a rotation of the body. However, it depends on the deformation; hence it depends on the strain tensor. Also, if the material is nonhomogeneous, U_0 depends on the coordinates (x, y, z). By considering that U_0 depends on (x, y, z), we admit the possibility that the configuration for which the displacement vector vanishes (i.e. the initial state) will be a state in which the stresses do not vanish. This condition is unavoidable if there are residual stresses due to previous inelastic deformation. Also, we may wish to regard a prescribed state of deformation as the zero configuration for potential energy. For example, in buckling problems, the zero configuration may be considered as the unbuckled state immediately before buckling.

Usually elastic deformations of solids do not cause significant changes of temperature. However, the strain energy density may depend on the prevailing temperature θ. Problems of thermal stress involving transient temperatures may often be analyzed with the assumption that the time rate of change of temperature is so slow that associated inertial effects may be disregarded. Then the stress distribution at any instant is the same as though the temperature distribution at that instant were maintained. Processes for which this type of approximation is admissible are said to be "quasi-stationary."

In view of the preceding remarks, the strain-energy-density function U_0 generally depends on the strain components, the coordinates, and the temperature; that is, $U_0 = f(\epsilon_x, \epsilon_y, \epsilon_z, \gamma_{yz}, \gamma_{zx}, \gamma_{xy}, x, y, z, \theta)$. If the displacement vector (u, v, w) receives variations $(\delta u, \delta v, \delta w)$, the strain tensor takes variations $\delta\epsilon_x, \delta\gamma_{yz}, \cdots$, and the corresponding variation of U_0 is

$$\delta U_0 = \frac{\partial U_0}{\partial \epsilon_x} \delta\epsilon_x + \frac{\partial U_0}{\partial \epsilon_y} \delta\epsilon_y + \frac{\partial U_0}{\partial \epsilon_z} \delta\epsilon_z + \frac{\partial U_0}{\partial \gamma_{yz}} \delta\gamma_{yz}$$
$$+ \frac{\partial U_0}{\partial \gamma_{zx}} \delta\gamma_{zx} + \frac{\partial U_0}{\partial \gamma_{xy}} \delta\gamma_{xy} \tag{4-34}$$

Since Eqs. (4-31) and (4-34) are valid for arbitrary variations $(\delta u, \delta v, \delta w)$,

$$\sigma_x = \frac{\partial U_0}{\partial \epsilon_x}, \qquad \sigma_y = \frac{\partial U_0}{\partial \epsilon_y}, \qquad \sigma_z = \frac{\partial U_0}{\partial \epsilon_z}$$
$$\tau_{yz} = \frac{\partial U_0}{\partial \gamma_{yz}}, \qquad \tau_{zx} = \frac{\partial U_0}{\partial \gamma_{zx}}, \qquad \tau_{xy} = \frac{\partial U_0}{\partial \gamma_{xy}} \tag{4-35}$$

Although Eqs. (4-35) have been derived for rectangular coordinates, they remain valid for any orthogonal coordinates (x, y, z). Furthermore, since rigid-body displacements have no effects on the stresses and strains, Eqs. (4-35) remain valid irrespective of the magnitudes of the translations and rotations of the particles, provided that the strains are small and the stresses are referred to the deformed coordinate system. For example, if a thin rod with axial coordinate x is bent into a loop, σ_x is to be interpreted as the normal stress on a cross-sectional element of the deformed rod. If the rotations are large, the derivatives of the displacement vector are no longer small, and the quadratic terms must be retained in the equations that define the strain tensor [Eqs. (4-3)]. However, this circumstance does not invalidate Eqs. (4-35) if the strains are small. If the strains are large, a generalization of Eqs. (4-35) is required (56, 61).

4-6. COMPLEMENTARY ENERGY DENSITY. * For convenience, the following notations are employed interchangeably with the preceding notations for stress and strain:

$$
\begin{aligned}
\epsilon_x &= \epsilon_1, & \epsilon_y &= \epsilon_2, & \epsilon_z &= \epsilon_3 \\
\gamma_{yz} &= \epsilon_4, & \gamma_{zx} &= \epsilon_5, & \gamma_{xy} &= \epsilon_6 \\
\sigma_x &= \sigma_1, & \sigma_y &= \sigma_2, & \sigma_z &= \sigma_3 \\
\tau_{yz} &= \sigma_4, & \tau_{zx} &= \sigma_5, & \tau_{xy} &= \sigma_6
\end{aligned}
\tag{4-36}
$$

The strain components ϵ_i may be regarded as the coordinates of a point in a six-dimensional space. The strain energy density U_0 of an elastic body is a point function in this space. Equations (4-35) are expressed as follows:

$$
\sigma_i = \frac{\partial U_0}{\partial \epsilon_i} ; \qquad i = 1, 2, \cdots, 6
\tag{4-37}
$$

According to the theory of real variables (28), Eq. (4-37) establishes a one-to-one correspondence between the stress tensor and the strain tensor in a neighborhood of the point $\epsilon_i = c_i$, provided that $\partial^2 U_0/\partial \epsilon_i\, \partial \epsilon_j$ is continuous in a neighborhood of point c_i, and the Hessian determinant, $\det (\partial^2 U_0/\partial \epsilon_i\, \partial \epsilon_j)$, does not vanish at point c_i. Then, in a neighborhood of point c_i the inverse relation exists; that is,

$$
\epsilon_i = f_i(\sigma_1, \sigma_2, \cdots, \sigma_6)
\tag{a}
$$

After this solution of Eq. (4-37) has been determined, U_0 may be expressed as a function of $(\sigma_1, \sigma_2, \cdots, \sigma_6)$.

* Through the courtesy of the Franklin Institute, the material in this article is extracted from the author's paper, "The Principle of Complementary Energy in Nonlinear Elasticity Theory," *Jour. Franklin Inst.*, **256**, 3, September 1953.

A. M. Legendre (1752–1833) showed that equations of the type in (4-37) with a nonvanishing Hessian determinant may be transformed into a conjugate form by the introduction of a new function $\Upsilon_0(\sigma_1, \sigma_2, \cdots, \sigma_6)$, defined by

$$\Upsilon_0 = -U_0 + \sum_{k=1}^{6} \sigma_k \epsilon_k \tag{4-38}$$

With the present physical interpretation of the variables, the function Υ_0 is called "complementary-energy density."

Differentiation of Eq. (4-38) yields

$$\frac{\partial \Upsilon_0}{\partial \sigma_i} = -\frac{\partial U_0}{\partial \sigma_i} + \sum_{k=1}^{6} \left(\epsilon_k \frac{\partial \sigma_k}{\partial \sigma_i} + \sigma_k \frac{\partial \epsilon_k}{\partial \sigma_i} \right)$$

Also, the chain rule of partial differentiation yields

$$\frac{\partial U_0}{\partial \sigma_i} = \sum_{k=1}^{6} \frac{\partial U_0}{\partial \epsilon_k} \frac{\partial \epsilon_k}{\partial \sigma_i}$$

Consequently,

$$\frac{\partial \Upsilon_0}{\partial \sigma_i} = \sum_{k=1}^{6} \epsilon_k \frac{\partial \sigma_k}{\partial \sigma_i} + \sum_{k=1}^{6} \left(\sigma_k - \frac{\partial U_0}{\partial \epsilon_k} \right) \frac{\partial \epsilon_k}{\partial \sigma_i} \tag{b}$$

By Eq. (4-37), the second sum in Eq. (b) is zero. Also,

$$\frac{\partial \sigma_k}{\partial \sigma_i} = 0 \quad \text{if} \quad k \neq i \quad \text{and} \quad \frac{\partial \sigma_k}{\partial \sigma_i} = 1 \quad \text{if} \quad k = i$$

Consequently, Eq. (b) reduces to

$$\epsilon_i = \frac{\partial \Upsilon_0}{\partial \sigma_i} ; \qquad i = 1, 2, \cdots, 6 \tag{4-39}$$

Equation (4-39) is conjugate to Eq. (4-37); it is known as the "Legendre transform" of Eq. (4-37). The derivation of Eq. (4-39) provides a proof that there is a function Υ_0 which satisfies Eq. (4-39). This function may be determined by Eq. (4-38), although more expedient methods are sometimes available. Temperature enters as a parameter in the formulas for U_0 and Υ_0.

For structures built of beams, struts, or torsional members, there may be only one significant component of stress, say σ_1. Then, in view of Eq. (4-35), the function U_0 depends on the single strain component ϵ_1. Consequently, by Eqs. (4-35), σ_1 depends on ϵ_1 only. Figure 4-1 illustrates a typical form of the (σ_1, ϵ_1) graph, called a "stress-strain curve." By Eqs. (4-35), $\sigma_1 = dU_0/d\epsilon_1$; therefore, $U_0 = \int \sigma_1 \, d\epsilon_1$. Consequently, U_0 is represented by the area under the stress-strain curve (cross hatched in

Fig. 4-1). This area may be measured from an arbitrary abscissa; for simplicity, the fixed abscissa is taken to be $\epsilon_1 = 0$. Since the product $\sigma_1 \epsilon_1$ is represented by the area of a rectangle in the (σ_1, ϵ_1) plane, Eq. (4-38) shows that the area above the stress-strain curve (shaded in Fig. 4-1) represents the function Υ_0. How-ever, this graphical interpretation of Υ_0 is applicable only if there is but one nonzero component of stress.

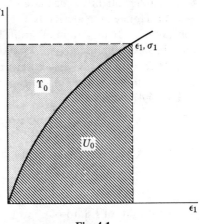

Fig. 4-1

It is assumed in some theories of plasticity that the increase of internal energy during an adiabatic deformation depends only on the final state of stress, and not on the path by which that state is attained, provided that the stresses increase monotonically during the loading (60, 76). These theories do not dis-tinguish between inelasticity and nonlinear elasticity unless unload-ing occurs. Consequently, the concept of complementary energy finds applications in analyses of plastically deformed bodies.

4-7. HOOKEAN MATERIALS. A Hookean material is defined by the condition that the strain energy density is a quadratic function of the strain components. Usually, the temperature θ, measured above an arbitrary zero (e.g. above ambient temperature), enters as a linear param-eter in the linear terms of the strain energy density. This condition is defined as a characteristic of a Hookean material; hence, with the notation of Eqs. (4-36), the strain energy density of a Hookean material is represented by

$$U_0 = \tfrac{1}{2}\sum_{i=1}^{6} \sum_{j=1}^{6} b_{ij}\epsilon_i\epsilon_j - \sum_{i=1}^{6}(c_i\theta + g_i)\epsilon_i + \text{constant} \qquad (4\text{-}40)$$

Equations (4-37) and (4-40) yield

$$\sigma_i = \sum_{j=1}^{6} b_{ij}\epsilon_j - (c_i\theta + g_i); \qquad i = 1, 2, \cdots, 6 \qquad (4\text{-}41)$$

These equations are valid only if the strains are small, since Eq. (4-37) is subject to this restriction. For a nonhomogeneous material, the quanti-ties b_{ij}, c_i, g_i are functions of x, y, z.

The state of zero strain may be designated arbitrarily; if this origin is changed, the strain components ϵ_i are changed by additive constants. Accordingly, for a homogeneous material, the state of zero strain can be chosen so that the constants g_i are eliminated. Then, by Eq. (4-41), the unstrained state also becomes the unstressed state when $\theta = 0$. The constants b_{ij} are independent of the state of zero strain. Since $b_{ij} = b_{ji}$, the most general Hookean material has 21 elastic constants b_{ij} and 6 thermal constants c_i. If the origin of strain is chosen so that $g_i = 0$ and if $\theta = 0$, the absolute minimum of U_0 occurs when $\epsilon_i = 0$. Consequently, the quadratic form $\Sigma\Sigma b_{ij}\epsilon_i\epsilon_j$ is always positive if all ϵ_i are not zero.

In what follows, the constants g_i are taken to be zero. Then inversion of Eq. (4-41) yields

$$\epsilon_i = \sum_{j=1}^{6} a_{ij}\sigma_j + k_i\theta; \qquad i = 1, 2, \cdots, 6 \qquad (4\text{-}42)$$

The matrix (a_{ij}) is the inverse of matrix (b_{ij}). Also,

$$k_i = \sum_{j=1}^{6} a_{ij}c_j, \qquad c_i = \sum_{j=1}^{6} b_{ij}k_j \qquad (4\text{-}43)$$

Substitution of Eq. (4-42) into Eq. (4-40) yields, with Eq. (4-43),

$$U_0 = \tfrac{1}{2}\sum_{i=1}^{6} \sum_{j=1}^{6} a_{ij}\sigma_i\sigma_j + \text{constant} \qquad (4\text{-}44)$$

The additive constant in Eq. (4-44) contains the temperature, but it does not contain the stress components; hence it is irrelevant.

Equations (4-39) and (4-42) yield

$$\Upsilon_0 = \tfrac{1}{2}\sum_{i=1}^{6} \sum_{j=1}^{6} a_{ij}\sigma_i\sigma_j + \sum_{i=1}^{6} k_i\sigma_i\theta \qquad (4\text{-}45)$$

If thermal strains are absent ($\theta = 0$), Eqs (4-44) and (4-45) yield $\Upsilon_0 = U_0$. Consequently, if there are no thermal strains, Eqs. (4-37) and (4-39) show that homogeneous Hookean materials obey the conjugate relations,

$$\sigma_i = \frac{\partial U_0}{\partial \epsilon_i}, \qquad \epsilon_i = \frac{\partial U_0}{\partial \sigma_i}; \qquad i = 1, 2, \cdots, 6 \qquad (4\text{-}46)$$

Crystals. Crystals conform most closely to the ideal behavior of a Hookean material. The atoms of crystals are arranged in regular arrays, called lattices. Crystals are classified according to the types of symmetry exhibited by their lattices (16).

A given straight line is called an "n-fold axis of symmetry" of a crystal if a rotation of the crystal through the angle $2\pi/n$ about that axis brings the lattice into coincidence with its original configuration. Any line is a 1-fold axis of symmetry, since a rotation through the angle 2π always brings the crystal back to its original configuration.

Crystals having only 1-fold axes of symmetry (that is, no symmetry at all) constitute the triclinic system. Crystals of this class have 21 elastic constants b_{ij} and 6 thermal constants c_i.

The coefficients b_{ij} in Eq. (4-40) depend on the orientation of the coordinate axes (x, y, z) with respect to the lattice of the crystal, since U_0 is invariant under coordinate transformations. The transformation of the coefficients b_{ij} under a given coordinate transformation may be determined by tensor theory. If a crystal is rotated through the angle $2\pi/n$ about an n-fold axis of symmetry, the coefficients b_{ij} are evidently unchanged. This condition determines the number of elastic constants for a crystal of a given class.

For example, a crystal that possesses only a single 2-fold axis of symmetry belongs to the monoclinic system. A rotation of the crystal through the angle π about this axis leaves the elastic coefficients unchanged. If the axis of symmetry is the z-axis, the coordinate transformation corresponding to the specified rotation is $x' = -x$, $y' = -y$, $z' = z$. Hence $u' = -u$, $v' = -v$, $w' = w$, and,

$$\epsilon_1' = \frac{\partial u'}{\partial x'} = \epsilon_1, \qquad \epsilon_2' = \frac{\partial v'}{\partial y'} = \epsilon_2, \qquad \epsilon_3' = \epsilon_3$$

$$\epsilon_4' = \frac{\partial w'}{\partial y'} + \frac{\partial v'}{\partial z'} = -\epsilon_4, \qquad \epsilon_5' = \frac{\partial u'}{\partial z'} + \frac{\partial w'}{\partial x'} = -\epsilon_5 \qquad \text{(a)}$$

$$\epsilon_6' = \frac{\partial v'}{\partial x'} + \frac{\partial u'}{\partial y'} = \epsilon_6$$

If the primed quantities ϵ_i' are substituted for ϵ_i in Eq. (4-40), U_0 is unchanged. In addition, the coefficients b_{ij} and c_i are unchanged. Hence

$$\tfrac{1}{2}\sum\sum b_{ij}\epsilon_i\epsilon_j - \theta\sum c_i\epsilon_i = \tfrac{1}{2}\sum\sum b_{ij}\epsilon_i'\epsilon_j' - \theta\sum c_i\epsilon_i' \qquad \text{(b)}$$

Replacing the primed quantities in Eq. (b) with unprimed quantities by means of Eq. (a) and observing that the resulting equation is true for all values of ϵ_i, we obtain $b_{14} = b_{15} = b_{24} = b_{25} = b_{34} = b_{35} = b_{46} = b_{56} = 0$ and $c_4 = c_5 = 0$. Accordingly, crystals of the monoclinic system have 13 elastic constants b_{ij} and 4 thermal constants c_i.

Crystals having face-centered or body-centered cubic lattices (51) possess three mutually perpendicular 4-fold axes of symmetry. This type of symmetry characterizes the regular system. For example, crystals of iron, copper, and aluminum belong to this class. By performing calculations similar to the preceding analysis of the monoclinic system we may show that crystals of the regular system possess only three elastic constants b_{ij} and one thermal constant c_i.

Isotropic Materials. A material is said to be "elastically isotropic" if its elastic constants are invariant under any rotation of coordinates; otherwise, it is "anisotropic." On a macroscopic scale, metals are isotropic if their crystals are oriented at random.

The strain energy density of an isotropic Hookean material depends only on the principal strains (ϵ_1, ϵ_2, ϵ_3). Accordingly, by Eq. (4-40),

$$U_0 = \tfrac{1}{2}\sum_{i=1}^{3}\sum_{j=1}^{3} b_{ij}\epsilon_i\epsilon_j - \sum_{i=1}^{3} c_i\epsilon_i\theta \qquad (c)$$

Since permutations of the quantities ϵ_i in this equation are irrelevant because of the isotropy,

$$b_{11} = b_{22} = b_{33} = \lambda + 2G$$
$$b_{23} = b_{31} = b_{12} = \lambda \qquad (4\text{-}47)$$
$$c_1 = c_2 = c_3 = c$$

The constants λ and G were introduced by Lamé. In view of Eqs. (c) and (4-47), the strain energy density of an isotropic Hookean material is

$$U_0 = \tfrac{1}{2}\lambda(\epsilon_1 + \epsilon_2 + \epsilon_3)^2 + G(\epsilon_1^2 + \epsilon_2^2 + \epsilon_3^2) - c(\epsilon_1 + \epsilon_2 + \epsilon_3)\theta \quad (4\text{-}48)$$

In terms of the strain invariants [Eq. (4-9)], this may be written as

$$U_0 = \tfrac{1}{2}\lambda I_1^2 + G(I_1^2 - 2I_2) - cI_1\theta \qquad (4\text{-}49)$$

Expressing I_1 and I_2 in terms of the strain components referred to any orthogonal coordinates (x, y, z) [see Eq. (4-7)], we obtain

$$U_0 = \tfrac{1}{2}\lambda(\epsilon_x + \epsilon_y + \epsilon_z)^2 + G(\epsilon_x^2 + \epsilon_y^2 + \epsilon_z^2)$$
$$+ \tfrac{1}{2}G(\gamma_{yz}^2 + \gamma_{zx}^2 + \gamma_{xy}^2) - c(\epsilon_x + \epsilon_y + \epsilon_z)\theta \quad (4\text{-}50)$$

Equations (4-35) and (4-50) yield the stress-strain-temperature relations for Hookean isotropic materials:

$$\sigma_x = \lambda e + 2G\epsilon_x - c\theta, \qquad \sigma_y = \lambda e + 2G\epsilon_y - c\theta$$
$$\sigma_z = \lambda e + 2G\epsilon_z - c\theta, \qquad \tau_{yz} = G\gamma_{yz}, \qquad \tau_{zx} = G\gamma_{zx} \quad (4\text{-}51)$$
$$\tau_{xy} = G\gamma_{xy}, \qquad e = \epsilon_x + \epsilon_y + \epsilon_z$$

Since small-deformation theory is considered, the volumetric strain e has been substituted for I_1 [see Eq. (4-13)].

For uniaxial tension of a prismatic bar, $\sigma_y = \sigma_z = 0, \tau_{yz} = \tau_{zx} = \tau_{xy} = 0$, $\theta = 0$. Then, by Eq. (4-51),

$$\epsilon_x = \frac{(\lambda + G)\sigma_x}{G(3\lambda + 2G)}, \qquad \epsilon_y = \epsilon_z = \frac{-\lambda\sigma_x}{2G(3\lambda + 2G)}$$
$$\gamma_{yz} = \gamma_{zx} = \gamma_{xy} = 0$$

In conventional engineering notations this result is written in the form,

$$\epsilon_x = \frac{\sigma_x}{E}, \qquad \epsilon_y = \epsilon_z = -\nu\epsilon_x, \qquad \gamma_{yz} = \gamma_{zx} = \gamma_{xy} = 0 \qquad (4\text{-}52)$$

where E is Young's modulus and ν is Poisson's ratio. Hence

$$E = \frac{G(3\lambda + 2G)}{\lambda + G}, \qquad \nu = \frac{\lambda}{2(\lambda + G)}$$

$$\lambda = \frac{2G\nu}{1 - 2\nu}, \qquad G = \frac{E}{2(1 + \nu)} \qquad (4\text{-}53)$$

Inverting Eqs. (4-51) and utilizing Eqs. (4-53), we obtain

$$\epsilon_x = \frac{1}{E}[\sigma_x - \nu(\sigma_y + \sigma_z)] + k\theta$$

$$\epsilon_y = \frac{1}{E}[\sigma_y - \nu(\sigma_z + \sigma_x)] + k\theta \qquad (4\text{-}54)$$

$$\epsilon_z = \frac{1}{E}[\sigma_z - \nu(\sigma_x + \sigma_y)] + k\theta$$

$$\gamma_{yz} = \frac{\tau_{yz}}{G}, \qquad \gamma_{zx} = \frac{\tau_{zx}}{G}, \qquad \gamma_{xy} = \frac{\tau_{xy}}{G}$$

where

$$k = \frac{(1 - 2\nu)c}{E} \qquad (4\text{-}55)$$

The constant k is known as the coefficient of thermal expansion. Equations (4-45) superpose strains due to stress on strains due to temperature.

Substituting Eqs. (4-54) into Eq. (4-50), we obtain the strain energy density of an isotropic Hookean body in terms of the stresses:

$$U_0 = \frac{1}{2E}[\sigma_x^2 + \sigma_y^2 + \sigma_z^2 - 2\nu(\sigma_y\sigma_z + \sigma_z\sigma_x + \sigma_x\sigma_y)$$

$$+ 2(1 + \nu)(\tau_{yz}^2 + \tau_{zx}^2 + \tau_{xy}^2)] \qquad (4\text{-}56)$$

Since Eq. (4-56) does not contain θ, the strain energy density is determined by the stresses alone. Of course, the temperature distribution may affect the stresses. Stress-free thermal expansion of a body contributes nothing to the strain energy.

Equations (4-39) and (4-54) yield

$$\Upsilon_0 = \frac{1}{2E}[\sigma_x^2 + \sigma_y^2 + \sigma_z^2 - 2\nu(\sigma_y\sigma_z + \sigma_z\sigma_x + \sigma_x\sigma_y)$$

$$+ 2(1 + \nu)(\tau_{yz}^2 + \tau_{zx}^2 + \tau_{xy}^2)] + k\theta(\sigma_x + \sigma_y + \sigma_z) \qquad (4\text{-}57)$$

Equations (4-56) and (4-57) show that if $\theta = 0$, $U_0 = \Upsilon_0$.

The foregoing equations must be generalized slightly if the origin of strain is not chosen so that the constants g_i in Eqs. (4-40) and (4-41) are zero.

4-8. GENERALIZATION OF CASTIGLIANO'S THEOREM OF LEAST WORK.

A variational formulation of boundary-value problems of elasticity in terms of the displacement components (u, v, w) is achieved directly by the principle of stationary potential energy. Because of its generality and because it evades the compatibility conditions, this method is particularly useful for studying large deflections or large strains of elastic systems. However, if approximation methods are used (e.g., the Ritz method, Sec. 3-11), the displacement vector must usually be determined with high accuracy if it is to be used to compute the stresses. Consequently, when stresses are of primary interest, a direct solution in terms of stresses may be preferable. Castigliano's theorem of least work then becomes a guiding variational principle, provided that small-displacement approximations [Eqs. (4-14)] may be used. Since Castigliano's researches were confined to Hookean bodies, he obtained only a special form of the theorem.

Rectangular coordinates (x, y, z) are adopted. Consider an elastic body or system that occupies a region R. Let the surface of the body be S, and let the direction cosines of the outward normal to S be (l, m, n). The body may have cavities or holes. On the surface of a cavity, the normal to S is directed into the cavity.

Let $(\sigma_x, \sigma_y, \cdots, \tau_{xy})$ be the state of stress that satisfies the equilibrium equations, the compatibility equations, and the boundary conditions. Let the stress components receive variations $(\delta\sigma_x, \delta\sigma_y, \cdots, \delta\tau_{xy})$ that satisfy the equilibrium equations. Then, by Eq. (4-24),

$$\frac{\partial}{\partial x}(\delta\sigma_x) + \frac{\partial}{\partial y}(\delta\tau_{yx}) + \frac{\partial}{\partial z}(\delta\tau_{zx}) = 0$$

$$\frac{\partial}{\partial x}(\delta\tau_{xy}) + \frac{\partial}{\partial y}(\delta\sigma_y) + \frac{\partial}{\partial z}(\delta\tau_{zy}) = 0 \qquad (a)$$

$$\frac{\partial}{\partial x}(\delta\tau_{xz}) + \frac{\partial}{\partial y}(\delta\tau_{yz}) + \frac{\partial}{\partial z}(\delta\sigma_z) = 0$$

The variation of the complementary-energy·density is

$$\delta\Upsilon_0 = \frac{\partial\Upsilon_0}{\partial\sigma_x}\delta\sigma_x + \frac{\partial\Upsilon_0}{\partial\sigma_y}\delta\sigma_y + \cdots + \frac{\partial\Upsilon_0}{\partial\tau_{xy}}\delta\tau_{xy} \qquad (b)$$

The complementary energy Υ of the body is defined as the volume integral of the complementary-energy-density function Υ_0; hence

$$\Upsilon = \iiint_R \Upsilon_0 \, dx \, dy \, dz \qquad (4\text{-}58)$$

For a Hookean body in the absence of residual stresses and thermal strains, Υ is identical to the strain energy, for then $\Upsilon_0 = U_0$. With Eq. (b), Eqs. (4-39) and (4-58) yield

$$\delta \Upsilon = \iiint_R (\epsilon_x \, \delta\sigma_x + \epsilon_y \, \delta\sigma_y + \epsilon_z \, \delta\sigma_z + \gamma_{yz} \, \delta\tau_{yz}$$
$$+ \gamma_{zx} \, \delta\tau_{zx} + \gamma_{xy} \, \delta\tau_{xy}) \, dx \, dy \, dz \qquad (c)$$

Since $(\sigma_x, \sigma_y, \cdots, \tau_{xy})$ is the actual equilibrium state of stress, the corresponding displacement vector exists. Consequently, by Eq. (c) and the strain displacement relations (4-14),

$$\delta \Upsilon = \iiint_R [u_x \, \delta\sigma_x + v_y \, \delta\sigma_y + w_z \, \delta\sigma_z + (w_y + v_z) \, \delta\tau_{yz}$$
$$+ (u_z + w_x) \, \delta\tau_{zx} + (v_x + u_y) \, \delta\tau_{xy}] \, dx \, dy \, dz$$

This may be expressed as follows:

$$\delta \Upsilon = \iiint_R \left[\frac{\partial}{\partial x} (u \, \delta\sigma_x + v \, \delta\tau_{xy} + w \, \delta\tau_{xz}) \right.$$
$$+ \frac{\partial}{\partial y} (u \, \delta\tau_{yx} + v \, \delta\sigma_y + w \, \delta\tau_{yz}) + \frac{\partial}{\partial z} (u \, \delta\tau_{zx} + v \, \delta\tau_{zy} + w \, \delta\sigma_z) \right] dx \, dy \, dz$$
$$- \iiint_R \left\{ u \left[\frac{\partial}{\partial x} (\delta\sigma_x) + \frac{\partial}{\partial y} (\delta\tau_{xy}) + \frac{\partial}{\partial z} (\delta\tau_{xz}) \right] \right.$$
$$+ v \left[\frac{\partial}{\partial x} (\delta\tau_{yx}) + \frac{\partial}{\partial y} (\delta\sigma_y) + \frac{\partial}{\partial z} (\delta\tau_{yz}) \right]$$
$$\left. + w \left[\frac{\partial}{\partial x} (\delta\tau_{zx}) + \frac{\partial}{\partial y} (\delta\tau_{zy}) + \frac{\partial}{\partial z} (\delta\sigma_z) \right] \right\} dx \, dy \, dz \qquad (d)$$

In view of Eq. (a), the second integral in Eq. (d) vanishes. Hence by the divergence theorem [Eq. (4-23)], Eq. (d) may be expressed as follows:

$$\delta \Upsilon = \iint_S [(u \, \delta\sigma_x + v \, \delta\tau_{xy} + w \, \delta\tau_{xz})l + (u \, \delta\tau_{yx} + v \, \delta\sigma_y + w \, \delta\tau_{yz})m$$
$$+ (u \, \delta\tau_{zx} + v \, \delta\tau_{zy} + w \, \delta\sigma_z)n] \, dS \qquad (e)$$

Let the stress vector on the surface S be $\mathbf{p} = \mathbf{i}p_x + \mathbf{j}p_y + \mathbf{k}p_z$. Then, by Eqs. (4-17), Eq. (e) may be expressed as follows:

$$\delta\Upsilon = \iint_S (u\,\delta p_x + v\,\delta p_y + w\,\delta p_z)\,dS \qquad (4\text{-}59)$$

Equation (4-59) expresses $\delta\Upsilon$ in terms of a variation of the boundary stresses that does not disturb equilibrium.

Suppose that the boundary consists of two parts, a part S_1 on which the stress vector is given, and a part S_2 on which the displacement vector is given. Then $(\delta p_x, \delta p_y, \delta p_z)$ vanish on S_1. Accordingly, the region of integration in Eq. (4-59) is reduced to S_2. In region S_2, $\delta u = \delta v = \delta w = 0$, since (u, v, w) are prescribed on S_2. Consequently, Eq. (4-59) may be expressed in the form $\delta\Psi = 0$, where

$$\Psi = \Upsilon - \iint_{S_2} (up_x + vp_y + wp_z)\,dS \qquad (4\text{-}60)$$

In forming $\delta\Psi$, we do not give variations to (u, v, w). The functional Ψ is commonly called "complementary energy," but, in accordance with the definition of the complementary-energy-density function, we shall refer to Υ as the complementary energy.

The preceding derivation establishes the fact that any variation of stress that satisfies the equilibrium equations (a) and the boundary conditions on S_1 yields $\delta\Psi = 0$. Consequently,

> Among all states of stress that satisfy the differential equations of equilibrium and the boundary conditions on S_1, that which represents the actual equilibrium state provides a stationary value to Ψ.

This is the general form of Castigliano's theorem of least work.

Sokolnikoff (75) has shown that under some conditions the state of stress that represents the actual equilibrium state provides an absolute minimum to Ψ. This is true if Υ_0 is a quadratic function of the stress components in which the second-degree terms are a positive definite quadratic form. This condition characterizes Hookean bodies. The increment of Υ_0 due to the variation of stress may be represented in the form,

$$\Delta\Upsilon_0 = \delta\Upsilon_0 + \frac{1}{2!}\delta^2\Upsilon_0 + \cdots$$

where $\delta\Upsilon_0$ is a linear form in $(\delta\sigma_x, \cdots, \delta\tau_{xy})$, $\delta^2\Upsilon_0$ is a quadratic form in $(\delta\sigma_x, \cdots, \delta\tau_{xy})$, etc. If Υ_0 is a quadratic function of the stress components, there are no variations beyond $\delta^2 \Upsilon_0$. By Eq. (4-45),

$$\delta^2 \Upsilon_0 = \sum_{i=1}^{6} \sum_{j=1}^{6} a_{ij} \, \delta\sigma_i \, \delta\sigma_j$$

By virtue of Eq. (4-44), $\Sigma\Sigma a_{ij} \, \delta\sigma_i \, \delta\sigma_j$ is positive definite (see Appendix A-1) Consequently, $\delta^2 \Upsilon_0$ is positive definite. Hence $\delta^2 \Upsilon$ is positive definite, since

$$\delta^2 \Upsilon = \iint \delta^2 \Upsilon_0 \, dx \, dy \, dz$$

Since $\delta\Psi$ vanishes for the equilibrium state of stress, $\Delta\Psi = \frac{1}{2}\delta^2 \Upsilon$. Accordingly, the following stronger but more special form of Castigliano's theorem is obtained:

> *Among all states of stress of a Hookean body which satisfy the differential equations of equilibrium and the boundary conditions on S_1, that which represents the actual equilibrium state provides an absolute minimum to Ψ.*

This theorem remains true even if there are thermal strains, since the linear terms in the complementary-energy density do not enter into the argument. For the same reason, the theorem does not require that the strains be measured from the unstressed state; any origin of strain is admissible. The theorem incidentally establishes the uniqueness of the equilibrium state of stress for a Hookean body. It is to be recalled, however, that small-displacement approximations have been used in the derivation of Castigliano's theorem.

Castigliano's principle of least work is useful for deriving compatibility equations for the stress components, particularly if the material has anisotropic or nonlinear stress-strain properties. If the boundary conditions are entirely of the stress type, the region S_2 vanishes, and, by Eq. (4-60), $\Psi = \Upsilon$. Then Castigliano's theorem is expressed by the variational equation, $\delta\Upsilon = 0$. In any case, the equations of compatibility for stress components are determined by the condition $\delta\Upsilon = 0$, since these equations are independent of the boundary conditions. Variations of the stresses are restricted to the class of functions that satisfy the equilibrium conditions. The equilibrium equations (4-24) may be treated by the Lagrange-multiplier method (Sec. 3-8), and the Euler equations for the modified complementary function $\overline{\Upsilon}$ may be identified as the compatibility equations for the stress components. This method has been used as an alternative derivation of Castigliano's principle (120). It is also possible to represent the general solution of the equilibrium equations by means of the stress functions of Maxwell or Morera or by a combination of the two.

Then the variational equation $\delta\Upsilon = 0$ yields the compatibility equations for the stress functions (126).

A simple illustration is provided by the theory of plane stress. In this case $\sigma_z = \tau_{xz} = \tau_{yz} = 0$, and the remaining stress components are independent of z. Then, if there are no body forces, the general solution of the equilibrium equations is

$$\sigma_x = F_{yy}, \qquad \sigma_y = F_{xx}, \qquad \tau_{xy} = -F_{xy} \qquad \text{(f)}$$

where $F(x, y)$ is the Airy stress function. If the material is isotropic and Hookean and the temperature terms are discarded, $U_0 = \Upsilon_0$, and Eq. (4-56) yields

$$\Upsilon = U = \frac{h}{2E} \iint [F_{xx}^2 + F_{yy}^2 - 2\nu F_{xx}F_{yy} + 2(1 + \nu)F_{xy}^2]\, dx\, dy \qquad \text{(g)}$$

where h is the thickness of the plate. The Euler equation for this integral is $\nabla^2\nabla^2 F = 0$. This is the known compatibility equation of plane-stress theory for isotropic Hookean bodies. The same method could be used to obtain the compatibility equation of plane-stress theory for an anisotropic Hookean body, but a generalization of the strain-energy formula (4-56) would be required.

Castigliano's principle of least work may be used to obtain approximate solutions of elasticity problems by the Ritz method. This application of the principle has received relatively few applications, but it has important potentialities for stress-type problems.

4-9. REISSNER'S VARIATIONAL THEOREM OF ELASTICITY.

The principle of stationary potential energy is well adapted to elasticity problems that are formulated in terms of displacements. On the other hand, Castigliano's theorem of least work is adapted to problems that are formulated in terms of stress. A variational theorem of elasticity that simultaneously provides the stress-displacement relations, the equilibrium equations, and the boundary conditions was developed by E. Reissner (138). In the form that the theorem is presented here, small-displacement approximations are required. A generalization of the theorem for finite deformations has been presented by Reissner (140).

An elastic body is considered to occupy a region R. The surface S of the body is considered to be separated into two parts, S_1 and S_2, such that the stress vector is prescribed on S_1 and the displacement vector is prescribed on S_2. If (l, m, n) is the outward unit normal of surface S, the stress vector (p_x, p_y, p_z) on S is expressed in terms of the stress tensor by Eqs. (4-17). The forced boundary conditions on S_1 are $p_x = \bar{p}_x$, $p_y = \bar{p}_y$, $p_z = \bar{p}_z$, where $(\bar{p}_x, \bar{p}_y, \bar{p}_z)$ are given functions on S_1. The forced boundary conditions on S_2 are $u = \bar{u}$, $v = \bar{v}$, $w = \bar{w}$, where $(\bar{u}, \bar{v}, \bar{w})$ are given functions on S_2. The complementary energy density Υ_0 is regarded as a function of the stress components. The body force, per unit volume, is $(\rho F_x, \rho F_y, \rho F_z)$, where ρ is the mass density. The following functional is introduced:

$$
\begin{aligned}
J = \iiint_R & [\sigma_x u_x + \sigma_y v_y + \sigma_z w_z + \tau_{yz}(w_y + v_z) + \tau_{zx}(u_z + w_x) \\
& + \tau_{xy}(v_x + u_y) - \rho(uF_x + vF_y + wF_z) - \Upsilon_0] \, dx \, dy \, dz \\
& - \iint_{S_1} (u\bar{p}_x + v\bar{p}_y + w\bar{p}_z) \, dS \\
& - \iint_{S_2} [(u - \bar{u})p_x + (v - \bar{v})p_y + (w - \bar{w})p_z] \, dS
\end{aligned}
\tag{4-61}
$$

Reissner's theorem asserts the following:

In the small-displacement theory of elasticity, the equilibrium state of a body is such that $\delta J = 0$ for arbitrary variations of u, v, w, σ_x, $\cdots$, τ_{xy}. The identical vanishing of δJ ensures the satisfaction of the differential equations of equilibrium, the stress-displacement relations, and the boundary conditions.

The fact that the equilibrium equations and the stress-displacement relations are satisfied follows immediately from the observation that these equations are simply the Euler equations for the triple integral in the formula for J. The double integrals in the equation for J are introduced to ensure the satisfying of the boundary conditions.

For proof of the theorem, the functions u, v, w, σ_x, σ_y, $\cdots$, τ_{xy} are given arbitrary variations δu, δv, δw, $\delta \sigma_x$, $\delta \sigma_y$, $\cdots$, $\delta \tau_{xy}$. Then the variation of J is

$$
\begin{aligned}
\delta J = \iiint_R & \left[\left(u_x - \frac{\partial \Upsilon_0}{\partial \sigma_x} \right) \delta \sigma_x + \cdots + \left(v_x + u_y - \frac{\partial \Upsilon_0}{\partial \tau_{xy}} \right) \delta \tau_{xy} \right. \\
& + \sigma_x \frac{\partial}{\partial x} \delta u + \cdots + \tau_{xy} \left(\frac{\partial}{\partial x} \delta v + \frac{\partial}{\partial y} \delta u \right) \\
& - \rho F_x \, \delta u - \rho F_y \, \delta v - \rho F_z \, \delta w] \, dx \, dy \, dz \\
& - \iint_{S_1} (\bar{p}_x \, \delta u + \bar{p}_y \, \delta v + \bar{p}_z \, \delta w) \, dS \\
& - \iint_{S_2} [p_x \, \delta u + p_y \, \delta v + p_z \, \delta w + (u - \bar{u}) \, \delta p_x + (v - \bar{v}) \, \delta p_y \\
& + (w - \bar{w}) \, \delta p_z] \, dS
\end{aligned}
\tag{a}
$$

Equation (a) may be transformed by integration by-parts. If $\mathbf{q} = \mathbf{i}u$, the divergence theorem [Eq. (4-23)] reduces to

$$\iiint_R u_x \, dx \, dy \, dz = \iint_S lu \, dS$$

Setting $u = PQ$, we obtain

$$\iiint_R PQ_x \, dx \, dy \, dz = - \iiint_R QP_x \, dx \, dy \, dz + \iint_S lPQ \, dS \qquad (4\text{-}62)$$

Analogous relations apply for the integrals of PQ_y and PQ_z. In view of Eqs. (4-17), Eqs. (a) and (4-62) yield

$$\delta J = \iiint_R \left[\left(u_x - \frac{\partial \Upsilon_0}{\partial \sigma_x} \right) \delta \sigma_x + \cdots + \left(v_x + u_y - \frac{\partial \Upsilon_0}{\partial \tau_{xy}} \right) \delta \tau_{xy} \right.$$

$$- \left(\frac{\partial \sigma_x}{\partial x} + \frac{\partial \tau_{xy}}{\partial y} + \frac{\partial \tau_{xz}}{\partial z} + \rho F_x \right) \delta u - \left(\frac{\partial \tau_{yx}}{\partial x} + \frac{\partial \sigma_y}{\partial y} + \frac{\partial \tau_{yz}}{\partial z} + \rho F_y \right) \delta v$$

$$\left. - \left(\frac{\partial \tau_{zx}}{\partial x} + \frac{\partial \tau_{zy}}{\partial y} + \frac{\partial \sigma_z}{\partial z} + \rho F_z \right) \delta w \right] dx \, dy \, dz$$

$$+ \iint_{S_1} [(l\sigma_x + m\tau_{yx} + n\tau_{zx} - \bar{p}_x) \delta u + (l\tau_{xy} + m\sigma_y + n\tau_{zy} - \bar{p}_y) \delta v$$

$$+ (l\tau_{xz} + m\tau_{yz} + n\sigma_z - \bar{p}_z) \delta w] \, dS$$

$$- \iint_{S_2} [(u - \bar{u}) \delta p_x + (v - \bar{v}) \delta p_y + (w - \bar{w}) \delta p_z] \, dS$$

Since δJ vanishes for arbitrary variations $\delta u, \delta v, \delta w, \delta \sigma_x, \delta \sigma_y, \cdots, \delta \tau_{xy}$, the following differential equations are obtained:

$$u_x = \frac{\partial \Upsilon_0}{\partial \sigma_x}, \quad v_y = \frac{\partial \Upsilon_0}{\partial \sigma_y}, \cdots, v_x + u_y = \frac{\partial \Upsilon_0}{\partial \tau_{xy}} \qquad (b)$$

$$\frac{\partial \sigma_x}{\partial x} + \frac{\partial \tau_{xy}}{\partial y} + \frac{\partial \tau_{xz}}{\partial z} + \rho F_x = 0$$

$$\frac{\partial \tau_{yx}}{\partial x} + \frac{\partial \sigma_y}{\partial y} + \frac{\partial \tau_{yz}}{\partial z} + \rho F_y = 0 \qquad (c)$$

$$\frac{\partial \tau_{zx}}{\partial x} + \frac{\partial \tau_{zy}}{\partial y} + \frac{\partial \sigma_z}{\partial z} + \rho F_z = 0$$

Equations (b) are the stress-displacement relations; Eqs. (c) are the equilibrium equations. The vanishing of the surface integrals yields the boundary conditions,

$$\left.\begin{array}{l} l\sigma_x + m\tau_{yx} + n\tau_{zx} = \bar{p}_x \\ l\tau_{xy} + m\sigma_y + n\tau_{zy} = \bar{p}_y \\ l\tau_{xz} + m\tau_{yz} + n\sigma_x = \bar{p}_z \end{array}\right\} \quad \text{on } S_1 \qquad (d)$$

$$u = \bar{u}, \qquad v = \bar{v}, \qquad w = \bar{w} \qquad \text{on } S_2 \qquad (e)$$

Thus the condition $\delta J \equiv 0$ formulates the elasticity problem completely.

An important practical advantage of a variational formulation of a problem is the opportunity that it provides to apply the Ritz approximation procedure (Sec. 3-11). Reissner has discussed such an application of his theorem in connection with the problem of torsion of a uniform bar with end constraints (139). Naghdi has applied Reissner's theorem to the general theory of elastic shells (129).

4-10. CASTIGLIANO'S THEOREM ON DEFLECTIONS. The theorem discussed in this section is often called the "principle of complementary energy." A restricted form of this theorem that applies to Hookean bodies was derived by A. Castigliano in a thesis in 1873. In 1889 it was generalized by Fr. Engesser, who substituted the complementary energy for strain energy. Engesser's concept attracted little attention, and it was rediscovered by H. M. Westergaard (150) in 1942. In recent years it has been discussed by several investigators (2, 95, 100, 101). The derivation presented here is taken from a paper by the author (120).

Consider an elastic body or system B that is mounted so that rigid-body displacements of the entire system are impossible (Fig. 4-2). Let the surface of the body be S, and let the outward unit normal to S be $\mathbf{n} = \mathbf{i}l + \mathbf{j}m + \mathbf{k}n$. Let the body be in equilibrium under the action of surface stresses and body forces. Certain point forces are included among the external loads. Contrary to the view that is adopted in the classical theory of elasticity, we consider a point force to be distributed on a small spot of surface S, so that excessive deformation, yielding, or rupture are precluded. The distribution of stress on the spot is unimportant; only the resultant force on the spot has significance. Consequently, the complementary energy Υ [see Eq. (4-58)] is regarded as a function of the concentrated loads $F_1, F_2, \cdots$. Also, Υ depends on the distributed loads on surface S and on the body forces, but the distributed forces will not enter explicitly into consideration.

Suppose that the boundary consists of two parts, a deformable part S_1 and a fixed part S_2. On S_2 the displacement vector vanishes because

of the constraints. The part S_2 may consist of a few small spots, commonly
called "point supports." Since the displacement vector vanishes on S_2,
the region of integration in Eq. (4-59) reduces to S_1. The variation of
boundary stresses on S_1 is not restricted by equilibrium conditions, since
equilibrium may be maintained by compensating reactions in S_2.

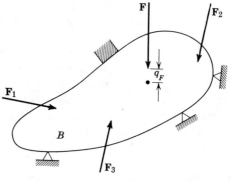

Fig. 4-2

Consider a concentration of load on a small area S_0 of region S_1—the
type of load that is ordinarily called a "point force." Consider a variation
of the surface stress that vanishes in the region $S_1 - S_0$. Then Eq. (4-59)
yields

$$\delta \Upsilon = \iint_{S_0} (u \, \delta p_x + v \, \delta p_y + w \, \delta p_z) \, dS \tag{a}$$

Equation (a) may be simplified by the theorem of the mean for integrals
(28). If $f(x, y)$ and $\phi(x, y)$ are continuous real functions in a region A of
the (x, y) plane and $\phi(x, y)$ retains a constant sign in A,

$$\iint_A f(x, y)\phi(x, y) \, dx \, dy = K \iint_A \phi(x, y) \, dx \, dy \tag{4-63}$$

where K is the value of $f(x, y)$ at some point in A. Accordingly, if
$(\delta p_x, \delta p_y, \delta p_z)$ all retain constant signs in S_0, Eq. (a) yields

$$\delta \Upsilon = u_0 \iint_{S_0} \delta p_x \, dS + v_0 \iint_{S_0} \delta p_y \, dS + w_0 \iint_{S_0} \delta p_z \, dS$$

where (u_0, v_0, w_0) are the values of (u, v, w) at certain points in S_0. Setting
$\mathbf{q}_0 = \mathbf{i}u_0 + \mathbf{j}v_0 + \mathbf{k}w_0$, we may write this result in vector notation:

$$\delta \Upsilon = \mathbf{q}_0 \cdot \iint_{S_0} \delta \mathbf{p} \, dS \tag{b}$$

The total load on the small area S_0 is

$$\mathbf{F} = \iint_{S_0} \mathbf{p} \, dS$$

Consequently, Eq. (b) yields

$$\delta \Upsilon = \mathbf{q}_0 \cdot \delta \mathbf{F} \qquad \text{(c)}$$

Since $\delta \mathbf{F}$ is arbitrary, we may let it be collinear with $\mathbf{F}$. Then $\delta \mathbf{F} = \mathbf{N} \, \delta F$, where $\mathbf{N}$ is a unit vector in the direction of $\mathbf{F}$. Hence, by Eq. (c),

$$\delta \Upsilon = \mathbf{q}_0 \cdot \mathbf{N} \, \delta F = q_F \, \delta F \qquad \text{(d)}$$

where q_F is the component of the displacement vector $\mathbf{q}_0$ in the direction of force $\mathbf{F}$. Also,

$$\delta \Upsilon = \frac{\partial \Upsilon}{\partial F} \delta F$$

Therefore, by Eq. (d),

$$\frac{\partial \Upsilon}{\partial F} = q_F$$

This conclusion signifies the following:

If an elastic system is mounted so that rigid-body displacements of the entire system are impossible and certain external point forces $F_1, F_2, \cdots$ act on the system, in addition to distributed loads and thermal strains, the displacement component q_i of the point of application of force F_i in the direction of force F_i is determined by the equation

$$q_i = \frac{\partial \Upsilon}{\partial F_i} \qquad \text{(4-64)}$$

This is the general form of Castigliano's theorem on deflections. It is to be emphasized that the derivation of this theorem employs small-displacement approximations.

Principle of Complementary Energy for Systems with Finite Degrees of Freedom. According to Eq. (1-13), the equilibrium configurations of an unchecked holonomic elastic system with generalized coordinates $(x_1, x_2, \cdots, x_n)$ are determined by the equations

$$\frac{\partial U}{\partial x_i} = P_i$$

where U is the potential energy of internal forces and P_i are the components of generalized external force. In this case the complementary energy is defined by

$$\Upsilon = -U + \sum_{k=1}^{n} P_k x_k \qquad \text{(4-65)}$$

The Legendre transformation (Sec. 4-6) yields

$$\frac{\partial \Upsilon}{\partial P_i} = x_i \qquad (4\text{-}66)$$

Equation (4-66) is a statement of Castigliano's theorem for systems with finite degrees of freedom. Its similarity to Eq. (4-64) is to be noted. This principle is not limited to small displacements, but, for large displacements, the complementary energy of the system cannot be regarded as the sum of the complementary energies of its parts. Then the concept of a complementary-energy-density function must be abandoned.

Castigliano limited his investigations to the technically important case for which U is a quadratic form in the generalized coordinates. Then Eqs. (1-14), (1-15), and (4-65) yield $U = \Upsilon$.

4-11. COROLLARIES TO CASTIGLIANO'S THEOREM. A theorem concerning the rotation produced by a couple follows immediately from Eq. (4-64). We again consider an arbitrary elastic body (which may represent an arch, a beam, a truss, etc.) that is mounted on rigid supports,*

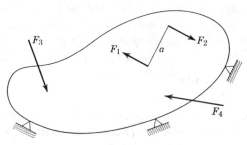

Fig. 4-3

and that is subjected to external forces $F_1, F_2, \cdots$ (Fig. 4-3). The body may also be subjected to distributed loads and temperature gradients. If two parallel concentrated forces, F_1 and F_2, with opposite senses, act perpendicular to the ends of a straight-line segment of length a (Fig. 4-3), Eq. (4-64) shows that the rotation of the line segment due to the deformation is

$$\theta = \frac{1}{a}\frac{\partial \Upsilon}{\partial F_1} + \frac{1}{a}\frac{\partial \Upsilon}{\partial F_2}$$

---

* If the body is supported by springs, the supports may be regarded as a part of the body.

If F_1 and F_2 are functions of a single scalar variable F, the chain rule of partial differentiation yields

$$\frac{\partial \Upsilon}{\partial F} = \frac{\partial \Upsilon}{\partial F_1}\frac{dF_1}{dF} + \frac{\partial \Upsilon}{\partial F_2}\frac{dF_2}{dF}$$

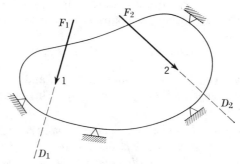

Fig. 4-4

In particular, if $F_1 = F_2 = F$, this yields

$$\frac{\partial \Upsilon}{\partial F} = \frac{\partial \Upsilon}{\partial F_1} + \frac{\partial \Upsilon}{\partial F_2}$$

Consequently,

$$\theta = \frac{1}{a}\frac{\partial \Upsilon}{\partial F}$$

Since the equal and opposite forces F_1 and F_2 constitute a couple of magnitude $M = Fa$, this equation may be written

$$\theta = \frac{\partial \Upsilon}{\partial M} \tag{4-67}$$

Equation (4-67) determines the angular displacement of the arm of a couple M that acts on an elastic body.

Maxwell's Law of Reciprocity. A reciprocal relationship between loads and deflections that was discovered by Clerk Maxwell is an immediate consequence of Castigliano's principle. It is restricted in its scope to Hookean bodies without thermal strains. In this case $U = \Upsilon$.

Consider any Hookean structure that is subjected to two point loads F_1 and F_2 (Fig. 4-4). Then the strain energy U is a function of F_1 and F_2. By Castigliano's principle, the deflection component at point 1 in the direction of force F_1 due to the action of force F_2 alone is

$$q_1 = \frac{\partial U}{\partial F_1}\bigg|^{F_1=0} \tag{a}$$

Similarly, the deflection component at point 2 in the direction of force F_2 due to the action of force F_1 alone is

$$q_2 = \frac{\partial U}{\partial F_2}\bigg|^{F_2=0} \tag{b}$$

For a Hookean body, the deflections are proportional to the loads, and the strain energy is accordingly a quadratic form in the loads. Therefore, in the present case

$$U = aF_1^2 + bF_1F_2 + cF_2^2 \tag{c}$$

where a, b, c are constants. Equations (a), (b), and (c) yield $q_1 = bF_2$, $q_2 = bF_1$. Consequently, if $F_1 = F_2$, $q_1 = q_2$. This conclusion may be expressed as follows:

> *For a Hookean body, the deflection component at point 1 in the direction D_1 due to a load F at point 2 in the direction D_2 equals the deflection component at point 2 in the direction D_2 due to a load F at point 1 in the direction D_1.*

This is Maxwell's law of reciprocity. It may be illustrated by the solutions of elementary problems of mechanics of materials. For example, in Sec. 2-2, the deflection of a simple beam of length L due to a concentrated lateral load Q at the point $x = c$ is shown to be

$$y = \frac{2QL^3}{\pi^4 EI} \sum_{n=1}^{\infty} n^{-4} \sin \frac{n\pi c}{L} \sin \frac{n\pi x}{L}$$

Since this equation remains unchanged if x and c are interchanged, the deflection of a simple beam at point x due to a lateral load Q at point c equals the deflection at point c due to a lateral load Q at point x. This statement may be regarded as a special case of Maxwell's theorem. Incidentally, Sec. 2-2 shows that Maxwell's law of reciprocity applies also for beam columns, since the only change in the above formula is the replacement of the term n^4 by $n^4 - rn^2$, where $r = P/P_e$.

It is apparent from Eq. (4-67) that there is also a reciprocal theorem for couples and rotations.

4-12. CASTIGLIANO'S THEOREM APPLIED TO TRUSSES. If a uniform elastic bar is subjected to tension or compression, the axial strain, by Eq. (4-39) is $\epsilon = d\Upsilon_0/d\sigma$, where σ is the tensile stress and Υ_0 is the complementary-energy density. This relationship remains valid if the bar is heated, since Υ_0 is then modified appropriately. The total complementary energy of the bar is $\Upsilon = AL\Upsilon_0$, where A is the cross-sectional area and L is the length. Also, the tension in the bar is $N = A\sigma$. Hence the total extension of the bar is

$$e = \frac{d\Upsilon}{dN} \qquad (4\text{-}68)$$

The members of a pin-jointed truss are subjected to tensions or compressions. The complementary energy of the truss is $\Upsilon = \Sigma\Upsilon_i$, where Υ_i is the complementary energy of the ith member, and the sum extends over all members. The truss may be subjected to various loads, and the members may be heated to different temperatures. If F is one of the loads, Eq. (4-64) shows that the component of displacement of the point of application of force F in the direction of force F is

$$q = \frac{\partial\Upsilon}{\partial F} = \Sigma \frac{\partial\Upsilon_i}{\partial F}$$

Since Υ_i depends only on the tension N_i in the ith member, this yields

$$q = \Sigma \frac{d\Upsilon_i}{dN_i} \frac{\partial N_i}{\partial F}$$

By Eq. (4-68) $d\Upsilon_i/dN_i$ is the extension e_i of the ith member due to the actual loads and thermal strains. Accordingly, if $\partial N_i/\partial F$ is denoted by n_i,

$$q = \Sigma e_i n_i \qquad (4\text{-}69)$$

In many cases N_i is a linear homogeneous function of the external loads; that is, the superposition principle applies for the loads. This condition always applies for a statically determinate truss, if the deflections are small. In addition, it is generally valid for Hookean trusses. If N_i is a linear function of the applied loads, n_i is a constant. Consequently, the relation $n_i = \partial N_i/\partial F$ admits the following interpretation: if N_i is a linear homogeneous function of the applied loads, n_i is the tension in the ith member if $F = 1$ and all other loads are removed from the structure.

Relative Displacements of Joints. The reduction q of the distance between two joints of a truss—say joints J' and J''—equals the displacement component of joint J' on the line $J'J''$ plus the displacement component of joint J'' on the line $J''J'$. Consequently, by Eq. (4-69)

$$q = \Sigma e_i(n_i' + n_i'') \qquad (a)$$

If the tensions in the members are linear homogeneous functions of the applied loads, n_i' (or n_i'') is the tension in the ith member due to a unit load applied at joint J' (or joint J'') in the direction $J'J''$ (or the direction $J''J'$). Let us set $n_i' + n_i'' = n_i$. Then, if the tensions in the members are linear homogeneous functions of the applied loads, n_i is the tension in the

ith member if the truss is loaded only by a unit tension between joints J' and J''. With this definition of n_i, Eq. (4-69) determines the reduction of the distance between joints J' and J'', provided that the superposition principle applies for the loads. This conclusion follows from a comparison of Eq. (4-69) with Eq. (a).

Symmetry Conditions. The foregoing theory is equally applicable to symmetrical trusses and unsymmetrical trusses. However, a slight modification of the theory is advantageous if the truss and the external loads possess symmetry properties.

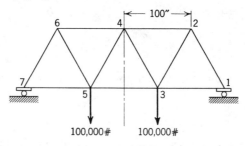

Fig. 4-5

Suppose, for example, that we wish to compute the reduction q of the distance between joints 2 and 5 of the truss shown in Fig. 4-5. By Eq. (4-69), $q = \Sigma e_i n_i'$, where e_i is the extension of the ith member due to the actual loads on the truss and n_i' is the tension in the ith member due to two external unit loads acting along the line 2–5 and tending to pull joints 2 and 5 together. This interpretation of n_i' is legitimate, since the tensions in the members are linear homogeneous functions of the applied loads. By symmetry the reduction of the distance between joints 2 and 5 equals the reduction of the distance between joints 3 and 6. Accordingly, q is also given by the equation $q = \Sigma e_i n_i''$, where n_i'' is the tension in the ith member due to a unit tension acting between joints 3 and 6. Therefore,

$$q = \Sigma e_i n_i' = \Sigma e_i n_i'' = \tfrac{1}{2}\Sigma e_i(n_i' + n_i'')$$

or

$$q = \Sigma e_i n_i, \quad \text{where} \quad n_i = \tfrac{1}{2}(n_i' + n_i'')$$

Since the superposition principle applies, n_i may be interpreted as the tension in the ith member caused by $\tfrac{1}{2}$-lb tensions acting simultaneously between joints (2, 5) and (3, 6) (see Fig. 4-7). Thus symmetry is preserved in the computation of the tensions n_i.

The extension of this idea to other symmetric trusses with symmetric loadings is self-evident. A truss may conceivably consist of several identical parts with identical deflections, say s parts. Then loads of magnitude $1/s$ should be applied to the parts for the computation of the tensions n_i.

Analysis of Statically Indeterminate Trusses by the Maxwell-Mohr Method. Any part of a truss may be regarded as an isolated mechanical system (free body). By the principles of momentum and moment of momentum, necessary conditions for statical equilibrium of the isolated part are that the vector sum of the external forces on it vanish and that the moment of these forces about any fixed point vanish. If these conditions determine the tensions in the members uniquely, the truss is said to be "statically determinate;" otherwise, "statically indeterminate." The expression "statical determinacy" implies that the deflections are so small that the angles between the members are appreciably constant; otherwise, the internal forces cannot be determined without consideration of the deformation.

To analyze a pin-jointed truss by the principle of stationary potential energy (Sec. 2-4), one need not distinguish between statical determinacy and statical indeterminacy. However, the tensions in the members of a statically determinate truss are obtained most directly by the balancing of forces and moments on isolated parts. If a statically determinate truss is analyzed by the principle of stationary potential energy, the tensions in the members are independent of the strain energy function.

A statically indeterminate truss is rendered statically determinate if certain members are removed. The deleted members are said to be "redundant."

Let us temporarily consider the tensions in the redundant members of a truss to be independent variables; that is, let us suppose that we may assign arbitrary values to these quantities. We may imagine that this situation is realized by installation of turnbuckles in the redundant members. When the external forces and the tensions in the redundant members are given, the tensions in the other members may be determined by the balancing of forces and moments on isolated parts.

We refer to the part of the truss that consists of nonredundant members as the "base structure." This, itself, is a statically determinate truss. The increase of the distance between any two joints of the base structure may be expressed in terms of the external loads and the tensions in the redundant members by the method of complementary energy already discussed in this article. Accordingly, the extensions of the redundant members corresponding to given tensions in these members may be computed. Now, if the extension e of a redundant member is also expressed in terms of its tension N by the load extension relation of that member, we obtain an equation among the tensions in the redundant members. By writing equations of this type for each of the redundant members, we obtain enough equations to determine all the tensions in the redundant members. This formulation of problems of statically indeterminate trusses was devised by Clerk Maxwell (1831–1879). It was later elaborated by Otto Mohr (1835–1918). The Maxwell-Mohr method introduces fewer unknowns and fewer simultaneous equations than the method of stationary potential energy.

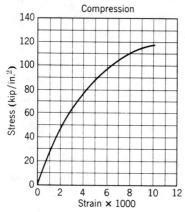
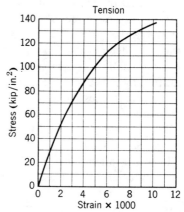

Fig. 4-6

Example.   Deflection of a Symmetrical Truss.   The members in the truss shown in Fig. 4-5 are each 100 in. long. Accordingly, the panels are equilateral triangles. The truss and the external loads are symmetrical about the vertical center line. The cross-sectional areas of the members are

$$A_{12} = A_{24} = A_{34} = 1.5 \text{ in.}^2$$
$$A_{23} = A_{13} = A_{35} = 1 \text{ in.}^2$$

The stress-strain curve for any member is represented by Fig. 4-6. Since there is no unloading, the distinction between inelasticity and nonlinear elasticity is irrelevant.

The reduction of the distance between joints 2 and 5 is required. The tensions in members 23, 24, and 35, as determined by the equations of equilibrium of the joints, are $N_{23} = 115,470$ lb, $N_{24} = -115,470$ lb, $N_{35} = 115,470$ lb. Consequently, the stresses are $\sigma_{23} = 115,470$ lb/in.^{2}, $\sigma_{24} = -76,980$ lb/in.^{2}, $\sigma_{35} = 115,470$ lb/in.^{2}. Hence, by Fig. 4-6, the strains are $\epsilon_{23} = 0.0065$, $\epsilon_{24} = -0.0041$, $\epsilon_{35} = 0.0065$. Accordingly, since the members are 100 in. long, their extensions are $e_{23} = 0.65$ in., $e_{24} = -0.41$ in., $e_{35} = 0.65$ in.

Since the truss is statically determinate, superposition of tensions is

legitimate. Since the truss consists of two identical parts with symmetrical loadings, we apply tensions of magnitude $\frac{1}{2}$ lb between joints 2 and 5 and between the corresponding joints 3 and 6 (Fig. 4-7). The tensions in the members due to these loads are
$n_{12} = n_{13} = n_{34} = 0$, $n_{23} = n_{24} =$ -0.2887 lb, $n_{35} = -0.5773$ lb. By Eq. (4-69) the reduction δ of the distance between joints 2 and 5 is

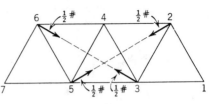

Fig. 4-7

$$\delta = 2n_{23}e_{23} + 2n_{24}e_{24} + n_{35}e_{35}$$

The factor 2 appears in this equation because of the symmetry relations, $n_{23}e_{23} = n_{56}e_{56}$ and $n_{24}e_{24} = n_{46}e_{46}$. With the preceding numerical values, we obtain

$$\delta = 2(-0.2887)(0.65) + 2(-0.2887)(-0.41) + (-0.5773)(0.65)$$

Therefore, $\delta = -0.51$ in. The negative sign signifies that the distance between joints 2 and 5 is increased 0.51 in.

Example of Maxwell-Mohr Method. In Fig. 4-8a the numbers adjacent to the members of the truss represent relative values of EA. Only relative values of EA are needed for computation of the tensions in the members. The truss has one redundant member—say the diagonal AC. This mem-

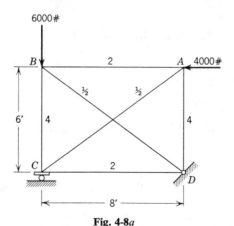

Fig. 4-8a

ber is conceived to be removed and replaced by its tension, $F = N_{AC}$ (Fig. 4-8b). By statics the tensions in the other members are

$$N_{AB} = -4000 - 0.8F, \qquad N_{BC} = -9000 - 0.6F$$
$$N_{BD} = F + 5000, \qquad N_{AD} = -0.6F, \qquad N_{CD} = -0.8F \tag{b}$$

The extension of a member with tension N and length L is NL/EA. Accordingly, by Eq. (b) the relative extensions of the members are

$$e_{AB} = -192{,}000 - 38.4F, \qquad e_{BC} = -162{,}000 - 10.8F$$
$$e_{BD} = 1{,}200{,}000 + 240F, \qquad e_{AD} = -10.8F, \qquad e_{CD} = -38.4F \qquad \text{(c)}$$

By statics the tensions in the members due to a unit tension acting between joints A and C (Fig. 4-8c) are

$$n_{AB} = -0.8, \qquad n_{BC} = -0.6, \qquad n_{BD} = 1, \qquad n_{AD} = -0.6,$$
$$n_{CD} = -0.8$$

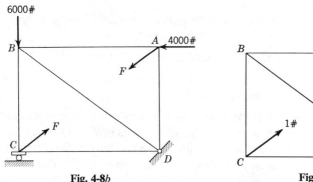

Fig. 4-8b   Fig. 4-8c

By Eq. (4-69) the reduction of the distance between joints A and C is

$$-e_{AC} = \sum n_i e_i = (-0.8)(-192{,}000 - 38.4F) + (-0.6)(-162{,}000 - 10.8F)$$
$$+ (1)(1{,}200{,}000 + 240F) + (-0.6)(-10.8F) + (-0.8)(-38.4F)$$
$$= 1{,}450{,}800 + 314.4F$$

In addition, the load-extension relation for member AC is $e_{AC} = 240F$. Hence,

$$-240F = 1{,}450{,}800 + 314.4F$$

Therefore, $F = N_{AC} = -2617$ lb. Equation (b) now yields

$$N_{AB} = -1906 \text{ lb}, \qquad N_{BC} = -7430 \text{ lb}, \qquad N_{BD} = 2383 \text{ lb},$$
$$N_{AD} = 1570 \text{ lb}, \qquad N_{CD} = 2094 \text{ lb}$$

4-13. COMPLEMENTARY ENERGY OF BEAMS. If a straight cantilever beam is subjected to an external couple M at the end, the rotation of the end, by Eq. (4-67), is $\theta = d\Upsilon/dM$. If the beam is uniform,

$\theta = L/R$, where R is the radius of curvature due to bending and L is the length of the beam. Hence

$$\Upsilon = L \int_0^M \frac{dM}{R}$$

The integral in this formula is determined by the moment-curvature relation of the beam. For convenience, the following notation is introduced:

$$f(M) = \int_0^M \frac{dM}{R} \tag{a}$$

Then, for a beam that is subjected to uniform bending, $\Upsilon = Lf(M)$. Consequently, for nonuniform bending, the complementary energy in an infinitesimal length ds is $f(M)\, ds$. Therefore, if any beam is subjected to a bending moment $M(s)$, where s is an axial coordinate,

$$\Upsilon = \int_0^L f(M)\, ds \tag{4-70}$$

This formula does not take account of shear deformation or axial forces.

By Eq. (4-70) the complementary energy of a structure that consists of flexural members (e.g., a frame) is

$$\Upsilon = \sum \int f(M_i)\, ds \tag{b}$$

where M_i is the bending moment at any point in the ith member, and the integral extends over the length of the ith member. The sum extends over all members. If P is a force applied to the structure, the component of deflection q at the point of application of force P and in the direction of force P is, by Eq. (4-64), $q = \partial \Upsilon / \partial P$. Hence, by Eq. (b)

$$q = \sum \int f'(M_i) \frac{\partial M_i}{\partial P}\, ds$$

By Eq. (a) $f'(M_i) = 1/R_i$, where R_i is the radius of curvature at an arbitrary point of the ith member due to the actual loads on the structure. Therefore, if $\partial M_i / \partial P = m_i$,

$$q = \sum \int \frac{m_i\, ds}{R_i} \tag{4-71}$$

If the bending moments vary linearly with the external forces, m_i is independent of the actual loads on the structure. We may then interpret m_i to be the bending moment in the ith member if $P = 1$, and all other loads are removed from the structure.

Likewise, if a couple M is applied to the structure, Eq. (4-67) shows that the rotation of the arm of the couple is $\theta = \partial \Upsilon / \partial M$. Hence, by Eq. (b)

$$\theta = \sum \int f'(M_i) \frac{\partial M_i}{\partial M} \, ds$$

Setting $\partial M_i / \partial M = m_i$, we obtain

$$\theta = \sum \int \frac{m_i \, ds}{R_i} \tag{4-72}$$

Example. Cantilever Beam with Parabolic Elasticity. A lateral load P acts at the end of a cantilever beam. The cross section of the beam is a rectangle of depth $2h$ and width b. The stress-strain relation is $\sigma = K\sqrt{|\epsilon|} \, \mathrm{sgn} \, \epsilon$, where K is a constant and $\mathrm{sgn} \, \epsilon$ (read "signum ϵ") is defined as $+1$ if $\epsilon > 0$, -1 if $\epsilon < 0$, and 0 if $\epsilon = 0$. By statics, the bending moment is

$$M = 2b \int_0^h \sigma \eta \, d\eta$$

where η is an ordinate in the cross section, measured from the neutral axis. Hence

$$M = 2bK \int_0^h \epsilon^{\frac{1}{2}} \eta \, d\eta$$

Since $\epsilon = \eta/R$, this yields

$$M = \frac{4bKh^{\frac{5}{2}}}{5R^{\frac{1}{2}}} \quad \text{or} \quad \frac{1}{R} = \frac{25M^2}{16 b^2 K^2 h^5}$$

By Eqs. (4-71) and (4-72)

$$\theta, q = \int_0^L \frac{m \, ds}{R} = \frac{25P^2}{16 b^2 K^2 h^5} \int_0^L s^2 m \, ds \tag{c}$$

since $M = Ps$. Applying a unit vertical load at the free end, we obtain $m = s$, and Eq. (c) yields

$$q = \frac{25P^2 L^4}{64 b^2 K^2 h^5}$$

By applying a unit bending moment at the free end we obtain $m = 1$ and Eq. (c) yields

$$\theta = \frac{25P^2 L^3}{48 b^2 K^2 h^5}$$

4-14. UNIT-DUMMY-LOAD METHOD. Let P denote an external point force (or an external couple) acting on a Hookean structure. Let the deflections be small, and let the deflection component of the point of application of load P along the line of action of P (or the angular displacement of the arm of the couple P) be denoted by q. Since the material is Hookean, $U = \Upsilon$. Consequently, by Castigliano's principle [Eq. (4-64) or (4-67)], $q = \partial U / \partial P$.

If the structure consists of beams, columns, ties, and torque bars, its strain energy is determined by the formulas in Sec. 2-1. Consequently, if the tension, bending moment, shear, and torque in a member are denoted by N, M, S, and T, respectively, and if x is an axial coordinate along a member,

$$q = \frac{\partial U}{\partial P} = \sum \int \left(\frac{N \, \partial N / \partial P}{EA} + \frac{M \, \partial M / \partial P}{EI} + \frac{\kappa S \, \partial S / \partial P}{GA} + \frac{T \, \partial T / \partial P}{GJ} \right) dx$$

where the sum extends over all members.

It is shown in the theory of elasticity that the effects of various loads may be superimposed, provided that the strains and the rotations are sufficiently small and the elasticity is linear. Under these conditions, the quantities N, M, S, and T are linear functions of the external loads and couples. Consequently, the derivatives $\partial N / \partial P$, $\partial M / \partial P$, $\partial S / \partial P$, and $\partial T / \partial P$ are constants. These constants are denoted by n, m, s, and t, respectively. Accordingly,

$$q = \sum \int \left(\frac{nN}{EA} + \frac{mM}{EI} + \frac{\kappa s S}{GA} + \frac{tT}{GJ} \right) dx \tag{4-73}$$

The significance of the constants is made clear by considering the load (or couple) P as acting alone. Then, since N, M, S, and T are linear homogeneous functions of the external loads and couples,

$$N = nP, \qquad M = mP, \qquad S = sP, \qquad T = tP$$

Accordingly, m, n, s, and t are, respectively, the moment, tension, shear, and torque due to a unit load (or unit couple) P acting alone. Consequently, we have the following useful theorem:

> To determine the displacement of a given particle in a given direction (or the angular deflection of a given line segment), let a unit load act at the given particle in the given direction (or let the given line segment be the arm of a unit couple). Let the moments, tensions, shears, and torques in the various members, due to the unit load (or couple) be denoted respectively by m, n, s, and t. Let the moments, tensions, shears, and torques, due to the real loads on the structure, be denoted by M, N, S, and T. The required deflection q is then given by Eq. (4-73).

This theorem is limited to small deflections of linearly elastic structures that consist of beams, ties, and struts. The structure may contain curved members, provided that the curvatures are not so great that the elementary formulas of mechanics of materials for straight members cease to be applicable. Many applications of the foregoing theorem are given in the book by Van Den Broek (88).

Example. **Angular Deflection of a Member in a Frame.** For the frame shown in Fig. 4-9, $L_1 = 40$, $L_2 = 20\sqrt{2}$, $L_3 = 20$, $L_4 = L_5 = 20$, $A_1 = A_2 = A_3 = A_4 = A_5 = 2$, $I_4 = I_5 = 3$. The angular deflection q of member 2 is required.

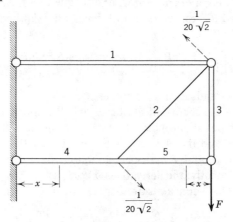

Fig. 4-9

The tensions, moments, and shears in the various members, due to load F, as determined by statics, are

$$N_1 = 2F \qquad M_1 = 0 \qquad S_1 = 0$$
$$N_2 = -2\sqrt{2}F \qquad M_2 = 0 \qquad S_2 = 0$$
$$N_3 = 2F \qquad M_3 = 0 \qquad S_3 = 0$$
$$N_4 = -2F \qquad M_4 = Fx \qquad S_4 = F$$
$$N_5 = 0 \qquad M_5 = Fx \qquad S_5 = -F$$

Since the angular deflection of member 2 is required, we place a unit couple on this member; that is, we impose equal and opposite forces of magnitude $\sqrt{2}/40$ at the extremities of the member. The tensions, moments, and shears in the various members, due to this couple, are

$$n_1 = -\tfrac{1}{20} \qquad m_1 = 0 \qquad s_1 = 0$$
$$n_2 = \sqrt{2}/40 \qquad m_2 = 0 \qquad s_2 = 0$$
$$n_3 = 0 \qquad m_3 = 0 \qquad s_3 = 0$$
$$n_4 = \tfrac{1}{20} \qquad m_4 = 0 \qquad s_4 = 0$$
$$n_5 = 0 \qquad m_5 = 0 \qquad s_5 = 0$$

Since the m's and s's are zero, Eq. (4-73) reduces to

$$q = \frac{n_1 N_1 L_1}{E_1 A_1} + \frac{n_2 N_2 L_2}{E_2 A_2} + \frac{n_4 N_4 L_4}{E_4 A_4}$$

Consequently, $q = -(F/E)(3 + \sqrt{2})$. The negative sign indicates that the angular deflection is clockwise, since the unit couple was taken counterclockwise.

4-15. ANALYSIS OF STATICALLY INDETERMINATE STRUCTURES BY THE UNIT-DUMMY-LOAD METHOD.

A structure is essentially unchanged if we remove the redundant members or redundant supports and replace them by external forces and moments that are identical to the forces and moments that were imposed by the deleted parts. These forces and moments are initially unknown, but we may denote them by $R_1, R_2, \cdots$. We may derive formulas for the linear and angular deflections of various parts of the simplified statically determinate structure that carries the prescribed loading and the statically indeterminate reactions, $R_1, R_2, \cdots$. These deflections, of course, are functions of $R_1, R_2, \cdots$. Accordingly, by setting the deflections equal to zero, we obtain equations that determine $R_1, R_2, \cdots$. The procedure is best illustrated by examples.

Example. Beam with a Statically Indeterminate Reaction. For the beam shown in Fig. 4-10a, the reaction R is regarded as a statically indeterminate quantity. The right-hand support is conceived to be removed, but the force R is retained. We compute the deflection of the free end of the beam for the loading shown in Fig. 4-10b. Subsequently, R is determined by the condition that this deflection is zero.

The bending moment for the loading shown in Fig. 4-10b is

$$M = Rx - \tfrac{1}{2}wx^2$$

To apply the unit-dummy-load method, we consider the moment due to a 1-lb load applied at the free end (Fig. 4-10c). This moment is $m = x$. Neglecting the effect of shear, we then have, by Eq. (4-73),

$$q = \int_0^L \frac{Mm}{EI}\,dx = \frac{1}{EI}\int_0^L (Rx^2 - \tfrac{1}{2}wx^3)\,dx$$

Hence

$$q = \frac{RL^3}{3EI} - \frac{wL^4}{8EI}$$

Since R is the reaction of a fixed support, $q = 0$. Therefore,

$$R = \tfrac{3}{8}wL$$

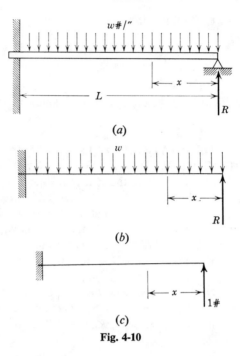

(a)

(b)

(c)

Fig. 4-10

Example. Internal Forces and Moments in a Ring. A large ring is loaded as shown in Fig. 4-11. The bending moment and the tension are required at the point where the load P is applied.

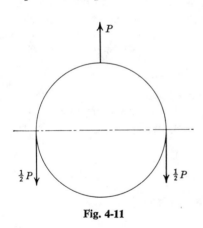

Fig. 4-11

Since the stress distribution is symmetrical about the vertical center line, we need consider only half the ring. In addition, since a rigid-body displacement has no effect on the stresses, we let the lowest point of the ring be clamped (Fig. 4-12a). It is clear that the angular deflection and the horizontal displacement at the topmost point of the ring are zero. By symmetry the shear at the topmost point is $\frac{1}{2}P$.

It is assumed that the strain energy is primarily due to bending. Since the radius of the ring is large compared to the thickness, the strain energy of bending is determined by the straight-beam formula.

Figure 4-12a shows that the bending moment due to the applied loads is

$$M = M_0 + N_0R(1 - \cos \theta) - \tfrac{1}{2}PR \sin \theta, \qquad 0 < \theta < \tfrac{1}{2}\pi$$
$$M = M_0 + N_0R(1 - \cos \theta) - \tfrac{1}{2}PR, \qquad \tfrac{1}{2}\pi < \theta < \pi \tag{a}$$

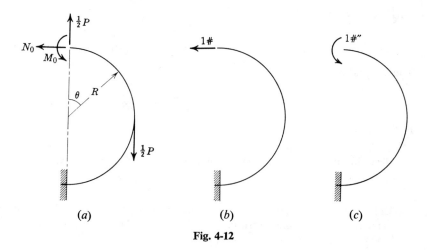

(a)   (b)   (c)

Fig. 4-12

To compute the horizontal deflection q at the top, we apply a unit dummy horizontal load at this point (Fig. 4-12b). The moment is then $m = R(1 - \cos \theta)$. Hence, since $q = 0$, Eq. (4-73) yields

$$q = \int \frac{Mm}{EI} ds = \frac{R^2}{EI} \int_0^{\pi} M(1 - \cos \theta) \, d\theta = 0$$

or

$$\int_0^{\pi} M(1 - \cos \theta) \, d\theta = 0 \tag{b}$$

To compute the angular deflection α at the top, we apply a unit moment at this point (Fig. 4-12c). The moment at any point is then $m = 1$. Hence, since $\alpha = 0$, Eq. (4-73) yields

$$\alpha = \int \frac{Mm}{EI} ds = \frac{R}{EI} \int_0^{\pi} M \, d\theta = 0$$

or

$$\int_0^{\pi} M \, d\theta = 0 \tag{c}$$

Equations (b) and (c) yield

$$\int_0^{\pi} M \cos \theta \, d\theta = 0 \tag{d}$$

Substitution of Eq. (a) into Eqs. (c) and (d) yields

$$M_0 \int_0^\pi d\theta + N_0 R \int_0^\pi (1 - \cos\theta)\, d\theta - \tfrac{1}{2} PR \int_0^{\pi/2} \sin\theta\, d\theta - \tfrac{1}{2} PR \int_{\pi/2}^\pi d\theta = 0$$

$$M_0 \int_0^\pi \cos\theta\, d\theta + N_0 R \int_0^\pi (1 - \cos\theta) \cos\theta\, d\theta$$

$$- \tfrac{1}{2} PR \int_0^{\pi/2} \sin\theta \cos\theta\, d\theta - \tfrac{1}{2} PR \int_{\pi/2}^\pi \cos\theta\, d\theta = 0$$

These equations yield

$$N_0 = \frac{P}{2\pi}, \qquad M_0 = \frac{PR}{4}$$

PROBLEMS

1. Determine the principal strains, the direction cosines of the principal axes of strain in the undeformed medium, and the volumetric strain corresponding to the strain components, $\epsilon_x = 5,\ \epsilon_y = 5,\ \epsilon_z = 6,\ \gamma_{yz} = -2\sqrt{2},\ \gamma_{zx} = 2\sqrt{2},$ $\gamma_{xy} = -2$.

2. Determine the principal strains, the principal axes of strain in the undeformed medium, and the volumetric strain corresponding to the displacement vector, $u = x - 2y,\ v = 3x + 2y,\ w = 5z$.

3. For plane stress, $\sigma_z = \tau_{xz} = \tau_{yz} = 0$. Also, $\sigma_x,\ \sigma_y,\ \tau_{xy}$ are independent of z. Let (x, y) and (ξ, η) be two sets of plane rectangular coordinates, and let the angle between the ξ- and x-axes be θ. For plane stress, express $\sigma_\xi,\ \sigma_\eta,\ \tau_{\xi\eta}$ in terms of $\sigma_x,\ \sigma_y,\ \tau_{xy}$ by balancing forces on a differential prismatic element. Hence prove that the following expressions are invariants:

$$\sigma_x + \sigma_y, \qquad \begin{vmatrix} \sigma_x & \tau_{xy} \\ \tau_{xy} & \sigma_y \end{vmatrix}$$

4. The stress components are $\sigma_x = 2,\ \sigma_y = 1,\ \sigma_z = -2,\ \tau_{yz} = 8,\ \tau_{zx} = -1,$ $\tau_{xy} = 3$. Determine the principal stresses and the direction cosines of the axes of principal stress.

5. The stress components are $\sigma_x = 0,\ \sigma_y = 0,\ \sigma_z = -10,000,\ \tau_{yz} = -5000,$ $\tau_{zx} = 5000,\ \tau_{xy} = 5000$. Determine the principal stresses and the direction cosines of the axes of principal stress. Hence determine the maximum shearing stress.

6. The stress components are $\sigma_x = 3000,\ \sigma_y = -6000,\ \sigma_z = 15,000,\ \tau_{yz} = 1000,$ $\tau_{zx} = 0,\ \tau_{xy} = 2000$. Determine the stress vector, the normal stress, and the shearing stress on a plane whose normal has direction cosines $(\tfrac{2}{3}, \tfrac{2}{3}, \tfrac{1}{3})$.

7. Specialize the strain-displacement relations and the equilibrium equations for any orthogonal coordinates to obtain the corresponding relations for cylindrical coordinates. For spherical coordinates.

8. By means of Eq. (4-50) express the potential energy of an isotropic Hookean body that is in equilibrium under the action of body force $(\rho F_x, \rho F_y, \rho F_z)$,

temperature $\theta(x, y, z)$, and surface forces in terms of the displacement vector (u, v, w), where (x, y, z) are rectangular coordinates. Hence derive the differential equations for (u, v, w) by the principle of stationary potential energy.

9. Prove that crystals of the regular system possess three elastic constants and one thermal constant. Derive the expression for U_0.

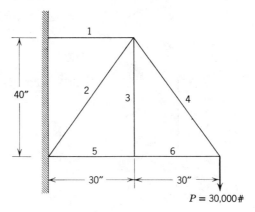

Fig. P4-10

10. The bars in the truss shown in Fig. P4-10 have the following cross-sectional areas: $A_1 = 1$ in.^{2}, $A_2 = A_4 = 2$ in.^{2}. $A_3 = \frac{1}{2}$ in.^{2}, $A_5 = A_6 = \frac{3}{2}$ in.^{2}. The elastic behavior of a bar is represented by the stress-strain relation $\sigma = 200,000\epsilon^{1/3}$ lb/in.^{2}. Compute the vertical deflection of the right-hand joint.

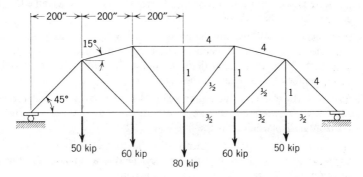

Fig. P4-11

11. The numbers adjacent to the members of the symmetric truss (Fig. P4-11) denote the cross-sectional areas (in.^{2}). Calculate the horizontal displacements of the end supports and the vertical displacement of the center joint, using the stress-strain diagram of Fig. 4-6.

12. Solve the problem of the pentagonal truss (Sec. 2-4) by the Maxwell-Mohr method.

13. Solve Prob. 21, Chap. 2, by the Maxwell-Mohr method.

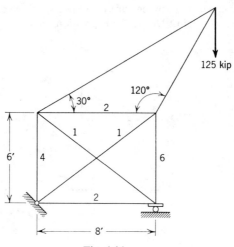

Fig. 4-14

14. The numbers adjacent to the members of the truss (Fig. P4-14) denote relative values of EA. Compute the tensions in the members by the Maxwell-Mohr method.

15. By Castigliano's theorem, derive a formula for the deflection at the center of a simple Hookean beam of rectangular cross section with a single load at the center, taking shear deformation into account.

16. A simple beam with rectangular cross section of depth h and width b carries a concentrated load P at its center. The stress-strain relation of the material is $\sigma = K\epsilon^{\frac{1}{3}}$. Neglecting shear deformation, derive a formula for the deflection at the center by Castigliano's theorem.

17. The uniform elastic semicircular beam is loaded by a vertical force P, as shown in Fig. P4-17. By the unit-dummy-load method, determine the horizontal displacement at the free end.

18. A beam that carries a uniformly distributed load has immovable hinge supports at the ends and at the center. Derive a formula for the center reaction, taking shear deformation into account.

19. The width of the spring is 2 in., and $E = 3 \times 10^7$ lb/in.^{2} (Fig. P4-19). Calculate the horizontal and vertical deflections of the free end and the angular deflection of the free end by the unit-dummy-load method. Consider only strain energy of bending and use the straight-beam formula for strain energy.

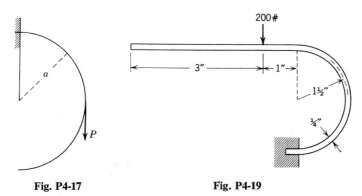

Fig. P4-17   Fig. P4-19

20. A steel piston ring with 3.3-in. mean diameter and 0.125-in. square cross section has 0.375-in. gap at the ends. What tangential force at the ends will close the gap?

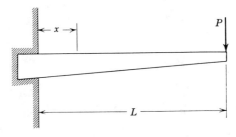

Fig. P4-21

21. For the tapered Hookean cantilever beam (Fig. P4-21) $I = (ax + b)^{-1}$, where a and b are constants. By Castigliano's theorem, derive a formula for the deflection of the free end, neglecting shear deformation.

22. Assuming that the shear q is proportional to $|\sin \theta|$ and that equilibrium exists (Fig. P4-22), derive formulas for the bending moment and the tension at the point at which the load is applied. Neglect shear deformation and suppose that the material is Hookean. Also, assume that the shear q acts at the centroidal axis.

23. A uniform semicircular arch that is hinged at the ends carries a vertical concentrated load P at the top. Calculate the horizontal reactions at the ends.

24. The uniform semicircular member (Fig. P4-24) is clamped at the left end and is loaded by a force P perpendicular to the plane of the semicircle at the mid-point. The right end is free. The torsional stiffness is GJ and the flexural stiffness is EI. Derive a formula for the deflection at point A in the direction of force P.

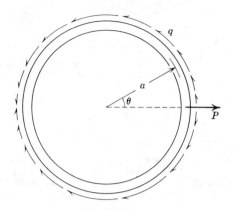

Fig. P4-22

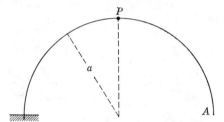

Fig. P4-24

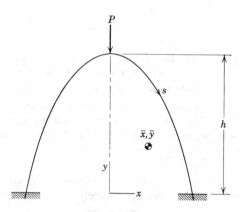

Fig. P4-25

25. The symmetrical arch is defined parametrically by $x = x(s)$, $y = y(s)$, where s is arc length measured from the top (Fig. P4-25). The arc length of the entire arch is $2L$. The stiffness EI is constant. By Castigliano's theorem or the unit-dummy-load method, determine the bending moment M_0 and the tension N_0 at the top. Consider only strain energy of bending. Express the integrals in terms of L, h, $\bar{x}$, $\bar{y}$, I_{xy}, and I_{xx}, where $(\bar{x}, \bar{y})$ are the coordinates of the centroid of the right half of the curve, and

$$I_{xy} = \int_0^L xy \, ds, \qquad I_{xx} = \int_0^L y^2 \, ds$$

26. The uniform semicircular Hookean bar is clamped at one end, and it is subjected to torque T_0 at the free end. (The vector T_0 in Fig. P4-26 represents

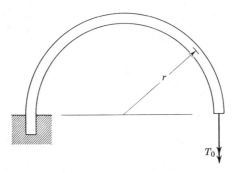

Fig. P4-26

torque in accordance with the right-hand-screw rule.) The flexural stiffness is EI and the torsional stiffness is GJ. By the unit-dummy-load method, determine the twist and the displacement at the free end.

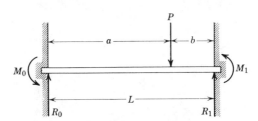

Fig. P4-27

27. By the unit-dummy-load method, determine the end moments M_0, M_1 and the end reactions R_0, R_1 for the clamped beam (Fig. P4-27). Neglect shear deformation and suppose that the material follows Hooke's law.

28. Using the strain-energy formula for a ring [Eq. (2-28)], derive the formula for curved beams corresponding to Eq. (4-73). Use the formula to determine the end deflection of the curved cantilever beam of Sec. 3-6 by the unit-dummy-load method.

29. $\sigma_x = \sigma_y = \sigma_z = 0, \qquad \tau_{yz} = \tau_{zx} = \tau_{xy} = \tau > 0$

Determine the principal stresses and the maximum shearing stress. Determine the direction cosines of the principal axis for which the corresponding principal stress is tensile.

30. Let (x, y, z) be curvilinear coordinates such that the coordinate lines coincide with the lines of principal stress. Obtain the differential equations of equilibrium for this case.

5 Theory of plates and shells

This function (strain energy) being known, we can immediately apply the general method given in the *Mécanique analytique*, and which appears to be more especially applicable to problems that relate to the motions of systems composed of an immense number of particles mutually acting upon each other. One of the advantages of this method, of great importance, is that we are necessarily led by the mere process of calculation, and with little care on our part, to all the equations and conditions which are requisite and sufficient for the complete solution of any problem to which it may be applied.

GEORGE GREEN

5-1. THE VON KÁRMÁN THEORY OF FLAT PLATES. In the initial state two of the surfaces of a flat plate are parallel planes, called the "faces" of the plate. The distance between the faces is called the thickness of the plate. A plate is distinguished from a block by the fact that the thickness is small compared to the dimensions parallel to the faces. The plane midway between the faces is called the "middle plane" of the plate. It is convenient to adopt rectangular coordinates (x, y, z), such that the x- and y-axes lie in the middle plane.

In the classical theory of flat plates the quadratic terms in the strain tensor are neglected. This approximation fails to account for an effect that often contributes materially to the resistive action of plates—the so-called "membrane effect," by which the tensions in a deflected plate provide components that react the applied lateral load. An improved theory was introduced by von Kármán (86) who suggested that quadratic terms in w_x and w_y be retained in the strain tensor [Eqs. (4-3)] but that other quadratic terms be dropped. It appears reasonable to drop $u_x{}^2$ from the formula for ϵ_x, since u_x occurs also to the first degree in this formula

[Eqs. (4-3)]. The same argument applies for $v_y{}^2$ and $w_z{}^2$ in the formulas for ϵ_y and ϵ_z and for $v_y v_z$, $w_y w_z$, $u_z u_x$, $w_z w_x$, $u_x u_y$, $v_x v_y$ in the formulas for γ_{yz}, γ_{zx}, γ_{xy}. The other quadratic terms in u and v are dropped because they have about the same magnitude as the squares of the strain components and the terms that have already been discarded. On the other hand, w_x and w_y are the slopes of cross sections of the deformed middle plane; apparently, they may be quite large compared to the strain components.

It will be convenient to denote the displacement of an arbitrary point (x, y, z) in the plate by (u, v, w) and to denote the displacement of the corresponding point $(x, y, 0)$ of the middle plane by $(\bar{u}, \bar{v}, \bar{w})$. Likewise, $(\epsilon_x, \epsilon_y, \cdots, \gamma_{xy})$ will denote the strain components at any point (x, y, z) and $(e_x, e_y, \cdots, e_{xy})$ will denote the corresponding strain components at the point $(x, y, 0)$. After discarding all quadratic terms from Eqs. (4-3), except $w_x{}^2$, $w_y{}^2$, $w_x w_y$, we obtain

$$\epsilon_x = u_x + \tfrac{1}{2}w_x{}^2, \qquad \epsilon_y = v_y + \tfrac{1}{2}w_y{}^2, \qquad \epsilon_z = w_z$$
$$\gamma_{yz} = w_y + v_z, \qquad \gamma_{zx} = u_z + w_x, \qquad \gamma_{xy} = v_x + u_y + w_x w_y \tag{a}$$

To determine how u and v vary through the thickness, we introduce the tentative assumption, $\epsilon_z = 0$. This assumption is obviously inadmissible in the stress-strain relations. However, the variation of w through the thickness has little effect on u and v. Consequently, the most expeditious assumption is used; namely, $\epsilon_z = 0$ or $w_z = 0$. This signifies that $w = \bar{w}$.

If there are no external shearing forces applied to the faces of the plate and the material is isotropic, γ_{yz} and γ_{zx} vanish at the faces, since $\tau_{yz} = G\gamma_{yz}$ and $\tau_{zx} = G\gamma_{zx}$. Since the plate is thin, γ_{yz} and γ_{zx} are then small everywhere. Accordingly, it is assumed that $\gamma_{yz} = \gamma_{zx} = 0$. The relations $\gamma_{yz} = \gamma_{zx} = \epsilon_z = 0$ signify that lines normal to the middle plane remain straight and normal under the deformation. This condition is analogous to Bernoulli's hypothesis that cross sections of a beam remain plane and normal to the centroidal axis; it is known as the "Kirchhoff assumption" in the theory of plates.

With Eq. (a), the assumptions $\gamma_{yz} = \gamma_{zx} = \epsilon_z = 0$ yield

$$u = \bar{u} - zw_x, \qquad v = \bar{v} - zw_y, \qquad w = \bar{w} \tag{5-1}$$

and

$$\epsilon_x = e_x - zw_{xx}, \qquad \epsilon_y = e_y - zw_{yy}, \qquad \gamma_{xy} = e_{xy} - 2zw_{xy} \tag{5-2}$$

where

$$e_x = \bar{u}_x + \tfrac{1}{2}w_x{}^2, \qquad e_y = \bar{v}_y + \tfrac{1}{2}w_y{}^2, \qquad e_{xy} = \bar{v}_x + \bar{u}_y + w_x w_y \tag{5-3}$$

The shearing stresses τ_{xz} and τ_{yz} have already been neglected. In the stress-strain relations σ_z is also neglected, since it has about the same magnitude as the external pressures applied to the faces of the plate. In other words, a state of plane stress is now assumed. Restricting attention to isotropic elastic plates, we obtain from Eqs. (4-54)

$$\epsilon_x = \frac{1}{E}(\sigma_x - \nu\sigma_y) + k\theta$$

$$\epsilon_y = \frac{1}{E}(\sigma_y - \nu\sigma_x) + k\theta, \qquad \gamma_{xy} = \frac{\tau_{xy}}{G}$$

(5-4)

Inversion of these equations yields

$$\sigma_x = \frac{E}{1-\nu^2}(\epsilon_x + \nu\epsilon_y) - \frac{Ek\theta}{1-\nu}$$

$$\sigma_y = \frac{E}{1-\nu^2}(\epsilon_y + \nu\epsilon_x) - \frac{Ek\theta}{1-\nu}, \qquad \tau_{xy} = G\gamma_{xy}$$

(5-5)

By Eq. (4-56), the strain-energy density is

$$U_0 = \frac{1}{2E}\left[\sigma_x^2 + \sigma_y^2 - 2\nu\sigma_x\sigma_y + 2(1+\nu)\tau_{xy}^2\right]$$

(5-6)

If an irrelevant additive constant containing θ is discarded, Eqs. (5-5) and (5-6) yield

$$U_0 = \frac{G}{1-\nu}\left[\epsilon_x^2 + \epsilon_y^2 + 2\nu\epsilon_x\epsilon_y + \tfrac{1}{2}(1-\nu)\gamma_{xy}^2 - 2(1+\nu)(\epsilon_x + \epsilon_y)k\theta\right]$$

(5-7)

The strain energy of the entire plate is

$$U = \iint \left(\int_{-h/2}^{h/2} U_0 \, dz\right) dx \, dy$$

where h is the thickness of the plate. Integration with respect to z is facilitated by the following notation:

$$\theta_0 = \int_{-h/2}^{h/2} k\theta \, dz, \qquad \theta_1 = \int_{-h/2}^{h/2} zk\theta \, dz$$

(5-8)

When Eqs. (5-2) and (5-7) are substituted into the preceding formula for U, the strain energy separates into a sum, $U = U_m + U_b + U_\theta$, where the membrane energy U_m is linear in h and the bending energy U_b is cubic in h. The term U_θ represents the part of the strain energy that results from heating. If G, h, and ν are considered to be constants, the result of integration with respect to z is

$$U_m = \frac{Gh}{1 - \nu} \iint [e_x^2 + e_y^2 + 2\nu e_x e_y + \tfrac{1}{2}(1 - \nu)e_{xy}^2]\, dx\, dy$$

$$U_b = \frac{Gh^3}{12(1 - \nu)} \iint [w_{xx}^2 + w_{yy}^2 + 2\nu w_{xx}w_{yy} + 2(1 - \nu)w_{xy}^2]\, dx\, dy \quad (5\text{-}9)$$

$$U_\theta = \frac{E}{1 - \nu} \iint [-(e_x + e_y)\theta_0 + (w_{xx} + w_{yy})\theta_1]\, dx\, dy$$

By means of Eq. (5-3), e_x, e_y, e_{xy} are expressed in terms of $\bar{u}$, $\bar{v}$, w.

If a lateral distributed load $p(x, y)$ is applied to the plate, the potential energy of the load is

$$\Omega = -\iint wp\, dx\, dy$$

The total potential energy of the plate is $V = U_m + U_b + U_\theta + \Omega$. The Euler equations for the integral V are the equilibrium equations of the plate, expressed in terms of $\bar{u}, \bar{v}, w$. By Eq. (3-20), these equations are

$$\frac{\partial}{\partial x}(e_x + \nu e_y) + \tfrac{1}{2}(1 - \nu)\frac{\partial e_{xy}}{\partial y} = \frac{1 + \nu}{h}\frac{\partial \theta_0}{\partial x}$$

$$\tfrac{1}{2}(1 - \nu)\frac{\partial e_{xy}}{\partial x} + \frac{\partial}{\partial y}(e_y + \nu e_x) = \frac{1 + \nu}{h}\frac{\partial \theta_0}{\partial y}$$

$$\nabla^2\nabla^2 w + \frac{12(1 + \nu)}{h^3}(\theta_0\nabla^2 w + \nabla^2\theta_1) = \frac{p}{D}$$

$$+ \frac{12}{h^2}[(e_x + \nu e_y)w_{xx} + (e_y + \nu e_x)w_{yy} + (1 - \nu)e_{xy}w_{xy}]$$

$(5\text{-}10)$

where

$$D = \frac{Eh^3}{12(1 - \nu^2)}$$

The constant D is called the "flexural rigidity" of the plate. The last of Eqs. (5-10) has been simplified with the aid of the first two equations.

A rectangular element of the plate with dimensions dx and dy is subjected to the forces $N_x\, dy$, $N_y\, dx$, $N_{xy}\, dy$, and $N_{yx}\, dx$ (Fig. 5-1). The tractions N_x, N_y, N_{xy}, N_{yx} are expressed in terms of the stresses by means of the statical relations

$$N_x = \int_{-h/2}^{h/2} \sigma_x\, dz, \qquad N_y = \int_{-h/2}^{h/2} \sigma_y\, dz, \qquad N_{xy} = N_{yx} = \int_{-h/2}^{h/2} \tau_{xy}\, dz$$

$(5\text{-}11)$

Hence, by Eqs. (5-2) and (5-5),

$$N_x = \frac{Eh}{1 - \nu^2} (e_x + \nu e_y) - \frac{E\theta_0}{1 - \nu}$$

$$N_y = \frac{Eh}{1 - \nu^2} (e_y + \nu e_x) - \frac{E\theta_0}{1 - \nu} \qquad (5\text{-}12)$$

$$N_{xy} = N_{yx} = Ghe_{xy}$$

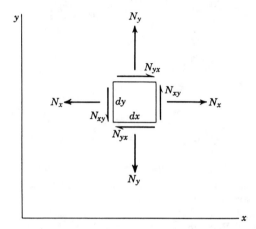

Fig. 5-1

Equations (5-10) and (5-12) yield

$$\frac{\partial N_x}{\partial x} + \frac{\partial N_{xy}}{\partial y} = 0, \qquad \frac{\partial N_{xy}}{\partial x} + \frac{\partial N_y}{\partial y} = 0 \qquad (5\text{-}13)$$

$$D\nabla^2\nabla^2 w + \frac{E}{1 - \nu} \nabla^2\theta_1 = p + N_x w_{xx} + N_y w_{yy} + 2N_{xy} w_{xy} \qquad (5\text{-}14)$$

The general solution of Eq. (5-13) is

$$N_x = F_{yy}, \qquad N_y = F_{xx}, \qquad N_{xy} = -F_{xy} \qquad (5\text{-}15)$$

where $F(x, y)$ is an arbitrary function known as the "Airy stress function" (57). Hence Eq. (5-14) yields

$$D\nabla^2\nabla^2 w + \frac{E}{1 - \nu} \nabla^2\theta_1 = p + F_{yy} w_{xx} + F_{xx} w_{yy} - 2F_{xy} w_{xy} \qquad (5\text{-}16)$$

A second relation between the functions F and w is obtained immediately. Equations (5-12) and (5-15) yield

$$he_x = \frac{1}{E}(N_x - \nu N_y) + \theta_0 = \frac{1}{E}(F_{yy} - \nu F_{xx}) + \theta_0$$

$$he_y = \frac{1}{E}(N_y - \nu N_x) + \theta_0 = \frac{1}{E}(F_{xx} - \nu F_{yy}) + \theta_0 \qquad (5\text{-}17)$$

$$he_{xy} = \frac{N_{xy}}{G} = -\frac{F_{xy}}{G}$$

Equation (5-3) yields

$$\frac{\partial^2 e_x}{\partial y^2} + \frac{\partial^2 e_y}{\partial x^2} - \frac{\partial^2 e_{xy}}{\partial x\,\partial y} = w_{xy}^2 - w_{xx}w_{yy}$$

Eliminating e_x, e_y, and e_{xy} by means of Eqs. (5-17), we obtain

$$\nabla^2\nabla^2 F + E\nabla^2\theta_0 + Eh(w_{xx}w_{yy} - w_{xy}^2) = 0 \qquad (5\text{-}18)$$

Equations (5-16) and (5-18) are the fundamental relations in the von Kármán theory of plates. The temperature term in Eq. (5-18) was introduced by Gossard, Seide, and Roberts (114).

Strain Energy of Bending.   The formula for the strain energy of bending [Eqs. (5-9)] may be written as follows:

$$U_b = \tfrac{1}{2}D\iint \left[(w_{xx} + w_{yy})^2 - 2(1 - \nu)(w_{xx}w_{yy} - w_{xy}^2)\right] dx\,dy \qquad (5\text{-}19)$$

The expression $K = w_{xx}w_{yy} - w_{xy}^2$ is an approximation for the Gaussian curvature of the deflected middle surface (Sec. 5-6). It satisfies Euler's equation of the calculus of variations automatically [Eq. (3-20)]; hence it contributes nothing to the differential equations of the plate [Eqs. (5-10)]. This circumstance reflects the fact that K may be transformed into a line integral on the boundary (121). However, K must not be discarded from Eq. (5-19) indiscriminately, since it may contribute to the natural boundary conditions. If the edge of the plate is clamped, there are no natural boundary conditions, and then K may be dropped. Furthermore, the Gauss-Bonnet theorem of differential geometry shows that K contributes nothing to the natural boundary conditions if the edges of the plate are geodesics on the deformed middle surface (121). In particular, if the plate is polygonal and the edges remain straight when the plate is deformed, K is irrelevant. This case was noted by Nadai (59).

5-2. SMALL-DEFLECTION THEORY OF PLATES. The classical small-deflection equations of plates result from von Kármán's theory if the quadratic terms in Eq. (5-3) are discarded. Then Eqs. (5-1), (5-2), (5-4), (5-5), (5-6), (5-7), (5-8), (5-9), (5-11), (5-12), (5-13), (5-15), (5-17), and (5-19) are unchanged. Also, the first two equations in (5-10) are unchanged. The third equation of (5-10) becomes

$$\nabla^2\nabla^2 w + \frac{12(1 + \nu)}{L^3}\nabla^2\theta_1 = \frac{p}{D} \tag{5-20}$$

Equation (5-18) becomes

$$\nabla^2\nabla^2 F + E\nabla^2\theta_0 = 0 \tag{5-21}$$

Accordingly, the coupling between the lateral deflection w and the in-plane displacements $(\bar{u}, \bar{v})$ is eliminated. The stresses due to the deflection w and the stresses due to the displacements $(\bar{u}, \bar{v})$ are simply additive. The use of complex variables for solving boundary-value problems for differential equations of the type in (5-20) and (5-21) has been developed extensively by Muskhelishvili and his school (57).

In addition to the tractions N_x, N_y, N_{xy}, defined by Eq. (5-11), the following integrals of the stresses are employed in plate theory:

$$Q_x = \int_{-h/2}^{h/2} \tau_{xz}\, dz, \qquad Q_y = \int_{-h/2}^{h/2} \tau_{yz}\, dz$$

$$M_x = \int_{-h/2}^{h/2} z\sigma_x\, dz, \qquad M_y = \int_{-h/2}^{h/2} z\sigma_y\, dz \tag{5-22}$$

$$M_{xy} = M_{yx} = \int_{-h/2}^{h/2} z\tau_{xy}\, dz$$

The quantities (Q_x, Q_y) represent intensities of transverse shearing forces [dimension (F/L)]; the quantities (M_x, M_y) represent intensities of bending moments [dimension $(FL/L) = (F)$]. The quantity $M_{xy} = M_{yx}$ represents intensity of twisting moment [dimension (F)]. The intensities of forces and moments on an infinitesimal element of the plate are represented by Figs. 5-2 and 5-3. Equations (5-11) and (5-22) are retained in the large-deflection theory. However, then the deformation of the element (Figs. 5-2 and 5-3) must be taken into account. In the large-deflection theory the obliquity of the forces $N_x\, dy$, $N_y\, dx$, etc., with respect to the undeformed middle surface is important.

In the small-deflection theory the deformation of the plate is considered to have a negligible effect on the equilibrium conditions. In other words, the equilibrium equations are derived with the supposition that the middle surface of the deformed plate is plane. These equations are obtained most

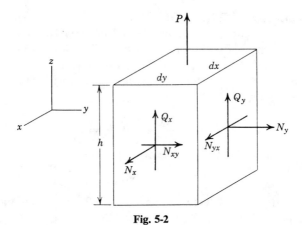

Fig. 5-2

simply by integration of the stress-equilibrium equations (4-24) with respect to z. Neglecting body force, we obtain, for example, from the first of equations in (4-24),

$$\frac{\partial}{\partial x} \int_{-h/2}^{h/2} \sigma_x \, dz + \frac{\partial}{\partial y} \int_{-h/2}^{h/2} \tau_{xy} \, dz + \int_{-h/2}^{h/2} \frac{\partial \tau_{xz}}{\partial z} \, dz = 0 \qquad (a)$$

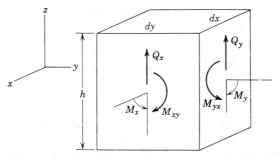

Fig. 5-3

The last integral is

$$\int_{-h/2}^{h/2} \frac{\partial \tau_{xz}}{\partial z} \, dz = \tau_{xz} \Big|_{-h/2}^{h/2} = 0$$

Hence, with Eqs. (5-22), Eq. (a) yields

$$\frac{\partial N_x}{\partial x} + \frac{\partial N_{xy}}{\partial y} = 0$$

Similar results are obtained from the second and third of the equations in (4-24). The moment-equilibrium equations are obtained if the first two equations in (4-24) are multiplied by z and then integrated through the thickness. Integration by parts must be used to evaluate two of the integrals. The complete set of results is

$$\frac{\partial N_x}{\partial x} + \frac{\partial N_{xy}}{\partial y} = 0, \qquad \frac{\partial N_{yx}}{\partial x} + \frac{\partial N_y}{\partial y} = 0$$

$$\frac{\partial Q_x}{\partial x} + \frac{\partial Q_y}{\partial y} + p = 0 \tag{5-23}$$

$$\frac{\partial M_x}{\partial x} + \frac{\partial M_{xy}}{\partial y} = Q_x, \qquad \frac{\partial M_{yx}}{\partial x} + \frac{\partial M_y}{\partial y} = Q_y$$

Equations (5-23) are purely statical; they are equally valid for elastic plates and plastic plates. The boundary conditions, $\tau_{xz} = \tau_{yz} = 0$ at $z = \pm h/2$ have been used for deriving Eqs. (5-23), but, by a slight generalization, this restriction could be eliminated. However, Eqs. (5-23) are not strictly valid unless the plate is flat in the deformed state. Consequently, as they are usually used, these equations are approximations.

Eliminating Q_x and Q_y from Eqs. (5-23), we obtain

$$\frac{\partial^2 M_x}{\partial x^2} + 2 \frac{\partial^2 M_{xy}}{\partial x \, \partial y} + \frac{\partial^2 M_y}{\partial y^2} + p = 0 \tag{5-24}$$

Equation (5-24) is the moment-equilibrium equation.

Equations (5-2) and (5-5) yield

$$\sigma_x = \frac{E}{1 - v^2}(e_x + ve_y) - \frac{Ek\theta}{1 - v} - \frac{Ez}{1 - v^2}(w_{xx} + vw_{yy})$$

$$\sigma_y = \frac{E}{1 - v^2}(e_y + ve_x) - \frac{Ek\theta}{1 - v} - \frac{Ez}{1 - v^2}(w_{yy} + vw_{xx}) \tag{5-25}$$

$$\tau_{xy} = Ge_{xy} - 2Gzw_{xy}$$

Hence by Eqs. (5-22)

$$M_x = -D(w_{xx} + vw_{yy}) - \frac{E\theta_1}{1 - v}$$

$$M_y = -D(w_{yy} + vw_{xx}) - \frac{E\theta_1}{1 - v} \tag{5-26}$$

$$M_{xy} = M_{yx} = -D(1 - v)w_{xy}$$

Substitution of Eqs. (5-26) into Eqs. (5-24) again yields Eq. (5-20).

If there are no boundary forces in the plane of the plate and $\theta = 0$, Eq. (5-21) yields $F = 0$. Then, by Eq. (5-15), $N_x = N_y = N_{xy} = 0$, and, by Eqs. (5-12), $e_x = e_y = e_{xy} = 0$. By Eqs. (5-25) the stresses are then determined if w is known. With the boundary conditions, Eq. (5-20) determines w.

5-3. BOUNDARY CONDITIONS IN THE CLASSICAL THEORY OF PLATES.

Suitable boundary conditions for a free edge of a plate were first derived by Kirchhoff in 1850 by a variational procedure. Earlier boundary conditions of Poisson were inconsistent with the fourth-order differential equation of the classical theory.

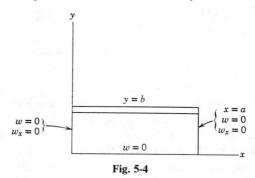

Fig. 5-4

Consider a rectangular plate of length a and width b (Fig. 5-4). For generality, the edge $y = b$ is considered to have a flange that is symmetrical with respect to the middle plane of the plate. Since, in the small-deflection theory, the effects of the membrane tractions N_x, N_y, N_{xy} are uncoupled from the effects of bending, the membrane energy U_m will be taken as zero. Also, the temperature θ is set equal to zero. Hence by Eq. (5-19) the total potential energy of the plate and the flange is

$$V = \int_0^b \int_0^a \left[\tfrac{1}{2} D(w_{xx} + w_{yy})^2 - D(1 - \nu)(w_{xx}w_{yy} - w_{xy}^2) - pw\right] dx\, dy$$

$$+ \tfrac{1}{2} EI \int_0^a w_{xx}^2(x, b)\, dx + \tfrac{1}{2} GJ \int_0^a w_{xy}^2(x, b)\, dx \qquad\qquad \text{(a)}$$

The last two integrals represent the strain energy of the flange [see Eqs. (2-8) and (2-17)]. Note that $w_{xy}(x, b)$ is the twist of the flange per unit length.

The terms in Eq. (a) that represent the strain energy of the flange do not enter into the part of the first variation of V that is expressed by a double integral. Consequently, the Euler equation of the problem is not affected by the presence of the flange. Accordingly, Eq. (5-20) determines the deflection of the plate (with $\theta_1 = 0$). The natural boundary conditions are determined by Eqs. (3-21) and Eq. (a). Let us suppose that the forced boundary conditions are

(a) Edges $x = 0$ and $x = a$; $w = w_x = 0$ (clamped edges).
(b) Edge $y = 0$; $w = 0$ (simply supported edge).

In view of these conditions, the variation η, with its derivatives η_x and η_y, vanish on the edges $x = 0$ and $x = a$. Also because of the forced boundary conditions, η and η_x vanish on the edge $y = 0$. Hence, by Eqs. (3-21), the natural boundary condition for the simply supported edge, $y = 0$, is $\partial F/\partial w_{yy} = 0$ or $w_{yy} + \nu w_{xx} = 0$. As might be expected, Eqs. (5-26) show that this is equivalent to the condition $M_y = 0$. Because of the forced boundary condition, w_{xx} vanishes on the edge $y = 0$. Consequently, the natural boundary condition for the edge $y = 0$ reduces to $w_{yy} = 0$.

To obtain the natural boundary conditions for the flanged edge, we must augment Eqs. (3-21) by terms that come from the line integrals in Eq. (a). By Eq. (a), the first variation of the strain energy of the flange is

$$\delta U_f = \epsilon \int_0^a (EI w_{xx} \eta_{xx} + GJ w_{xy} \eta_{xy})\, dx$$

Since the flange is attached to the plate, it conforms to the boundary conditions of the plate at $x = 0$ and $x = a$. Hence η, η_x, and η_y vanish at the ends of the flange. Consequently, integration by parts yields

$$\delta U_f = \epsilon EI \int_0^a \eta w_{xxxx}\, dx - \epsilon GJ \int_0^a \eta_y w_{xxy}\, dx \tag{b}$$

Appending terms from Eq. (b) to Eqs. (3-21) and noting that the integrals in Eq. (d) of Sec. 3-9 are taken in the opposite sense to the integrals in Eq. (b) above, we obtain

$$\eta \left(EI w_{xxxx} + \frac{\partial F}{\partial w_y} - \frac{\partial}{\partial y} \frac{\partial F}{\partial w_{yy}} - \frac{\partial}{\partial x} \frac{\partial F}{\partial w_{xy}} \right) = 0$$

$$\eta_y \left(-GJ w_{xxy} + \frac{\partial F}{\partial w_{yy}} \right) = 0$$

where F denotes the integrand of the double integral in Eq. (a). Since η and η_y are arbitrary, these equations yield the natural boundary conditions for the flanged edge:

$$w_{yyy} + (2 - \nu)w_{xxy} - \frac{EI}{D} w_{xxxx} = 0 \quad \text{for} \quad y = b$$

$$w_{yy} + \nu w_{xx} - \frac{GJ}{D} w_{xxy} = 0 \quad \text{for} \quad y = b \tag{5-27}$$

If the flange is absent, $EI = GJ = 0$, and Eqs. (5-27) reduce to the natural boundary conditions for a free edge of a plate:

$$w_{yyy} + (2 - \nu)w_{xxy} = 0 \quad \text{for} \quad y = b$$

$$w_{yy} + \nu w_{xx} = 0 \quad \text{for} \quad y = b \tag{5-28}$$

Equations (5-28) were derived by Kirchhoff by the present method.

It is natural to assume, as Poisson did, that the boundary conditions at a free edge $y = b$ are $M_y = 0$, $Q_y = 0$, and $M_{yx} = 0$. However, in general, it is impossible to adapt the solution of Eq. (5-20) to three boundary conditions. The fact that Kirchhoff obtained only two boundary conditions for a free edge is consistent with the differential equation of the classical theory of plates. In view of Eqs. (5-26), the second of the equations in (5-28) means that $M_y = 0$ at a free edge, $y = b$. This condition might have been anticipated. Similarly, the second of the equations in (5-27) means that torque in the flange is "fed in" by bending moments in the plate. However, the first of the equations in (5-28) is not a result that might have been easily anticipated.

By Eqs. (5-23) and (5-26),

$$Q_x = -D(w_{xxx} + w_{xyy}), \qquad Q_y = -D(w_{xxy} + w_{yyy}) \qquad (5\text{-}29)$$

Quantities V_x and V_y are defined as follows:

$$V_x = Q_x + \frac{\partial M_{xy}}{\partial y}, \qquad V_y = Q_y + \frac{\partial M_{yx}}{\partial x} \qquad (5\text{-}30)$$

By Eqs. (5-26) and (5-29), $V_y = -D[w_{yyy} + (2 - \nu)w_{xxy}]$. Accordingly, the first of the equations in (5-28) means that V_y vanishes on the free edge, $y = b$.

By an ingenious mechanistic argument, Thomson and Tait (86) concluded that the true transverse shear intensity on an edge of the plate parallel to the x-axis is V_y rather than Q_y. The boundary condition for a free edge, $V_y = 0$, is thus rendered plausible. However, the transverse shear intensities on cross sections of the plate perpendicular to the x- or y-axes are still to be regarded as Q_x or Q_y. The argument of Thomson and Tait generally requires that lateral point loads be introduced at the corners of the plate to react the edge shears. Actually, the theory of transverse shear in plates lies outside the scope of the classical theory, since the initial assumption $\tau_{xz} = \tau_{yz} = 0$ signifies that $Q_x = Q_y = 0$. It is reasonable to reintroduce the shears Q_x and Q_y through the equilibrium equations (5-23), since these equations were derived without the assumption $\tau_{xz} = \tau_{yz} = 0$. Nevertheless, puzzling inconsistencies arise from this procedure, as shown in Sec. 5-4.

5-4. SIMPLY SUPPORTED RECTANGULAR PLATES. According to the discussion at the end of Sec. 5-1, the strain energy of bending of an isotropic polygonal plate with fixed edges is

$$U_b = \tfrac{1}{2} D \iint (w_{xx} + w_{yy})^2 \, dx \, dy \qquad (5\text{-}31)$$

The term $(w_{xx}w_{yy} - w_{xy}^2)$ is dropped from Eq. (5-19) because it neither contributes to the Euler equation nor to the natural boundary conditions, provided that the edges of the plate remain straight.

Let the edges of a simply supported rectangular plate lie on the lines $x = 0$, $x = a$, $y = 0$, and $y = b$. The deflection of the plate may be represented by a double sine series:

$$w = \sum_{m=1}^{\infty} \sum_{n=1}^{\infty} a_{mn} \sin \frac{m\pi x}{a} \sin \frac{n\pi y}{b} \tag{5-32}$$

This series automatically satisfies the boundary condition $w = 0$. Also, it satisfies the condition that the bending moments vanish on the edges ($w_{xx} = 0$ for $x = 0$ or $x = a$ and $w_{yy} = 0$ for $y = 0$ or $y = b$).

Introducing Eq. (5-32) into Eq. (5-31), we obtain the integral of a quadruple series. However, the integrals of the cross products are all zero, and only the integrals of the squared terms need be considered. Integration yields

$$U_b = \tfrac{1}{8}\pi^4 ab D \sum_{m=1}^{\infty} \sum_{n=1}^{\infty} \left(\frac{m^2}{a^2} + \frac{n^2}{b^2}\right)^2 a_{mn}^2 \tag{5-33}$$

The potential energy of the external forces is

$$\Omega = -\iint pw \, dx \, dy$$

where $p(x, y)$ is the lateral distributed load. Hence, by Eq. (5-32),

$$\Omega = -\sum_{m=1}^{\infty} \sum_{n=1}^{\infty} a_{mn} \iint p \sin \frac{m\pi x}{a} \sin \frac{n\pi y}{b} \, dx \, dy \tag{5-34}$$

In the small-deflection theory forces in the plane of the plate have no effect on bending. Consequently, if there is no bending caused by heating, the deflection w may be computed with the understanding that the total potential energy is $V = U_b + \Omega$. The coefficients are determined by the principle of stationary potential energy, $\partial V / \partial a_{ij} = 0$. Differentiating, and subsequently replacing (i, j) by (m, n), we obtain

$$a_{mn} = \frac{4 \iint p \sin (m\pi x/a) \sin (n\pi y/b) \, dx \, dy}{\pi^4 ab \, D(m^2/a^2 + n^2/b^2)^2} \tag{5-35}$$

Equations (5-32) and (5-35) are a general solution of the deflection problem for simply supported elastic isotropic rectangular plates with arbitrary distributed loads, since $p(x, y)$ is arbitrary. This solution was first derived by Navier. Many solutions of special problems of simply supported rectangular plates are discussed by Timoshenko and Woinowsky-Krieger (86).

For example, a particularly simple solution is obtained if

$$p = p_0 \sin \frac{\pi x}{a} \sin \frac{\pi y}{b}$$

Then Eqs. (5-32) and (5-35) yield

$$w = \frac{p_0 a^4 b^4}{\pi^4 D(a^2 + b^2)^2} \sin \frac{\pi x}{a} \sin \frac{\pi y}{b} \tag{5-36}$$

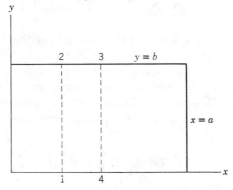

Fig. 5-5

Equations (5-26), (5-29), (5-30), and (5-36) yield

$$Q_x = \frac{p_0 a b^2}{\pi(a^2 + b^2)} \cos \frac{\pi x}{a} \sin \frac{\pi y}{b} \tag{5-37}$$

$$V_y = \frac{-p_0 a^2 b}{\pi(a^2 + b^2)^2} [a^2 + (2 - v)b^2] \sin \frac{\pi x}{a} \quad \text{at} \quad y = b$$

Equation (5-37) is inconsistent with the equilibrium requirement for a strip of the plate.* For example, if the strip 1234 is regarded as a free body (Fig. 5-5), the integral of the load p over the face of the strip must be balanced by the shears V_y on the edges 23 and 14 and by the shears Q_x on the cross sections 12 and 34. By symmetry, the shears on the edges 23 and 14 are equal. Accordingly, without performing the integration, we can see that the equilibrium condition for the strip is not generally satisfied, since the integral of V_y over the segment 23 contains the factor $[a^2 + (2 - v)b^2]$, whereas Q_x and p are independent of v. The inconsistency must be attributed to the initial assumption, $\tau_{xz} = \tau_{yz} = 0$, upon which the classical theory of plates is based. In many cases, the classical theory is adequate for determining the deflection w and the stresses $\sigma_x, \sigma_y, \tau_{xy}$. However, calculations of the shears V_x, V_y, Q_x, Q_y and the shearing stresses τ_{xz}, τ_{yz} must be viewed with caution.

* This fact was called to the author's attention by M. C. Stippes.

5-5. SHEAR DEFORMATION OF PLATES.

The example in Sec. 5-4 demonstrates that the classical theory is not entirely satisfactory for determining the transverse shears in a plate. An improved small-deflection theory for moderately thick plates that yields consistent results for the transverse shears was developed by E. Reissner (136, 137). This theory permits Poisson's three boundary conditions, $M_y = 0$, $M_{yx} = 0$, $Q_y = 0$, on a free edge, $y = b$. Since it is a small-deflection theory, it employs the linear strain-displacement relations [Eqs. (4-14)]. The plate is accordingly conceived to be flat in the deformed state.

The boundary conditions on the faces of the plate are taken to be $\sigma_z = 0$ at $z = h/2$ and $\sigma_z = -p(x, y)$ at $z = -h/2$. Also, $\tau_{xz} = \tau_{yz} = 0$ on either face. It is assumed that σ_x, σ_y, and τ_{xy} are linear functions of z. This approximation is also used in the classical theory. By Eqs. (5-11) and (5-22),

$$\sigma_x = \frac{N_x}{h} + \left(\frac{12M_x}{h^3}\right)z, \qquad \sigma_y = \frac{N_y}{h} + \left(\frac{12M_y}{h^3}\right)z, \qquad \tau_{xy} = \frac{N_{xy}}{h} + \left(\frac{12M_{xy}}{h^3}\right)z$$

$$(5\text{-}38)$$

Body forces will be neglected. Then, by substituting Eqs. (5-38) into the equilibrium equations (4-24) and simplifying the results by Eqs. (5-23) and the boundary conditions, we obtain

$$\tau_{xz} = \frac{3Q_x}{2h}\left[1 - \left(\frac{2z}{h}\right)^2\right], \qquad \tau_{yz} = \frac{3Q_y}{2h}\left[1 - \left(\frac{2z}{h}\right)^2\right]$$

$$\sigma_z = -\frac{3p}{4}\left[\frac{2}{3} - \frac{2z}{h} + \frac{1}{3}\left(\frac{2z}{h}\right)^3\right]$$

$$(5\text{-}39)$$

These equations satisfy the boundary conditions stated above.

By means of Eqs. (5-38) and (5-39), the complementary energy of the plate may be expressed in terms of N_x, N_y, etc. Temperature terms will be included, since they introduce no significant complications. Equations (4-57), (4-58), (5-38), and (5-39) yield, after integration with respect to z,

$$\Upsilon = \frac{1}{2Eh} \iint [N_x^2 + N_y^2 - 2\nu N_x N_y + 2(1 + \nu)N_{xy}^2$$

$$+ (\nu p h + 2E\theta_0)(N_x + N_y)]\, dx\, dy$$

$$+ \frac{6}{Eh^3} \iint [M_x^2 + M_y^2 - 2\nu M_x M_y + 2(1 + \nu)M_{xy}^2 + \tfrac{1}{5}(1 + \nu)h^2$$

$$\times (Q_x^2 + Q_y^2) - (\tfrac{1}{5}\nu p h^2 - 2E\theta_1)(M_x + M_y)]\, dx\, dy \qquad (5\text{-}40)$$

Additive terms that do not contain N_x, N_y, N_{xy}, Q_x, Q_y, M_x, M_y, or M_{xy} are not included in Eqs. (5-40), since they are irrelevant. The terms θ_0 and θ_1 are defined by Eqs. (5-8).

According to the generalized form of Castigliano's theorem of least work (Sec. 4-8), the quantity Ψ that is to be rendered stationary is obtained by appending to the complementary energy Υ a certain line integral on the boundary of the plate. However, this line integral does not affect the Euler equations for the double integrals. Consequently, it will be disregarded.* The first and second integrals in Eq. (5-40) may be treated separately.

When N_x, N_y, N_{xy} are expressed in terms of an Airy function $F(x, y)$ by Eq. (5-15), the first integral in Eq. (5-40) becomes

$$\iint [F_{xx}{}^2 + F_{yy}{}^2 - 2\nu F_{xx}F_{yy} + 2(1 + \nu)F_{xy}{}^2$$
$$+ (\nu ph + 2E\theta_0)(F_{xx} + F_{yy})] \, dx \, dy$$

The Airy function ensures that the equilibrium equations for N_x, N_y, N_{xy} are satisfied. The Euler equation [Eq. (3-20)] for the preceding integral is

$$\nabla^2\nabla^2 F + \nabla^2(\tfrac{1}{2}\nu ph + E\theta_0) = 0 \qquad (5\text{-}41)$$

or

$$\nabla^2(N_x + N_y + \tfrac{1}{2}\nu ph + E\theta_0) = 0 \qquad (5\text{-}42)$$

Because p appears in Eq. (5-41), there is a slight coupling between the lateral load and the tractions N_x, N_y, N_{xy}. This coupling does not appear in the classical small-deflection theory.

The shears Q_x and Q_y in Eq. (5-40) may be eliminated by the equilibrium equations [Eqs. (5-23)]. The moment-equilibrium equation [Eq. (5-24)] is treated by means of a Lagrange multiplier $\lambda(x, y)$ (see Sec. 3-8). Hence the second integral in Eq. (5-40) is replaced by

$$\Upsilon = \iint \left\{ M_x{}^2 + M_y{}^2 - 2\nu M_x M_y + 2(1 + \nu)M_{xy}{}^2 - (\tfrac{1}{5}\nu ph^2 - 2E\theta_1) \right.$$
$$\times (M_x + M_y) + \tfrac{1}{5}(1 + \nu)h^2 \left[\left(\frac{\partial M_x}{\partial x} + \frac{\partial M_{xy}}{\partial y} \right)^2 + \left(\frac{\partial M_{xy}}{\partial x} + \frac{\partial M_y}{\partial y} \right)^2 \right]$$
$$+ \left. \lambda \left(\frac{\partial^2 M_x}{\partial x^2} + 2\frac{\partial^2 M_{xy}}{\partial x \, \partial y} + \frac{\partial^2 M_y}{\partial y^2} \right) \right\} dx \, dy$$

Writing the Euler equations [Eq. (3-20)] for this integral and simplifying

---

* Reissner included the line integral in order to obtain natural boundary conditions. He also used the line integral to interpret a Lagrange multiplier. However, certain approximations (136, 137, 110) concerning the displacement vector at the edge of the plate are required in this procedure. Reissner did not include N_x, N_y, and N_{xy} in his theory. Also, he did not include temperature terms.

the results by means of the equilibrium equations (5-23) and (5-24), we obtain

$$M_x - \nu M_y + E\theta_1 - \tfrac{1}{10}\nu ph^2 - \tfrac{1}{5}(1+\nu)h^2\frac{\partial Q_x}{\partial x} + \tfrac{1}{2}\lambda_{xx} = 0$$

$$M_y - \nu M_x + E\theta_1 - \tfrac{1}{10}\nu ph^2 - \tfrac{1}{5}(1+\nu)h^2\frac{\partial Q_y}{\partial y} + \tfrac{1}{2}\lambda_{yy} = 0 \quad (5\text{-}43)$$

$$(1+\nu)M_{xy} - \tfrac{1}{10}(1+\nu)h^2\left(\frac{\partial Q_y}{\partial x} + \frac{\partial Q_x}{\partial y}\right) + \tfrac{1}{2}\lambda_{xy} = 0$$

Equations (5-23) and (5-43) yield

$$M_x + \frac{E\theta_1}{1-\nu} + \frac{\nu ph^2}{10(1-\nu)} - \frac{h^2}{5}\frac{\partial Q_x}{\partial x} = \frac{-(\lambda_{xx} + \nu\lambda_{yy})}{2(1-\nu^2)}$$

$$M_y + \frac{E\theta_1}{1-\nu} + \frac{\nu ph^2}{10(1-\nu)} - \frac{h^2}{5}\frac{\partial Q_y}{\partial y} = \frac{-(\lambda_{yy} + \nu\lambda_{xx})}{2(1-\nu^2)} \qquad (5\text{-}44)$$

$$M_{xy} - \frac{h^2}{10}\left(\frac{\partial Q_y}{\partial x} + \frac{\partial Q_x}{\partial y}\right) = \frac{-\lambda_{xy}}{2(1+\nu)}$$

With the boundary conditions, the six differential equations (5-23) and (5-44) determine the functions M_x, M_y, M_{xy}, Q_x, Q_y, λ.

By Eqs. (4-54), (5-38), and (5-39), the strains are given by

$$E\epsilon_x = \left(\frac{N_x}{h} - \frac{\nu N_y}{h} + \frac{\nu p}{2}\right) + \left(\frac{12M_x}{h^3} - \frac{12\nu M_y}{h^3} - \frac{3\nu p}{2h}\right)z$$

$$+ 2\nu p\frac{z^3}{h^3} + Ek\theta$$

$$E\epsilon_y = \left(\frac{N_y}{h} - \frac{\nu N_x}{h} + \frac{\nu p}{2}\right) + \left(\frac{12M_y}{h^3} - \frac{12\nu M_x}{h^3} - \frac{3\nu p}{2h}\right)z$$

$$+ 2\nu p\frac{z^3}{h^3} + Ek\theta \qquad (5\text{-}45)$$

$$G\gamma_{xz} = \frac{3Q_x}{2h}\left(1 - \frac{4z^2}{h^2}\right), \qquad G\gamma_{yz} = \frac{3Q_y}{2h}\left(1 - \frac{4z^2}{h^2}\right)$$

$$G\gamma_{xy} = \frac{N_{xy}}{h} + \left(\frac{12M_{xy}}{h^3}\right)z$$

By the strain-displacement relations [Eqs. (4-14)],

$$w_{xx} = \frac{\partial \gamma_{xz}}{\partial x} - \frac{\partial \epsilon_x}{\partial z}, \qquad w_{yy} = \frac{\partial \gamma_{yz}}{\partial y} - \frac{\partial \epsilon_y}{\partial z}$$

$$2w_{xy} = \frac{\partial \gamma_{yz}}{\partial x} + \frac{\partial \gamma_{xz}}{\partial y} - \frac{\partial \gamma_{xy}}{\partial z}$$

Hence, by Eqs. (5-45),

$$Ew_{xx} = \frac{3(1+v)}{h} \frac{\partial Q_x}{\partial x} - \frac{12M_x}{h^3} + \frac{12vM_y}{h^3} + \frac{3vp}{2h}$$

$$- \left[\frac{12(1+v)}{h^3} \frac{\partial Q_x}{\partial x} + \frac{6vp}{h^3} \right] z^2 - Ek \frac{\partial \theta}{\partial z}$$

$$Ew_{yy} = \frac{3(1+v)}{h} \frac{\partial Q_y}{\partial y} - \frac{12M_y}{h^3} + \frac{12vM_x}{h^3} + \frac{3vp}{2h}$$

$$- \left[\frac{12(1+v)}{h^3} \frac{\partial Q_y}{\partial y} + \frac{6vp}{h^3} \right] z^2 - Ek \frac{\partial \theta}{\partial z}$$

$$Ew_{xy} = \frac{3(1+v)}{2h} \left(1 - \frac{4z^2}{h^2} \right) \left(\frac{\partial Q_y}{\partial x} + \frac{\partial Q_x}{\partial y} \right) - \frac{12(1+v)}{h^3} M_{xy}$$

Defining the flexural rigidity D as in Eq. (5-10), and noting Eqs. (5-23), we may write these equations as follows:

$$M_x + D(1+v)k \frac{\partial \theta}{\partial z} + \frac{vph^2}{8(1-v)} \left(1 - \frac{4z^2}{h^2} \right) - \frac{h^2}{4} \left(1 - \frac{4z^2}{h^2} \right) \frac{\partial Q_x}{\partial x}$$

$$= -D(w_{xx} + vw_{yy})$$

$$M_y + D(1+v)k \frac{\partial \theta}{\partial z} + \frac{vph^2}{8(1-v)} \left(1 - \frac{4z^2}{h^2} \right) - \frac{h^2}{4} \left(1 - \frac{4z^2}{h^2} \right) \frac{\partial Q_y}{\partial y} \qquad (5\text{-}46)$$

$$= -D(w_{yy} + vw_{xx})$$

$$M_{xy} - \frac{h^2}{8} \left(1 - \frac{4z^2}{h^2} \right) \left(\frac{\partial Q_y}{\partial x} + \frac{\partial Q_x}{\partial y} \right) = -D(1-v)w_{xy}$$

Reissner (136, 137) identified the quantity $6\lambda/Eh^3$ as the mean value of the deflection w. This relationship reconciles Eqs. (5-44) and (5-46), except for the temperature term, provided that the mean value of $1 - 4z^2/h^2$ is taken to be $\frac{4}{5}$. However, according to the usual definition of the mean, this value should be $\frac{2}{3}$ instead of $\frac{4}{5}$. The temperature terms in Eqs. (5-44) and (5-46) are in agreement if the temperature is a linear function of z.

Letting w denote the deflection of the middle plane, we obtain from Eqs. (5-46),

$$M_x + D(1 + v)k\theta' + \frac{vph^2}{8(1 - v)} - \frac{h^2}{4}\frac{\partial Q_x}{\partial x} = -D(w_{xx} + vw_{yy})$$

$$M_y + D(1 + v)k\theta' + \frac{vph^2}{8(1 - v)} - \frac{h^2}{4}\frac{\partial Q_y}{\partial y} = -D(w_{yy} + vw_{xx}) \quad (5\text{-}47)$$

$$M_{xy} - \frac{h^2}{8}\left(\frac{\partial Q_y}{\partial x} + \frac{\partial Q_x}{\partial y}\right) = -D(1 - v)w_{xy}$$

where θ' denotes the value of $\partial\theta/\partial z$ at $z = 0$.

Equations (5-47) were derived by Goodier (110). They are consistent with the initial assumptions, and they may be used instead of Eqs. (5-44). With the boundary conditions, Eqs. (5-23) and (5-47) determine the functions M_x, M_y, M_{xy}, Q_x, Q_y, w. Goodier remarked that if $z = \pm h/2$ Eqs. (5-46) reduce to the classical equations (5-26), except for the temperature term, which he did not discuss.

5-6. GEOMETRY OF SHELLS. The theory of curved shells has become important for the design of pressure vessels, submarine hulls, ship hulls, airplane structures, concrete roofs, containers for liquids, and many other structures. It includes the theories of flat plates and curved beams as special cases. Shell analysis entails problems of stress concentration, elasticity, buckling, creep, and plasticity. Practical results can be obtained only with the aid of approximations, yet the subject has proved to be very sensitive in this respect, particularly in problems of buckling.

The theory of shells rests on geometry and mechanics of materials. Some preliminary geometrical relations are presented in this article. Most of the proofs are omitted; derivations may be found in texts on differential geometry (79).

A surface is defined by equations of the type $X = X(x, y)$, $Y = Y(x, y)$, $Z = Z(x, y)$, in which (X, Y, Z) are rectangular coordinates and (x, y) are parameters called "surface coordinates." If $\mathbf{i}, \mathbf{j}, \mathbf{k}$ are unit vectors along the X-, Y-, Z-axes, the point (X, Y, Z) is located by the vector $\mathbf{r} = \mathbf{i}X + \mathbf{j}Y + \mathbf{k}Z$. Accordingly, the surface is defined by the vector equation, $\mathbf{r} = \mathbf{r}(x, y)$. A line on the surface on which only x varies (or only y varies) is called an x-coordinate line (or a y-coordinate line). Since $d\mathbf{r} = \mathbf{r}_x\, dx + \mathbf{r}_y\, dy$, the vectors $\mathbf{r}_x$ and $\mathbf{r}_y$ are tangent to the x and y coordinate lines, respectively. Accordingly, the x- and y-coordinate lines are orthogonal to each other if, and only if, $\mathbf{r}_x \cdot \mathbf{r}_y = 0$. In this case the surface coordinates are said to be "orthogonal."

The distance ds between points with the surface coordinates (x, y) and $(x + dx, y + dy)$ is determined by $ds^2 = d\mathbf{r} \cdot d\mathbf{r}$. Hence, for orthogonal surface coordinates,

$$ds^2 = A^2\, dx^2 + B^2\, dy^2 \tag{5-48}$$

where

$$\begin{aligned}
A^2 &= \mathbf{r}_x \cdot \mathbf{r}_x = X_x^{\,2} + Y_x^{\,2} + Z_x^{\,2} \\
B^2 &= \mathbf{r}_y \cdot \mathbf{r}_y = X_y^{\,2} + Y_y^{\,2} + Z_y^{\,2}
\end{aligned} \tag{5-49}$$

The area of any part of the surface is evidently determined by

$$\text{area} = \iint AB\, dx\, dy$$

For orthogonal surface coordinates, the magnitudes of the vectors $\mathbf{r}_x$ and $\mathbf{r}_y$ are A and B, respectively. Therefore, the unit vector normal to the surface is

$$\mathbf{n} = \frac{\mathbf{r}_x \times \mathbf{r}_y}{AB} \tag{5-50}$$

The positive sense of the surface normal is determined by this equation; it naturally depends on the identification of the parameters x and y.

The following differential expression, known as the "second fundamental form," is important in surface theory:

$$e\, dx^2 + 2f\, dx\, dy + g\, dy^2 = -d\mathbf{r} \cdot d\mathbf{n} \tag{5-51}$$

For orthogonal surface coordinates, the coefficients e, f, g are determined by

$$e = \frac{1}{AB} \begin{vmatrix} X_{xx} & Y_{xx} & Z_{xx} \\ X_x & Y_x & Z_x \\ X_y & Y_y & Z_y \end{vmatrix} \qquad f = \frac{1}{AB} \begin{vmatrix} X_{xy} & Y_{xy} & Z_{xy} \\ X_x & Y_x & Z_x \\ X_y & Y_y & Z_y \end{vmatrix}$$

$$g = \frac{1}{AB} \begin{vmatrix} X_{yy} & Y_{yy} & Z_{yy} \\ X_x & Y_x & Z_x \\ X_y & Y_y & Z_y \end{vmatrix} \tag{5-52}$$

The extreme values of the curvatures of normal plane cross sections of the surface at any point are denoted by $1/r_1$ and $1/r_2$. These quantities are known as the "principal curvatures" of the surface at the given point. The directions of the cross-sectional curves corresponding to r_1 and r_2 are called the principal directions at the given point. It is shown in differential geometry that the principal directions are orthogonal. Consequently, the surface may be covered by two orthogonal families of curves which everywhere have the principal directions. These curves are called "lines of principal curvature." For a surface of revolution they are the meridians and the circles of latitude.

It is shown in differential geometry that the lines of principal curvature coincide with the coordinate lines if, and only if, the coordinates are orthogonal and $f = 0$. In this case

$$\frac{1}{r_1} = \frac{-e}{A^2}, \qquad \frac{1}{r_2} = \frac{-g}{B^2} \tag{5-53}$$

These equations affix signs to the principal radii of curvature, r_1 and r_2. They signify that r_1 or r_2 is positive if the corresponding center of curvature lies on the negative side of the tangent plane of the surface, the positive side of the tangent plane being the side toward which the surface normal $\mathbf{n}$ is directed. For example, for a sphere of radius a, $r_1 = r_2 = a$ if $\mathbf{n}$ is directed outward, but $r_1 = r_2 = -a$ if $\mathbf{n}$ is directed inward.

If the lines of principal curvature are coordinate lines (that is, if $\mathbf{r}_x \cdot \mathbf{r}_y = f = 0$), a theorem of Rodrigues is expressed as follows:

$$\frac{\partial \mathbf{n}}{\partial x} = \frac{1}{r_1} \frac{\partial \mathbf{r}}{\partial x}, \qquad \frac{\partial \mathbf{n}}{\partial y} = \frac{1}{r_2} \frac{\partial \mathbf{r}}{\partial y} \tag{5-54}$$

The product $K = 1/(r_1 r_2)$ is known as the "Gaussian curvature" of the surface. If the surface coordinates are orthogonal, K satisfies the following differential equation of Gauss:

$$-KAB = \frac{\partial}{\partial x}\left(\frac{B_x}{A}\right) + \frac{\partial}{\partial y}\left(\frac{A_y}{B}\right) \tag{5-55}$$

This equation shows that K is determined by A and B. The functions A and B do not change if the surface is bent without straining, since ds does not change [see Eq. (5-48)]. Therefore, by Eq. (5-55), K does not change if the surface is bent without straining. In particular, $K = 0$ for any developable surface, since a developable surface may be formed by bending a plane.

The functions A, B, r_1, r_2 also satisfy two differential equations of Codazzi. If the coordinate lines coincide with the lines of principal curvature, the Codazzi equations take the following form:

$$\frac{\partial}{\partial y}\left(\frac{A}{r_1}\right) = \frac{1}{r_2}\frac{\partial A}{\partial y}, \qquad \frac{\partial}{\partial x}\left(\frac{B}{r_2}\right) = \frac{1}{r_1}\frac{\partial B}{\partial x} \tag{5-56}$$

Dupin proved that the surfaces of a triply orthogonal system intersect each other on their lines of principal curvature. This theorem shows that if two orthogonal families of surfaces are given we generally cannot construct a third family that is orthogonal to both of them; the construction is impossible unless the original two families intersect on their lines of principal curvature.

In shell theory a special type of curvilinear coordinate system is usually employed. The middle surface of the shell is defined by $X = X(x, y)$,

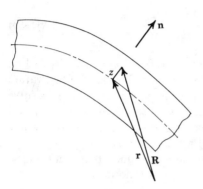

Fig. 5-6

$Y = Y(x, y)$, and $Z = Z(x, y)$, where (X, Y, Z) are rectangular coordinates and (x, y) are surface coordinates. The normal distance from the middle surface is denoted by $\pm z$. Positive z is measured in the sense of the positive normal $\mathbf{n}$ of the middle surface. To any set of values of (x, y, z) there corresponds a point in the shell. Hence (x, y, z) are curvilinear space coordinates; they will be called "shell coordinates." The surface $z = $ constant is parallel to the middle surface in the sense that it lies at a constant distance from it. The exterior surfaces of the shell are represented by $z = \pm h/2$, where h is the thickness of the shell. If $h = $ constant, the exterior surfaces are coordinate surfaces.

If the shell coordinates are orthogonal, the coordinate lines on the middle surface must be lines of principal curvature, as we can see from the aforementioned theorem of Dupin. Conversely, if the coordinate lines on the middle surface are lines of principal curvature, the shell coordinates are orthogonal, as the following argument shows.

Figure 5-6 represents a cross section of a shell. The position vector of a point on the middle surface is $\mathbf{r}$, and the position vector of the corresponding point at distance z from the middle surface is $\mathbf{R}$. From Fig. 5-6,

$$\mathbf{R} = \mathbf{r} + \mathbf{n}z \qquad \text{(a)}$$

Hence, by differentiation,

$$\mathbf{R}_x = \mathbf{r}_x + \mathbf{n}_x z, \qquad \mathbf{R}_y = \mathbf{r}_y + \mathbf{n}_y z \qquad \text{(b)}$$

Since the coordinate lines are lines of principal curvature, the Rodrigues formulas apply [Eq. (5-54)]. By Eq. (b),

$$\mathbf{R}_x = \left(1 + \frac{z}{r_1}\right)\mathbf{r}_x, \qquad \mathbf{R}_y = \left(1 + \frac{z}{r_2}\right)\mathbf{r}_y, \qquad \mathbf{R}_z = \mathbf{n} \qquad \text{(c)}$$

By Eqs. (5-49), $\mathbf{r}_x \cdot \mathbf{r}_x = A^2$ and $\mathbf{r}_y \cdot \mathbf{r}_y = B^2$. Also, $\mathbf{r}_x \cdot \mathbf{n} = \mathbf{r}_y \cdot \mathbf{n} = \mathbf{r}_x \cdot \mathbf{r}_y$ $= 0$. Consequently, since $ds^2 = d\mathbf{R} \cdot d\mathbf{R} = (\mathbf{R}_x \, dx + \mathbf{R}_y \, dy + \mathbf{R}_z \, dz)^2$, Eq. (c) yields

$$ds^2 = \alpha^2 \, dx^2 + \beta^2 \, dy^2 + \gamma^2 \, dz^2 \tag{5-57}$$

where

$$\alpha = A\left(1 + \frac{z}{r_1}\right), \qquad \beta = B\left(1 + \frac{z}{r_2}\right), \qquad \gamma = 1 \tag{5-58}$$

The factors α, β, γ are the Lamé coefficients [see Eq. (4-27)]. Equations (5-56), (5-57), and (5-58) are basic geometric relations in the theory of shells.

Equation (c) shows that the vectors $\mathbf{R}_x$ and $\mathbf{R}_y$ are parallel to the vectors $\mathbf{r}_x$ and $\mathbf{r}_y$, respectively. Accordingly, all x-coordinate lines (or all y-coordinate lines) that intersect a given straight normal N to the middle surface have parallel tangents at their intercepts with line N.

5-7. EQUILIBRIUM OF SHELLS. Figure 5-7a represents a differential element of a shell, cut out by surfaces $x = $ constant and $y = $ constant. The variables (x, y, z) are orthogonal shell coordinates. Accordingly, the coordinate lines on the middle surface are lines of principal curvature. By Eqs. (5-58) and (5-57), the elements of area of the cross sections (Fig. 5-7a) are

$$dA_x = \alpha \, dx \, dz = A\left(1 + \frac{z}{r_1}\right) dx \, dz$$

$$dA_y = \beta \, dy \, dz = B\left(1 + \frac{z}{r_2}\right) dy \, dz$$

where r_1 and r_2 are the principal radii of curvature of the middle surface.

Let N_x be the tensile force on a cross section per unit length of a y-coordinate line (Fig. 5-7b). Then the total tensile force on the differential element in the x-direction is $N_x B \, dy$. Hence

$$N_x B \, dy = dy \int_{-h/2}^{h/2} \beta \sigma_x \, dz$$

where h is the thickness of the shell. Therefore,

$$N_x = \frac{1}{B} \int_{-h/2}^{h/2} \beta \sigma_x \, dz = \int_{-h/2}^{h/2} \sigma_x \left(1 + \frac{z}{r_2}\right) dz$$

Similarly, the tension N_y, the shears N_{xy} and N_{yx}, the transverse shears Q_x and Q_y, the bending moments M_x and M_y, and the twisting moments M_{xy}

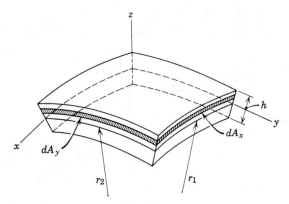

Fig. 5-7a

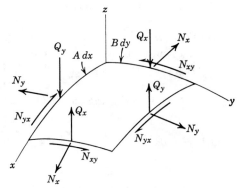

Fig. 5-7b

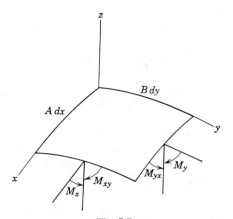

Fig. 5-7c

and M_{yx} are defined (see Figs. 5-7b and 5-7c). The complete set of relations is

$$N_x = \frac{1}{B} \int_{-h/2}^{h/2} \beta \sigma_x \, dz = \int_{-h/2}^{h/2} \sigma_x \left(1 + \frac{z}{r_2}\right) dz$$

$$N_y = \frac{1}{A} \int_{-h/2}^{h/2} \alpha \sigma_y \, dz = \int_{-h/2}^{h/2} \sigma_y \left(1 + \frac{z}{r_1}\right) dz$$

$$N_{xy} = \frac{1}{B} \int_{-h/2}^{h/2} \beta \tau_{xy} \, dz = \int_{-h/2}^{h/2} \tau_{xy} \left(1 + \frac{z}{r_2}\right) dz$$

$$N_{yx} = \frac{1}{A} \int_{-h/2}^{h/2} \alpha \tau_{xy} \, dz = \int_{-h/2}^{h/2} \tau_{xy} \left(1 + \frac{z}{r_1}\right) dz$$

$$Q_x = \frac{1}{B} \int_{-h/2}^{h/2} \beta \tau_{xz} \, dz = \int_{-h/2}^{h/2} \tau_{xz} \left(1 + \frac{z}{r_2}\right) dz \qquad (5\text{-}59)$$

$$Q_y = \frac{1}{A} \int_{-h/2}^{h/2} \alpha \tau_{yz} \, dz = \int_{-h/2}^{h/2} \tau_{yz} \left(1 + \frac{z}{r_1}\right) dz$$

$$M_x = \frac{1}{B} \int_{-h/2}^{h/2} \beta z \, \sigma_x \, dz = \int_{-h/2}^{h/2} z \left(1 + \frac{z}{r_2}\right) \sigma_x \, dz$$

$$M_y = \frac{1}{A} \int_{-h/2}^{h/2} \alpha z \, \sigma_y \, dz = \int_{-h/2}^{h/2} z \left(1 + \frac{z}{r_1}\right) \sigma_y \, dz$$

$$M_{xy} = \frac{1}{B} \int_{-h/2}^{h/2} \beta z \tau_{xy} \, dz = \int_{-h/2}^{h/2} z \left(1 + \frac{z}{r_2}\right) \tau_{xy} \, dz$$

$$M_{yx} = \frac{1}{A} \int_{-h/2}^{h/2} \alpha z \tau_{xy} \, dz = \int_{-h/2}^{h/2} z \left(1 + \frac{z}{r_1}\right) \tau_{xy} \, dz$$

The positive senses of forces and moments are shown in Fig. 5-7. Equations (5-59) are also valid for flat plates, with $1/r_1 = 1/r_2 = 0$.

Equation (5-59) shows that M_{xy} is not exactly equal to M_{yx}, unless $r_1 = r_2$. Equation (5-59) yields the following general relationship:

$$\frac{M_{xy}}{r_1} - \frac{M_{yx}}{r_2} = N_{yx} - N_{xy} \qquad (5\text{-}60)$$

The element of the shell shown in Fig. 5-7a may be subjected to external forces caused by gravity and by external pressures and shears applied to the outer and inner surfaces of the shell. Since the area of the element $dx \, dy$ of the middle surface is $AB \, dx \, dy$, the resultant external force on the element of the shell will be denoted by $\mathbf{P} AB \, dx \, dy$. The vector $\mathbf{P}$ is the resultant external force per unit area of the middle surface. It is a function of the coordinates (x, y) of the middle surface. The vector $\mathbf{P}$ is considered to act at the middle surface of the shell, and it is resolved into components (P_x, P_y, P_z) along the (x, y, z) coordinate lines. Often the component P_z is denoted by p or q, since usually it results from normal pressures on the faces of the shell.

In addition to the external force $\mathbf{P}AB\,dx\,dy$, an external couple $\mathbf{R}AB\,dx\,dy$ may act on an element of the shell. It is assumed that the couple results only from the shearing stresses on the external surfaces of the shell. Hence $R_z = 0$. Also,

$$ABR_x = -\alpha\beta z\tau_{yz}\Big|_{-h/2}^{h/2}, \qquad ABR_y = \alpha\beta z\tau_{xz}\Big|_{-h/2}^{h/2}$$

Likewise,

$$ABP_x = \alpha\beta\tau_{xz}\Big|_{-h/2}^{h/2}, \qquad ABP_y = \alpha\beta\tau_{yz}\Big|_{-h/2}^{h/2}$$

We obtain the equilibrium equations for N_x, N_y, N_{xy}, N_{yx}, Q_x, Q_y by integrating the differential equations of equilibrium through the thickness. Consider, for example, the first of the equations in (4-29). The first term is $\partial/\partial x(\beta\sigma_x)$. Integrating this expression with respect to z between the limits $-h/2$ and $h/2$, and utilizing Eqs. (5-59), we obtain

$$\int_{-h/2}^{h/2} \frac{\partial}{\partial x}(\beta\sigma_x)\,dz = \frac{\partial}{\partial x}\int_{-h/2}^{h/2}\beta\sigma_x\,dz = \frac{\partial}{\partial x}(BN_x)$$

Similarly, for the second term,

$$\int_{-h/2}^{h/2} \frac{\partial}{\partial y}(\alpha\tau_{xy})\,dz = \frac{\partial}{\partial y}\int_{-h/2}^{h/2}\alpha\tau_{xy}\,dz = \frac{\partial}{\partial y}(AN_{yx})$$

For the integral of the third term, we obtain

$$\int_{-h/2}^{h/2} \frac{\partial}{\partial z}(\alpha\beta\tau_{xz})\,dz = \alpha\beta\tau_{xz}\Big|_{-h/2}^{h/2} = ABP_x$$

The fourth integral obtained from Eqs. (4-29) is

$$\int_{-h/2}^{h/2} \frac{\partial\alpha}{\partial y}\tau_{xy}\,dz$$

By Eq. (5-58), this is equal to

$$\int_{-h/2}^{h/2} \frac{\partial}{\partial y}\left[A\left(1 + \frac{z}{r_1}\right)\right]\tau_{xy}\,dz$$

Hence

$$\int_{-h/2}^{h/2} \frac{\partial \alpha}{\partial y} \tau_{xy}\, dz = \int_{-h/2}^{h/2} \left[\frac{\partial A}{\partial y} + z\, \frac{\partial}{\partial y}\left(\frac{A}{r_1}\right) \right] \tau_{xy}\, dz$$

With the Codazzi equation (5-56), this yields

$$\int_{-h/2}^{h/2} \frac{\partial \alpha}{\partial y} \tau_{xy}\, dz = \int_{-h/2}^{h/2} \left(\frac{\partial A}{\partial y} + \frac{z}{r_2}\, \frac{\partial A}{\partial y} \right) \tau_{xy}\, dz$$

$$= \frac{\partial A}{\partial y} \int_{-h/2}^{h/2} \left(1 + \frac{z}{r_2} \right) \tau_{xy}\, dz$$

Hence by Eqs. (5-59)

$$\int_{-h/2}^{h/2} \frac{\partial \alpha}{\partial y} \tau_{xy}\, dz = N_{xy}\, \frac{\partial A}{\partial y}$$

Applying Eqs. (5-58) and (5-59) to the fifth term in Eqs. (4-29), we obtain

$$\int_{-h/2}^{h/2} \beta\, \frac{\partial \alpha}{\partial z}\, \tau_{xz}\, dz = \int_{-h/2}^{h/2} \beta\, \frac{A}{r_1}\, \tau_{xz}\, dz = \frac{A}{r_1} \int_{-h/2}^{h/2} \beta \tau_{xz}\, dz$$

$$= \frac{ABQ_x}{r_1}$$

Finally, by Eqs. (5-56), (5-58), and (5-59), the integral of the sixth term in Eqs. (4-29) is

$$\int_{-h/2}^{h/2} \frac{\partial \beta}{\partial x}\, \sigma_y\, dz = \int_{-h/2}^{h/2} \frac{\partial}{\partial x}\left[B\left(1 + \frac{z}{r_2} \right) \right] \sigma_y\, dz$$

$$= \int_{-h/2}^{h/2} \left[\frac{\partial B}{\partial x} + z\, \frac{\partial}{\partial x}\left(\frac{B}{r_2}\right) \right] \sigma_y\, dz = \int_{-h/2}^{h/2} \left(\frac{\partial B}{\partial x} + \frac{z}{r_1}\, \frac{\partial B}{\partial x} \right) \sigma_y\, dz$$

$$= \frac{\partial B}{\partial x} \int_{-h/2}^{h/2} \left(1 + \frac{z}{r_1} \right) \sigma_y\, dz = N_y\, \frac{\partial B}{\partial x}$$

With the foregoing results, the first of the equations in (4-29) yields

$$\frac{\partial}{\partial x}(BN_x) + \frac{\partial}{\partial y}(AN_{yx}) + N_{xy}\, \frac{\partial A}{\partial y} - N_y\, \frac{\partial B}{\partial x} + \frac{ABQ_x}{r_1} + ABP_x = 0$$

Similarly, the second and third equations in (4-29) yield the equilibrium equations for forces in the y- and z-directions.

The equilibrium equations for moments are obtained by multiplying the first and second equations in (4-29) by z and integrating through the thickness. Considering the first of these equations and noting Eqs. (5-59), we obtain the following integrals:

$$\int_{-h/2}^{h/2} z \frac{\partial}{\partial x} (\beta \sigma_x) \, dz = \frac{\partial}{\partial x} \int_{-h/2}^{h/2} z \beta \sigma_x \, dz = \frac{\partial}{\partial x} (BM_x)$$

$$\int_{-h/2}^{h/2} z \frac{\partial}{\partial y} (\alpha \tau_{xy}) \, dz = \frac{\partial}{\partial y} \int_{-h/2}^{h/2} z \alpha \tau_{xy} \, dz = \frac{\partial}{\partial y} (AM_{yx})$$

$$\int_{-h/2}^{h/2} z \frac{\partial}{\partial z} (\alpha \beta \tau_{xz}) \, dz = - \int_{-h/2}^{h/2} \alpha \beta \tau_{xz} \, dz + ABR_y$$

The last result is obtained by integration by parts. It yields

$$\int_{-h/2}^{h/2} z \frac{\partial}{\partial z} (\alpha \beta \tau_{xz}) \, dz + \int_{-h/2}^{h/2} z \beta \frac{\partial \alpha}{\partial z} \tau_{xz} \, dz$$

$$= \int_{-h/2}^{h/2} \left(-\alpha + \frac{zA}{r_1} \right) \beta \tau_{xz} \, dz + ABR_y = -A \int_{-h/2}^{h/2} \beta \tau_{xz} \, dz + ABR_y$$

$$= -ABQ_x + ABR_y$$

Also,

$$\int_{-h/2}^{h/2} z \frac{\partial \alpha}{\partial y} \tau_{xy} \, dz = \int_{-h/2}^{h/2} z \left[\frac{\partial A}{\partial y} + z \frac{\partial}{\partial y} \left(\frac{A}{r_1} \right) \right] \tau_{xy} \, dz$$

$$= \int_{-h/2}^{h/2} z \left(\frac{\partial A}{\partial y} + \frac{z}{r_2} \frac{\partial A}{\partial y} \right) \tau_{xy} \, dz = \frac{\partial A}{\partial y} \int_{-h/2}^{h/2} z \left(1 + \frac{z}{r_2} \right) \tau_{xy} \, dz = M_{xy} \frac{\partial A}{\partial y}$$

and

$$\int_{-h/2}^{h/2} z \frac{\partial \beta}{\partial x} \sigma_y \, dz = \int_{-h/2}^{h/2} z \left[\frac{\partial B}{\partial x} + z \frac{\partial}{\partial x} \left(\frac{B}{r_2} \right) \right] \sigma_y \, dz$$

$$= \int_{-h/2}^{h/2} z \left(\frac{\partial B}{\partial x} + \frac{z}{r_1} \frac{\partial B}{\partial x} \right) \sigma_y \, dz = \frac{\partial B}{\partial x} \int_{-h/2}^{h/2} z \left(1 + \frac{z}{r_1} \right) \sigma_y \, dz = M_y \frac{\partial B}{\partial x}$$

Hence by Eqs. (4-29) the first equilibrium equation for moments is

$$\frac{\partial}{\partial x} (BM_x) + \frac{\partial}{\partial y} (AM_{yx}) + M_{xy} \frac{\partial A}{\partial y} - M_y \frac{\partial B}{\partial x} - ABQ_x + ABR_y = 0$$

Similarly, the second of the equations in (4-29) yields another moment equation. The complete set of equilibrium equations is

$$\frac{\partial}{\partial x}(BN_x) + \frac{\partial}{\partial y}(AN_{yx}) + N_{xy}\frac{\partial A}{\partial y} - N_y\frac{\partial B}{\partial x} + \frac{ABQ_x}{r_1} + ABP_x = 0 \quad (5\text{-}61)$$

$$\frac{\partial}{\partial x}(BN_{xy}) + \frac{\partial}{\partial y}(AN_y) + N_{yx}\frac{\partial B}{\partial x} - N_x\frac{\partial A}{\partial y} + \frac{ABQ_y}{r_2} + ABP_y = 0 \quad (5\text{-}62)$$

$$\frac{\partial}{\partial x}(BQ_x) + \frac{\partial}{\partial y}(AQ_y) - \frac{AB}{r_1}N_x - \frac{AB}{r_2}N_y + ABP_z = 0 \quad (5\text{-}63)$$

$$\frac{\partial}{\partial x}(BM_x) + \frac{\partial}{\partial y}(AM_{yx}) + M_{xy}\frac{\partial A}{\partial y} - M_y\frac{\partial B}{\partial x} - ABQ_x + ABR_y = 0 \quad (5\text{-}64)$$

$$\frac{\partial}{\partial x}(BM_{xy}) + \frac{\partial}{\partial y}(AM_y) + M_{yx}\frac{\partial B}{\partial x} - M_x\frac{\partial A}{\partial y} - ABQ_y - ABR_x = 0 \quad (5\text{-}65)$$

These equations are supplemented by Eq. (5-60), which is an identity. The shears Q_x and Q_y may be eliminated by Eqs. (5-64) and (5-65). Equations (5-60) to (5-65) apply also in the theory of flat plates. If curvature due to bending is negligible, $1/r_1 = 1/r_2 = 0$ for a flat plate and (x, y) may be any orthogonal coordinates on the middle plane. Equations (5-61) and (5-62) are, accordingly, generalizations of Eq. (5-13). Relations equivalent to the preceding equilibrium equations were derived by Love (51). Various specializations of the equilibrium equations, with applications to engineering structures, are discussed in the book by Flügge (25).

5-8. STRAIN ENERGY OF SHELLS. For shell coordinates ($\gamma = 1$), the strain-displacement relations [Eqs. (4-28)] are approximated as follows:

$$\epsilon_x = \frac{1}{\alpha}\left(u_x + \frac{\alpha_y v}{\beta} + \alpha_z w\right) + \frac{w_x^2}{2A^2}$$

$$\epsilon_y = \frac{1}{\beta}\left(\frac{\beta_x u}{\alpha} + v_y + \beta_z w\right) + \frac{w_y^2}{2B^2}$$

$$\gamma_{yz} = \frac{w_y}{\beta} + v_z - \frac{\beta_z v}{\beta} \qquad\qquad (5\text{-}66)$$

$$\gamma_{zx} = u_z + \frac{w_x}{\alpha} - \frac{\alpha_z u}{\alpha}$$

$$\gamma_{xy} = \frac{u_y}{\beta} + \frac{v_x}{\alpha} - \frac{\beta_x v}{\alpha\beta} - \frac{\alpha_y u}{\alpha\beta} + \frac{w_x w_y}{AB}$$

The discarding of quadratic terms except those involving w_x and w_y is not always a legitimate approximation. Nevertheless, it has been done in nearly all investigations of large deflections of shells. In small-deflection theories all quadratic terms in the strain-displacement relations are neglected.

As in the classical theory of flat plates (Sec. 5-1), we assume that the transverse shearing stresses, τ_{xz} and τ_{yz}, vanish. The assumption $\tau_{xz} = \tau_{yz} = 0$ yields $\gamma_{xz} = \gamma_{yz} = 0$.

In comparison to u and v, the normal displacement w does not vary much with z. Consequently, we assume that $w = w(x, y)$. By Eqs. (5-66), the equations $\gamma_{xz} = \gamma_{yz} = 0$ may be expressed as follows:

$$\frac{\partial}{\partial z}\left(\frac{u}{\alpha}\right) + \frac{w_x}{\alpha^2} = 0, \qquad \frac{\partial}{\partial z}\left(\frac{v}{\beta}\right) + \frac{w_y}{\beta^2} = 0$$

With Eqs. (5-58), these equations yield

$$\frac{u}{\alpha} = \frac{-w_x}{A^2}\int\frac{dz}{\left(1 + \dfrac{z}{r_1}\right)^2}, \qquad \frac{v}{\beta} = \frac{-w_y}{B^2}\int\frac{dz}{\left(1 + \dfrac{z}{r_2}\right)^2}$$

Hence by integration

$$u = \frac{w_x r_1}{A} + \alpha f(x, y), \qquad v = \frac{w_y r_2}{B} + \beta g(x, y)$$

The additive functions, $f(x, y)$ and $g(x, y)$, are determined by the conditions $u = \bar{u}$ and $v = \bar{v}$ if $z = 0$. Accordingly, with Eq. (5-58), there results

$$u = A^{-1}(\alpha\bar{u} - zw_x), \qquad v = B^{-1}(\beta\bar{v} - zw_y), \qquad w = \bar{w} \qquad (5\text{-}67)$$

Equation (5-67) determines how the displacement vector varies through the thickness.

The strain components are now determined by substituting Eq. (5-67) o Eqs. (5-66). Thus we obtain

$$\epsilon_x = \frac{1}{\alpha}\frac{\partial}{\partial x}\left(\frac{\alpha\bar{u} - zw_x}{A}\right) + \frac{\alpha_y}{\alpha}\left(\frac{\beta\bar{v} - zw_y}{\beta B}\right) + \frac{\alpha_z w}{\alpha} + \frac{w_x^2}{2A^2}$$

$$\epsilon_y = \frac{1}{\beta}\frac{\partial}{\partial y}\left(\frac{\beta\bar{v} - zw_y}{B}\right) + \frac{\beta_x}{\beta}\left(\frac{\alpha\bar{u} - zw_x}{\alpha A}\right) + \frac{\beta_z w}{\beta} + \frac{w_y^2}{2B^2} \qquad (5\text{-}68)$$

$$\gamma_{xy} = \frac{\beta}{\alpha}\frac{\partial}{\partial x}\left(\frac{\beta\bar{v} - zw_y}{\beta B}\right) + \frac{\alpha}{\beta}\frac{\partial}{\partial y}\left(\frac{\alpha\bar{u} - zw_x}{\alpha A}\right) + \frac{w_x w_y}{AB}$$

These equations, which express the strain components at any point in terms of the displacements of the middle surface, are geometrical; they are not contingent on elastic or isotropic behavior of the material. Equations equivalent to those in (5-68), with quadratic terms omitted, were derived by Love (51). He also proposed that Eqs. (5-68) be linearized in z. Then,

$$\epsilon_x = e_x + z\kappa_x, \qquad \epsilon_y = e_y + z\kappa_y, \qquad \gamma_{xy} = e_{xy} + z\kappa_{xy} \qquad (5\text{-}69)$$

Here, (e_x, e_y, e_{xy}) are the values of $(\epsilon_x, \epsilon_y, \gamma_{xy})$ on the middle surface of the shell. The factors $(\kappa_x, \kappa_y, \kappa_{xy})$ are closely related to the changes of curvature of the middle surface caused by bending. Setting $z = 0$ in Eqs. (5-68), we obtain

$$e_x = \frac{\bar{u}_x}{A} + \frac{\bar{v}A_y}{AB} + \frac{w}{r_1} + \frac{w_x^2}{2A^2}$$

$$e_y = \frac{\bar{v}_y}{B} + \frac{\bar{u}B_x}{AB} + \frac{w}{r_2} + \frac{w_y^2}{2B^2} \qquad (5\text{-}70)$$

$$e_{xy} = \frac{\bar{v}_x}{A} + \frac{\bar{u}_y}{B} - \frac{A_y\bar{u}}{AB} - \frac{B_x\bar{v}}{AB} + \frac{w_x w_y}{AB}$$

Equations (5-70) is a generalization of Eq. (5-3) which was derived for flat plates.

The factors $(\kappa_x, \kappa_y, \kappa_{xy})$ are obtained from Eqs. (5-68) and (5-69) if the z-terms are expanded to first powers by the binomial series. The first-order effects of $\bar{u}$ and $\bar{v}$ on $\kappa_x, \kappa_y, \kappa_{xy}$ represent changes of curvature due to the circumstance that the elements of a surface are usually bent if the surface is deformed into itself. For example, if a surface element of a cone is displaced along a generator, it must be bent to remain in the original conical surface. Also, if a surface element of a cylinder is rotated about its normal, it must be bent to remain in the original cylindrical surface. These effects are quite small compared to the bending that results from the normal displacement w. Consequently, $\bar{u}$ and $\bar{v}$ will be discarded from the formulas for $\kappa_x, \kappa_y, \kappa_{xy}$. Also, the bending effect of w results mainly through the derivatives $w_x, w_y, w_{xx}, w_{xy}, w_{yy}$. There is a slight bending that depends explicitly on w; it remains if w is constant. For example, a spherical shell experiences changes of curvature if w is a nonzero constant, since the radius of the sphere is changed. However, this effect is quite small. Consequently, w will be discarded from the formulas for $\kappa_x, \kappa_y, \kappa_{xy}$, although the derivatives of w will be retained. Accordingly,

$$\kappa_x = -\frac{1}{A}\frac{\partial}{\partial x}\left(\frac{w_x}{A}\right) - \frac{A_y w_y}{AB^2}$$

$$\kappa_y = -\frac{B_x w_x}{A^2 B} - \frac{1}{B}\frac{\partial}{\partial y}\left(\frac{w_y}{B}\right) \qquad (5\text{-}71)$$

$$\kappa_{xy} = \frac{2}{AB}\left(\frac{A_y w_x}{A} + \frac{B_x w_y}{B} - w_{xy}\right)$$

The stress-strain relations for a flat plate [Eqs. (5-5)] remain valid for a curved shell. Accordingly, if the terms z/r_1 and z/r_2 in Eqs. (5-59) are disregarded, Eqs. (5-12) expresses the quantities N_x, N_y, N_{xy} in terms of e_x, e_y, e_{xy}. For a curved shell, e_x, e_y, e_{xy} are determined by Eqs. (5-70). Likewise, when z/r_1 and z/r_2 are discarded from Eqs. (5-59), we obtain with Eqs. (5-5), (5-8), and (5-69)

$$M_x = D(\kappa_x + \nu\kappa_y) - \frac{E\theta_1}{1 - \nu}$$

$$M_y = D(\kappa_y + \nu\kappa_x) - \frac{E\theta_1}{1 - \nu} \qquad (5\text{-}72)$$

$$M_{xy} = M_{yx} = \tfrac{1}{2}D(1 - \nu)\kappa_{xy}$$

The quantities κ_x, κ_y, κ_{xy} are determined by Eqs. (5-71). Equations (5-72) is a generalization of Eqs. (5-26) which was derived for flat plates. After M_x, M_y, M_{xy} are determined by Eqs. (5-72), the shears Q_x, Q_y may be obtained from the equilibrium equations (5-64) and (5-65).

The formula for the strain-energy density U_0 that was used for flat plates [Eq. (5-7)] remains valid for curved shells. The volume element of the shell is $\alpha\beta \, dx \, dy \, dz$. In view of Eq. (5-58), this will be approximated by $AB \, dx \, dy \, dz$. Then,

$$U = \iint AB \, dx \, dy \int_{-h/2}^{h/2} U_0 \, dz$$

When Eqs. (5-7) and (5-69) are substituted into the preceding equation, the strain energy separates into a sum of three terms, $U = U_m + U_b + U_\theta$, where the *membrane energy* U_m is linear in h and the *bending energy* U_b is cubic in h. The term U_θ represents the part of the strain energy that results from heating. Again, the notation in Eq. (5-8) is used. Then, if G and ν are considered to be constants, integration with respect to z yields

$$U_m = \frac{G}{1 - \nu} \iint [e_x{}^2 + e_y{}^2 + 2\nu e_x e_y + \tfrac{1}{2}(1 - \nu)e_{xy}{}^2]hAB \, dx \, dy$$

$$U_b = \frac{G}{12(1 - \nu)} \iint [\kappa_x{}^2 + \kappa_y{}^2 + 2\nu\kappa_x\kappa_y + \tfrac{1}{2}(1 - \nu)\kappa_{xy}{}^2]h^3AB \, dx \, dy$$

$$U_\theta = \frac{-E}{1 - \nu} \iint [(e_x + e_y)\theta_0 + (\kappa_x + \kappa_y)\theta_1]AB \, dx \, dy \qquad (5\text{-}73)$$

By means of Eqs. (5-70), (5-71), and (5-73), the strain energy is expressed as a functional of $\bar{u}$, $\bar{v}$, w.

The stress-strain-temperature relations [Eqs. (5-5)] remain valid for curved shells. Also, Eqs. (5-12) remain approximately valid. If there are no tangential external forces and equilibrium exists, the principle of stationary potential energy requires that the strain energy U satisfy the Euler equations of the calculus of variations for $\bar{u}$ and $\bar{v}$. With Eqs. (5-12), these conditions yield the equilibrium equations for tangential tractions. The results agree with Eqs. (5-61) and (5-62), except for the terms containing Q_x and Q_y, which have been lost because of the approximations.

Various refinements of the theory of shells have been developed (102, 124, 129), and many of them possess particular merits, but the foregoing theory, which is essentially that of Love (51) is attractive because of its comparative simplicity and adaptability. D. O. Brush (96) has shown that many recent studies of elastic shells are based on equations that result immediately from Eqs. (5-73) by variational methods. The von Kármán theory of large deflections of flat plates (Sec. 5-1), the eigenvalue theory of buckling of plates (84), the Donnell equations for large deflections of cylindrical shells, the equations of Reiss, Greenberg, and Keller for snap-through of shallow spherical caps (135), and the equations of Seide (144) for buckling of conical shells may all be obtained in this way. It is apparent that the equations of this article reduce to the flat-plate equations of Sec. 5-1 if $1/r_1 = 1/r_2 = 0$; in fact, the present theory is a natural generalization of the theory of large deflections of flat plates. When the equations in Secs. 5-7 and 5-8 are applied to flat plates, the coordinates are arbitrary, except for the requirement of orthogonality.

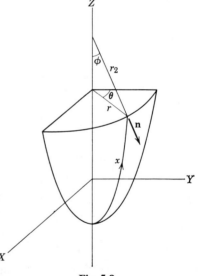

Fig. 5-8

5-9. AXIALLY SYMMETRIC SHELLS. The middle surface of an axially symmetric shell may be generated by rotating a plane curve about an axis in its plane. A cross section of the surface by a plane perpendicular to the axis of symmetry is a circle. A cross section of the surface by a plane that contains the axis of symmetry is called a "meridian." The cross-sectional circles and the meridians are the lines of principal curvature on the surface.

Figure 5-8 represents an axially symmetric surface. The Z-axis is the axis of symmetry. Any plane that contains the Z-axis intersects the surface in a meridian. The angle θ that this plane forms with the YZ-plane is called the longitude of the meridian. As surface coordinates, we adopt the longitude θ and any coordinate x that locates a point on a meridian. The radius of the cross-sectional circle with coordinate x is denoted by r. Accordingly, the equation of a meridian is $r = r(x)$, $Z = Z(x)$.

It is seen from Fig. 5-8 that the equations of the surface are

$$X = r \sin \theta, \qquad Y = r \cos \theta, \qquad Z = Z(x) \qquad (5\text{-}74)$$

Therefore, $ds^2 = A^2 \, dx^2 + B^2 \, d\theta^2$, in which

$$A = A(x), \qquad B = r \qquad (5\text{-}75)$$

The positive normal $\mathbf{n}$ of the surface is directed outward.

One principal curvature $1/r_1$ is the curvature of a meridian. The center of curvature corresponding to r_2 lies on the Z-axis. Consequently, with the notation of Fig. 5-8,

$$r_2 = r \csc \phi \qquad (5\text{-}76)$$

Also, $ds = r_1 \, d\phi$, where ds is an element of arc length on a meridian. Sometimes, coordinates may be chosen conveniently so that $\phi = x$; then $A = r_1$. However, for generality, we shall not fix the coordinate x. From consideration of an infinitesimal triangle with hypoteneuse on a meridian, it is apparent that

$$\frac{dr}{dx} = A \cos \phi \qquad (5\text{-}77)$$

Accordingly, since $B = r$, the equilibrium equations for tangential tractions, obtained by specialization of Eqs. (5-61) and (5-62), are

$$\frac{\partial}{\partial x}(rN_x) + A\frac{\partial N_{\theta x}}{\partial \theta} - AN_\theta \cos \phi + \frac{ArQ_x}{r_1} + ArP_x = 0$$

$$\frac{\partial}{\partial x}(rN_{x\theta}) + A\frac{\partial N_\theta}{\partial \theta} + AN_{\theta x} \cos \phi + AQ_\theta \sin \phi + ArP_\theta = 0 \qquad (5\text{-}78)$$

If $x = \phi$, Eqs. (5-78) are equivalent to those derived by Flügge (25), except for the terms containing Q_x and Q_θ, which he neglected.

If the loading is axially symmetrical, $N_{x\theta} = N_{\theta x} = 0$, $Q_\theta = P_\theta = 0$, and $\partial N_0/\partial \theta = 0$, Then the second of Eqs. (5-78) is satisfied automatically.

Cylindrical Shells. Circular cylindrical shells are included among axially symmetric shells. The coordinate x is conveniently taken to be distance along a generator. Then $A = 1$ and $B = a$, where a is the radius of the middle surface. In addition, $1/r_1 = 0$, $r_2 = r = a$, and $\phi = 90°$. If there are no tangential external forces and Q_x and Q_θ are neglected, Eqs. (5-78) reduce to

$$\frac{\partial N_x}{\partial x} + \frac{1}{a}\frac{\partial N_{x\theta}}{\partial \theta} = 0, \qquad \frac{\partial N_{x\theta}}{\partial x} + \frac{1}{a}\frac{\partial N_\theta}{\partial \theta} = 0 \qquad (5\text{-}79)$$

Here, the distinction between $N_{x\theta}$ and $N_{\theta x}$ is disregarded. The general solution of Eq. (5-79) is

$$N_x = \frac{\partial^2 F}{\partial \theta^2}, \qquad N_\theta = a^2 \frac{\partial^2 F}{\partial x^2}, \qquad N_{x\theta} = -a \frac{\partial^2 F}{\partial x\, \partial \theta} \qquad (5\text{-}80)$$

where $F(x, \theta)$ is an arbitrary function, called the "Airy stress function."
   By Eqs. (5-70)

$$e_x = \bar{u}_x + \tfrac{1}{2} w_x^2, \qquad e_\theta = \frac{\bar{v}_\theta + w}{a} + \frac{w_\theta^2}{2a^2}$$

$$e_{x\theta} = \bar{v}_x + \frac{\bar{u}_\theta}{a} + \frac{w_x w_\theta}{a} \qquad (5\text{-}81)$$

By Eqs. (5-71)

$$\kappa_x = -w_{xx}, \qquad \kappa_\theta = -\frac{w_{\theta\theta}}{a^2}, \qquad \kappa_{x\theta} = -\frac{2}{a} w_{x\theta} \qquad (5\text{-}82)$$

Accordingly, if thermal effects are disregarded and there is an internal pressure $p(x, \theta)$, the total potential energy, determined by Eqs. (5-73), is

$$V = a \iint \Bigg\{ \frac{Gh}{1-\nu} \Bigg[(\bar{u}_x + \tfrac{1}{2} w_x^2)^2 + \left(\frac{\bar{v}_\theta + w}{a} + \frac{w_\theta^2}{2a^2} \right)^2$$

$$+ 2\nu(\bar{u}_x + \tfrac{1}{2} w_x^2)\left(\frac{\bar{v}_\theta + w}{a} + \frac{w_\theta^2}{2a^2} \right) + \tfrac{1}{2}(1-\nu)\left(\bar{v}_x + \frac{\bar{u}_\theta}{a} + \frac{w_x w_\theta}{a} \right)^2 \Bigg]$$

$$+ \frac{Gh^3}{12(1-\nu)} \Bigg[w_{xx}^2 + \frac{w_{\theta\theta}^2}{a^4} + 2\nu w_{xx}\frac{w_{\theta\theta}}{a^2} + 2(1-\nu)\frac{w_{x\theta}^2}{a^2} \Bigg] - pw \Bigg\} dx\, d\theta$$

$$(5\text{-}83)$$

If the integrand in Eq. (5-83) is denoted by J, Eqs. (5-12) and (5-81) yield

$$\frac{\partial J}{\partial w} = \frac{N_\theta}{a} - p, \qquad \frac{\partial J}{\partial w_x} = N_x w_x + N_{x\theta}\frac{w_\theta}{a}$$

$$\frac{\partial J}{\partial w_\theta} = \frac{N_\theta w_\theta}{a^2} + N_{x\theta}\frac{w_x}{a}, \qquad \frac{\partial J}{\partial w_{xx}} = \frac{Gh^3}{6(1-\nu)}\left(w_{xx} + \nu\frac{w_{\theta\theta}}{a^2} \right)$$

$$\frac{\partial J}{\partial w_{x\theta}} = \frac{Gh^3}{3}\frac{w_{x\theta}}{a^2}, \qquad \frac{\partial J}{\partial w_{\theta\theta}} = \frac{Gh^3}{6(1-\nu)a^2}\left(\frac{w_{\theta\theta}}{a^2} + \nu w_{xx} \right)$$

Consequently, in view of Eq. (5-79), the Euler equation for w is

$$D\left(w_{xxxx} + \frac{2}{a^2} w_{xx\theta\theta} + \frac{1}{a^4} w_{\theta\theta\theta\theta} \right) + \frac{N_\theta}{a} = p + N_x w_{xx}$$

$$+ \frac{2}{a} N_{x\theta} w_{x\theta} + \frac{1}{a^2} N_\theta w_{\theta\theta}; \qquad D = \frac{Eh^3}{12(1-\nu^2)} \qquad (5\text{-}84)$$

Equation (5-84) was derived originally by Donnell (103) by balancing forces
and moments on an element of a deformed shell. It has played an important role in theories of buckling of cylindrical shells. The terms N_x, $N_{x\theta}$,
and N_θ on the right side of Eq. (5-84) result from the nonlinear terms in
Eqs. (5-81). Consequently, these terms do not appear in the small-
deflection theory of cylindrical shells.

If w is constant, Eq. (5-84) yields $N_\theta/a = p$. This result agrees with the
elementary theory of a thin cylindrical shell with uniform internal pressure
p. Since N_θ/a is a tensile stress, the positive sense of p must be outward.

If an infinitely long cylindrical shell is subjected to uniform external
pressure q, the buckling pattern is independent of x. Consequently, since
$N_\theta/a = p = -q$, Eq. (5-84) yields

$$w_{\theta\theta\theta\theta} + k^2 w_{\theta\theta} = 0, \qquad k^2 = \frac{qa^3}{D}$$

The general solution of this equation is

$$w = A \sin k\theta + B \cos k\theta + C\theta + C_1$$

Since w has period 2π, $C = 0$ and k is an integer. The value $k = 1$
corresponds to a rigid-body displacement. Consequently, the buckling
pressure is determined by $k = 2$. This yields

$$q_{\mathrm{cr}} = \frac{4D}{a^3}$$

Actually, this result is too high. The correct buckling pressure, determined
by the theory of buckling of rings (84), is $q_{\mathrm{cr}} = 3D/a^3$. The error occurs
because of the approximations (125). However, there is reason to believe
that Donnell's equation gives more
accurate results when multiple wave
patterns occur in the buckled form, as
in the case of short cylindrical shells
buckled by hydrostatic pressure.

Example—Cylindrical Tank. A vertical cylindrical tank with constant
wall thickness h is filled with liquid
with specific weight γ (Fig. 5-9). The
height of the tank is L. The axial
coordinate x is measured from the
bottom. If the tank is open at the
top, the pressure at ordinate x is
$p = \gamma(L - x)$.

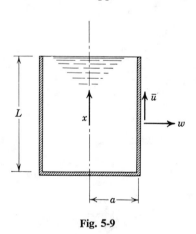

Fig. 5-9

Because of axial symmetry, $v = 0$, $N_{x\theta} = 0$, and all partial derivatives with respect to θ vanish. The first of Eqs. (5-79) reduces to $dN_x/dx = 0$. Since the boundary condition at the top is $N_x = 0$, this equation shows that N_x vanishes everywhere. Consequently, by Eqs. (5-12), $e_x + \nu e_\theta = 0$. If quadratic terms in Eqs. (5-81) are neglected, this yields $\bar{u}_x = -\nu w/a$. Hence by Eqs. (5-12)

$$N_\theta = \frac{Ehw}{a} \tag{a}$$

Accordingly, when quadratic terms are neglected, Eq. (5-84) yields

$$w_{xxxx} + \frac{Eh}{a^2 D} w = \frac{p}{D} \tag{b}$$

Equation (b) may be integrated readily when the preceding linear expression for p is introduced. The solutions of this equation are discussed in detail in the book by Timoshenko and Woinowsky-Krieger (86). The four arbitrary constants in the general solution may be adapted to the two forced boundary conditions ($w = w_x = 0$ at $x = 0$) and the two natural boundary conditions ($w_{xx} = w_{xxx} = 0$ at $x = L$). The natural boundary conditions signify that the bending moment and the shear are zero at the top edge. They may be derived by applying the variational procedure to the potential energy integral which results from Eqs. (5-83) when simplifications due to axial symmetry are introduced.

5-10. CIRCULAR PLATES. A circular plate may be regarded as a special case of an axially symmetric shell. If polar coordinates (r, θ) are adopted, $A = 1$ and $B = r$. In addition, $1/r_1 = 1/r_2 = 0$ and $\phi = 0$. By Eqs. (5-59), $N_{r\theta} = N_{\theta r}$ and $M_{r\theta} = M_{\theta r}$. The equilibrium equations for tractions in the plane of the plate are obtained immediately from Eqs. (5-78):

$$\frac{\partial N_r}{\partial r} + \frac{1}{r}\frac{\partial N_{r\theta}}{\partial \theta} + \frac{N_r - N_\theta}{r} + P_r = 0$$

$$\frac{\partial N_{r\theta}}{\partial r} + \frac{1}{r}\frac{\partial N_\theta}{\partial \theta} + \frac{2}{r} N_{r\theta} + P_\theta = 0 \tag{5-85}$$

Only axially symmetric loading is considered. Then $\bar{v} = 0$, $P_\theta = 0$, and partial derivatives with respect to θ vanish. By Eqs. (5-70),

$$e_r = \bar{u}_r + \tfrac{1}{2} w_r{}^2, \qquad e_\theta = \frac{\bar{u}}{r}, \qquad e_{r\theta} = 0$$

By Eqs. (4-71)

$$\kappa_r = -w_{rr}, \qquad \kappa_\theta = -\frac{w_r}{r}, \qquad \kappa_{r\theta} = 0$$

Accordingly, if h is constant, Eqs. (5-73) yield

$$U_m = \frac{\pi E h}{1 - \nu^2} \int \left[r(\bar{u}_r + \tfrac{1}{2}w_r^2)^2 + \frac{\bar{u}^2}{r} + 2\nu \bar{u}(\bar{u}_r + \tfrac{1}{2}w_r^2) \right] dr$$

$$U_b = \pi D \int \left(r w_{rr}^2 + \frac{w_r^2}{r} + 2\nu w_r w_{rr} \right) dr; \qquad D = \frac{E h^3}{12(1 - \nu^2)} \qquad (5\text{-}86)$$

$$U_\theta = -\frac{2\pi E}{1 - \nu} \int \left[\left(\bar{u}_r + \tfrac{1}{2}w_r^2 + \frac{\bar{u}}{r} \right) \theta_0 - \left(w_{rr} + \frac{w_r}{r} \right) \theta_1 \right] r \, dr$$

If thermal strains are absent, Eqs. (5-72) yield

$$M_r = -D\left(w_{rr} + \frac{\nu w_r}{r} \right)$$

$$M_\theta = -D\left(\frac{w_r}{r} + \nu w_{rr} \right)$$

$$(5\text{-}87)$$

If R_θ is neglected, Eqs. (5-64) and (5-87) yield

$$Q_r = -D \frac{d}{dr}\left[\frac{1}{r} \frac{d}{dr}\left(r \frac{dw}{dr} \right) \right] \qquad (5\text{-}88)$$

As an illustrative application of Eqs. (5-86), we consider a simply supported circular plate that carries an axially symmetric distributed load $p(r)$ and a concentrated load F at the center. If small-deflection approximations are used, $U_m = 0$. In addition, if there is no heating, $U_\theta = 0$. By Eqs. (5-86) the total potential energy is

$$V = \pi \int_0^a \{ D[r(w'')^2 + r^{-1}(w')^2 + 2\nu w' w''] - 2pwr \} \, dr - Fw_0$$

in which primes denote derivatives with respect to r. The deflection at the center is denoted by w_0.

The term $2\nu w' w''$ may be integrated directly. Accordingly, the preceding equation takes the following simpler form:

$$V = -Fw_0 + \pi \nu D(w')^2 \Big|_0^a + \pi \int_0^a \{ D[r(w'')^2 + r^{-1}(w')^2] - 2pwr \} \, dr \qquad (a)$$

It is known from the general theory of plates that w' is defined at a point where a concentrated load is applied but that w'' has a logarithmic singularity. Accordingly, in the present case, $w'(0) = 0$. Furthermore, w'/r has a logarithmic singularity. Accordingly, the integrand in Eq. (a) remains finite; that is, the integral is proper. The variation $\eta(r)$ is restricted so that the integrand in δV is also proper. Then, $\eta'(0) = 0$.

By adding a variation η to w in Eq. (a), deducting V, and retaining only the part of ΔV that is linear in η, we obtain

$$\delta V = -F\eta_0 + 2\pi v Dw'\eta' \Big|_0^a + 2\pi \int_0^a [D(rw''\eta'' + r^{-1}w'\eta') - pr\eta]\, dr$$

Integration by parts yields

$$\delta V = -F\eta_0 + 2\pi D(vw' + rw'')\eta' \Big|_0^a - 2\pi D(rw''' + w'' - r^{-1}w')\eta \Big|_0^a$$

$$+ 2\pi \int_0^a [D(rw'''' + 2w''' - r^{-1}w'' + r^{-2}w') - pr]\eta\, dr = 0$$

Accordingly, the Euler equation is

$$w'''' + \frac{2}{r} w''' - \frac{1}{r^2} w'' + \frac{1}{r^3} w' = \frac{p}{D} \tag{b}$$

Integration of Eq. (b) is facilitated if the equation is written in the following form:

$$\frac{1}{r}\frac{d}{dr}\left\{ r\frac{d}{dr}\left[\frac{1}{r}\frac{d}{dr}\left(r\frac{dw}{dr}\right)\right]\right\} = \frac{p}{D} \tag{5-89}$$

If the plate is simply supported at the edge, the forced boundary condition for the edge is $w = 0$ at $r = a$. The condition $w'(0) = 0$ may also be regarded as a forced boundary condition. The preceding expression for δV yields the following natural boundary conditions:

$$vw' + rw'' = 0 \quad \text{at} \quad r = a \tag{c}$$

$$\lim_{r \to 0}\left(rw''' + w'' - \frac{1}{r}w'\right) = \frac{F}{2\pi D} \tag{d}$$

In view of Eqs. (5-87) and (5-88), these conditions mean that the bending moment at the edge is zero and the shear balances the force F.

For example, suppose that the distributed load p is zero. Then the general solution of Eq. (5-89) is

$$w = A + Br^2 + Cr^2 \log r + C' \log r \tag{e}$$

in which A, B, C, C' are constants of integration. Differentiation of Eq. (e) yields

$$w' = (2B + C)r + 2Cr \log r + \frac{C'}{r}$$

$$w'' = 2B + 3C + 2C \log r - \frac{C'}{r^2}$$

$$w''' = \frac{2C}{r} + \frac{2C'}{r^3}$$

Hence

$$rw''' + w'' - \frac{1}{r} w' = 4C$$

Accordingly, Eq. (d) yields $C = F/(8\pi D)$. The condition $w'(0) = 0$ yields $C' = 0$. This condition might also be derived from the fact that w remains finite when $r = 0$.

Equation (c) now yields

$$B = \frac{-F}{16\pi D} \left(\frac{3 + \nu}{1 + \nu} + 2 \log a \right)$$

The forced boundary condition ($w = 0$ at $r = a$) yields accordingly

$$A = \frac{Fa^2(3 + \nu)}{16\pi D(1 + \nu)}$$

Consequently, by Eq. (e)

$$w = \frac{F}{16\pi D} \left[\frac{3 + \nu}{1 + \nu} (a^2 - r^2) - 2r^2 \log \frac{a}{r} \right] \qquad (5\text{-}90)$$

It is apparent that $\lim_{r \to 0} w'' = -\infty$; that is, the radius of curvature at the center of the plate is zero. This circumstance illustrates a characteristic difference between plate theory and beam theory. In plate theory the second derivative of the deflection function is discontinuous at a point of concentrated load. In beam theory the second derivative remains continuous, but the third derivative is discontinuous at a point where a concentrated load is applied.

PROBLEMS

1. A rectangular elastic isotropic plate of constant thickness is heated. The temperature is $\theta(x, y)$. By the principle of stationary potential energy, derive the differential equations for the displacement components $(\bar{u}, \bar{v})$. The edge $x = a$ is free; derive the natural boundary conditions for that edge by the variational method.

2. A rectangular elastic plate has edges simply supported at $x = 0$ and $x = \pi$. The edges $y = \pm b$ are free. The load is $p = p_0 \sin x$. Set $w = f(y) \sin x$ and obtain the solution of Eq. (5-20) that satisfies the boundary conditions. Let $\theta = 0$.

3. Suppose that the flange (Fig. 5-4) carries a distributed lateral line load $q(x)$. Determine the natural boundary conditions for the flanged edge.

4. A simply supported rectangular elastic plate carries a constant lateral pressure p on a square region of width $2c$ that has sides parallel to the edges of the plate and center at the center of the plate. Outside this square, there is no load. Determine the deflection $w(x, y)$ in the form of a double sine series. Use small-deflection theory.

5. A circular elastic disk of constant thickness is subjected to a temperature distribution $\theta(r)$. There are no external loads. The edge of the disk is free. Derive the differential equation and the natural boundary condition for the radial displacement $\bar{u}(r)$. Note that $\bar{u}(0) = 0$.

6. A circular elastic membrane of constant thickness h and radius a is subjected to a lateral pressure $p(r)$. The membrane is so thin that bending energy is negligible. The membrane has no initial stress. Using Eqs. (5-86) (large-deflection theory), derive the differential equations for $\bar{u}$ and w.

7. An isotropic elastic conical shell with constant thickness h and total vertex angle $2a$ is subjected to axially symmetric pressure $p(x)$, where x is distance along a generator from the vertex. Using small deflection theory, derive the strain-energy equation.

8. A circular elastic plate is supported by a post at the center, and the edge of the plate is free. The plate carries an axially symmetric load $p(r)$. Using small-deflection theory, derive the natural boundary conditions for the free edge.

9. A rectangular elastic plate with simply supported edges is subjected to the temperature distribution $\theta = \beta z$, where $\beta = $ constant. There are no external loads. Using small-deflection theory and employing Eqs. (5-9), (5-32), and (5-33), determine the coefficients a_{mn} by the principle of stationary potential energy.

10. Derive the strain formula for curved beams [Eq. (c), Sec. 2-3] from the corresponding formula for curved shells [Eqs. (5-68)].

11. Show by means of the Codazzi equations that the Lamé coefficients for shell coordinates satisfy the following relations of H. Lamb:

$$A\beta_x = \alpha B_x, \ B\alpha_y = \beta A_y, \quad \text{and} \quad \beta_x \alpha_z = \alpha \beta_{xz}, \ \alpha_y \beta_z = \beta \alpha_{yz}$$

(Subscripts denote partial derivatives).

12. Determine A, B, α, β, $1/r_1$, $1/r_2$ for a circular conical shell with total vertex angle $2a$. Let x be distance along a generator from the vertex.

13. Determine A, B, α, β, $1/r_1$, $1/r_2$ for a spherical shell of radius a. Let $x = \phi = $ colatitude.

14. Determine A, B, α, β, $1/r_1$, $1/r_2$, n for an elliptic cylindrical shell with middle surface defined by $X = a \cos \theta$, $Y = b \sin \theta$, $Z = x$. Let $x = x$, $y = \theta$.

15. Derive Eqs. (5-62), (5-63), and (5-65).

16. The terms Q_x and Q_y in Eqs. (5-61) and (5-62) are frequently neglected. With the aid of the Gauss equation (5-55), show that when $P_x = P_y = Q_x = Q_y = 0$ and the Gaussian curvature K is constant Eqs. (5-61) and (5-62) are satisfied automatically by

$$N_x = B^{-2}F_{yy} + A^{-2}B^{-1}B_xF_x - B^{-3}B_yF_y + KF$$
$$N_y = A^{-2}F_{xx} - A^{-3}A_xF_x + A^{-1}B^{-2}A_yF_y + KF$$
$$N_{xy} = -A^{-1}B^{-1}F_{xy} + A^{-2}B^{-1}A_yF_x + A^{-1}B^{-2}B_xF_y$$

where $F(x, y)$ is an arbitrary function (generalized Airy stress function).

17. Specialize the equations given in Prob. 16 for a spherical shell of radius a referred to spherical coordinates ($x = \phi, y = \theta$).

18. Let (X, Y) be rectangular coordinates in the middle plane of a flat plate. Set $X = c \cosh x \cos y$, $Y = c \sinh x \sin y$. Prove that the coordinates (x, y) are orthogonal and that the x- and y-coordinate lines are hyperbolas and ellipses, respectively, with foci at the points ($\pm c, 0$). Determine A and B and write the general solution of the equilibrium equations for N_x, N_y, N_{xy} by means of the equations given in Prob. 16.

19. Specialize Eq. (5-68) for spherical shells. Let $x = \phi$, $y = \theta$, where ϕ is colatitude.

20. Specialize Eqs. (5-68) for conical shells. Let x be distance along a generator from the vertex and let $y = \theta$ be longitude.

21. Specialize the strain-energy expression for cylindrical shells [Eqs. (5-83)] for the case of axially symmetric loading and small deflections (quadratic terms neglected). Hence derive the differential equations and the boundary conditions for a semi-infinite cylindrical shell that is subjected to constant shear $Q_x = Q_0$ and constant bending moment $M_x = M_0$ at the free end, $x = 0$

22. A simply supported isotropic elastic circular plate of radius a carries a uniform lateral pressure p. Determine the deflection $w(r)$. Compare with the approximate solution obtained in Prob. 20, Chap. 3.

23. Solve Prob. 22 for a plate with a clamped edge.

6 Theory of buckling

If the system is influenced only by internal forces, or if the applied forces follow the law of doing always the same amount of work upon the system passing from one configuration to another by all possible paths, the whole potential energy must be constant, in all positions, for neutral equilibrium; must be a minimum for positions of thoroughly stable equilibrium; must be either an absolute maximum, or a maximum for some displacements and a minimum for others when there is unstable equilibrium.

KELVIN AND TAIT

6-1. INTRODUCTION. The theory of buckling deals principally with conditions under which equilibrium ceases to be stable. The theory of stability (Sec. 1-11) is accordingly an essential preliminary to this chapter.

In the theory of buckling we consider a class Γ of unbuckled configurations, corresponding to a range of values of a real parameter p. Ordinarily, unbuckled configurations are characterized by geometric symmetry. Occasionally, it happens that there is only one unbuckled configuration, particularly if the mechanical system consists of rigid members. However, the class Γ need not be defined explicitly; the essential thing is that to each value of p in the range of interest there corresponds a single configuration in class Γ. Usually p designates the external load on the system. However, this is not essential. For example, in problems of thermal buckling, p may denote temperature.

In the classical problem of buckling the configuration in class Γ is stable if p is small, but it is not stable when p is large. The problem is to determine the value of p for which the configuration in class Γ ceases to be stable. This value is called the "critical value" of p; it is denoted by p_{cr}. Friedrichs

(106) suggested that p_{cr} be called the "Euler buckling load" or "Euler critical value," since it is analogous to the critical load that Euler determined for columns. Modern buckling theory treats many practical questions besides the determination of p_{cr}; for example, the degree of stability before p_{cr} is attained, effects of geometric imperfections, effects of inelastic action or other anomalies in behavior of materials, and nonlinear response of imperfect systems to impulsive loads (119).

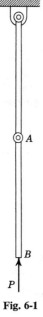

Fig. 6-1

Euler's classic investigation of columns in 1759 stimulated the study of numerous special column problems. Poincaré (131) developed the fundamentals of a general theory of buckling. He observed that the equilibrium points corresponding to all values of the parameter p constitute a path in configuration space, since the equilibrium configuration **X** corresponding to any value of p is a function of p alone. It may happen that the path $\mathbf{X} = \mathbf{X}(p)$ possesses "forks" or "bifurcation points." Ordinarily, the unbuckled form, represented by the stem of the path, becomes unstable at a bifurcation point. Accordingly, the critical value of p usually corresponds to the first bifurcation point on the path of equilibrium configurations.

However, not all buckling is of the bifurcation type. This fact is illustrated by a mechanism devised by Ziegler (152). The two bars shown in Fig. 6-1 are rigid. The joints are frictionless, but they contain springs, so that the restoring moment is proportional to the angular deflection of a hinge. The force P always remains collinear with member AB. It is easily seen that the only possible equilibrium configuration is the straight vertical position shown in Fig. 6-1. Furthermore, this form is an equilibrium configuration irrespective of the magnitude of force P. Nevertheless, Ziegler has shown that there exists a critical load P_{cr}, dependent on the spring constants and the lengths and mass distributions of the bars, such that the slightest lateral push will cause the system to execute large nonlinear oscillations if P exceeds P_{cr}. If P is less than P_{cr}, infinitesimal initial velocities cause only infinitesimal oscillations about the vertical configuration.

The peculiar nature of Ziegler's mechanism stems from the fact that the external force is nonconservative; that is, the work performed by the constant load P depends on the path described by point B. For a conservative system, the theory of stability developed in Sec. 1-11 may be used to determine the buckling load.

In practice, the mere existence of stability provides no assurance of safe design. What is important in engineering is the degree of stability. Von Kármán and Tsien (118) noted that some structures—particularly, shell-like structures—may experience states of weak stability such that small blows or other disturbances cause them to snap into badly deformed shapes. This condition, known as "snap-through," is characterized by a dip in the load-deflection curve. The theory of snap-through is essentially nonlinear, since it entails the study of buckled forms. Degree of stability may be measured by the amount of external work of lateral forces required to produce snap-through. The fact that snap-through can often be produced with very little supplementary work suggests that the Euler critical value is almost unattainable in some cases.

6-2. POSTBUCKLING BEHAVIOR OF A SIMPLE COLUMN.

Poincaré's bifurcation concept usually serves to determine the buckling load of a conservative holonomic system. If the equilibrium configurations corresponding to all values of the load parameter p are determined, the

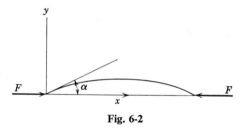

Fig. 6-2

bifurcation points appear automatically. All equilibrium configurations may be determined by the principle of stationary potential energy. Consequently, there is no fundamental need for a special theory of buckling of conservative systems. In fact, recognition of the importance of snap-through and imperfections has led to greatly increased emphasis on the practical significance of postbuckling behavior. However, with the exception of a few cases in which axial symmetry exists, buckled forms of plates and shells do not submit to mathematical analysis without the use of drastic approximations that reduce the systems to a few degrees of freedom. However, the study of buckled forms of elastic columns is tractable. This theory, known as the problem of the "elastica," was introduced by Euler. Love (51) devoted a chapter to the problem, and Schleusner (73) has written a book on the subject. In this article a simple case is discussed from the standpoint of the principle of stationary potential energy.

Suppose that a column is hinged at its ends, that one end remains at the origin, and that the other end lies on the x-axis (Fig. 6-2). The form of the buckled column is represented parametrically by the equations $x = x(s)$,

$y = y(s)$, in which s denotes arc length on the centroidal axis. Since $dx^2 + dy^2 = ds^2$,

$$(x')^2 + (y')^2 - 1 = 0 \tag{a}$$

in which primes denote derivatives with respect to s.

The forced boundary conditions are

$$x(0) = y(0) = 0, \qquad y(L) = 0 \tag{b}$$

The potential energy of the external forces is

$$\Omega = -F[L - x(L)] \tag{c}$$

The strain energy is given by Eq. (2-7). Consequently, aside from an additive constant and a constant factor, the total potential energy is

$$V = k^2 x(L) + \tfrac{1}{2}\int_0^L [(x'')^2 + (y'')^2] \, ds \tag{d}$$

where $k^2 = F/EI$. It is necessary to minimize V in the class of regular functions that satisfy the forced boundary conditions and the auxiliary differential equation (a).

Since $x(0) = 0$,

$$x(L) = \int_0^L x' \, ds$$

Consequently, Eq. (d) may be written:

$$V = \tfrac{1}{2}\int_0^L [(x'')^2 + (y'')^2 + 2k^2 x'] \, ds \tag{e}$$

The general solution of Eq. (a) is

$$x' = \cos\theta, \qquad y' = \sin\theta \tag{f}$$

where θ is a function of s. The variable θ may be identified as the angle between the tangent to the deflection curve and the x-axis. Equations (e) and (f) yield

$$V = \tfrac{1}{2}\int_0^L [(\theta')^2 + 2k^2 \cos\theta] \, ds \tag{g}$$

Giving θ a variation $\delta\theta$ and integrating by parts, we obtain

$$\delta V = \theta' \, \delta\theta \Big|_0^L - \int_0^L (\theta'' + k^2 \sin\theta) \, \delta\theta \, ds \tag{h}$$

Since the variations of θ at the end points are arbitrary, Eq. (h) yields the natural boundary conditions,

$$\theta'(0) = c'(L) = 0 \tag{i}$$

These conditions mean that there is no bending moment at either end.

Since $\delta\theta$ is an arbitrary function of s, Eq. (h) yields the following Euler equation:

$$\theta'' + k^2 \sin\theta = 0 \tag{j}$$

Thus the problem is reduced to the determination of the solution of Eq. (j) that satisfies the natural boundary conditions [Eq. (i)]. One solution is $\theta = 0$. However, if k is large enough, there are other solutions.

To integrate Eq. (j), set $\theta' = u$. Then

$$\theta'' = \frac{du}{ds} = \frac{du}{d\theta}\frac{d\theta}{ds} = u\frac{du}{d\theta}$$

Consequently, Eq. (j) yields

$$u^2 - 2k^2 \cos\theta = C$$

By Eq. (i), $C = -2k^2 \cos\alpha$, where α is the value of θ at the end $s = 0$. Hence

$$\frac{d\theta}{ds} = -k\sqrt{2(\cos\theta - \cos\alpha)} \tag{k}$$

The negative sign of the square root is chosen because $d\theta/ds$ is obviously negative. Equation (k) yields

$$s = \frac{-1}{k}\int_0^\theta \frac{d\theta}{\sqrt{2(\cos\theta - \cos\alpha)}} + C_1$$

The constant C_1 is determined by the condition $\theta = \alpha$ when $s = 0$. Hence, since $s = L/2$ when $\theta = 0$,

$$kL = \sqrt{2}\int_0^\alpha \frac{d\theta}{\sqrt{\cos\theta - \cos\alpha}} = \int_0^\alpha \frac{d\theta}{\sqrt{\sin^2\alpha/2 - \sin^2\theta/2}} \tag{l}$$

The integral in Eq. (l) is transformed to the standard form of an elliptic integral by the substitutions, $\sin\alpha/2 = a$, $\sin\theta/2 = a\sin\phi$. Thus we obtain

$$kL = 2\int_0^{\pi/2} \frac{d\phi}{\sqrt{1 - a^2 \sin^2\phi}} \tag{m}$$

The integral in Eq. (m) is tabulated for various values of a in tables of elliptic integrals. However, for small values of a, the solution is obtained readily by a binomial expansion of the integrand. Thus we obtain

$$kL = 2\int_0^{\pi/2}(1 + \tfrac{1}{2}a^2 \sin^2\phi + \tfrac{3}{8}a^4 \sin^4\phi + \tfrac{5}{16}a^6 \sin^6\phi + \cdots)\,d\phi$$

Hence

$$kL = \pi[1 + (\tfrac{1}{2})^2 a^2 + (\tfrac{3}{8})^2 a^4 + (\tfrac{5}{16})^2 a^6 + \cdots] \tag{n}$$

Equation (n) shows that if $kL < \pi$ there is no solution other than $\theta = 0$. However, if $kL > \pi$, there are deflected equilibrium forms of the column. Consequently, a bifurcation point occurs at $kL = \pi$, that is, at $F = F_{cr} = \pi^2 EI/L^2$. Thus the Euler buckling load is obtained as a bifurcation point.

If $\alpha = 30°$, $a = \sin 15° = 0.25882$. Then Eq. (n) yields $kL = 1.0174\pi$. Therefore, if $\alpha = 30°$, $F/F_{cr} = 1.035$. This result indicates that a column will sustain only a very small increment of load above the Euler buckling load, particularly if yielding results from severe bending.

6-3. BUCKLING OF CONSERVATIVE SYSTEMS WITH ENUMERABLE DEGREES OF FREEDOM.

The theory of stability presented in Sec. 1-11 may be used to determine the buckling load of a conservative holonomic system with finite degrees of freedom. The class Γ of unbuckled configurations lies on the path $X = X(p)$ that represents all equilibrium configurations. For small values of p, the configuration in class Γ is stable, but for large values of p it is unstable. Consequently, if $\delta^2 V$ is positive definite for $p < p_{cr}$ and $\delta^2 V$ is indefinite, negative definite, or negative semidefinite* for $p > p_{cr}$, the buckling load is p_{cr}.

By Eq. (1-27), $\delta^2 V = \Sigma \Sigma a_{ij} h_i h_j$. The coefficients a_{ij} are functions of p. A necessary and sufficient condition that $\delta^2 V$ be positive definite is that the determinant D of matrix (a_{ij}) and all its principal minors be positive (Appendix, Sec. A-1). Consequently, we may expect that the buckling load is the smallest load for which D or a principal minor of D vanishes. Usually it is unnecessary to examine a sequence of minors of D, since D itself vanishes before any of its principal minors.

To be more precise, we require that the functions $a_{ij}(p)$ be analytic in a range π of the variable p that covers the range of interest. Analyticity signifies that the functions $a_{ij}(p)$ admit power series expansions with nonzero intervals of convergence about any point in π. Analytic functions have the following pertinent properties: (a) They are continuous in π. (b) The sum or product of any two of them is again an analytic function. (c) An analytic function that is not identically zero vanishes at only a finite number of points in any closed subinterval of π. Any branch of an algebraic function that is continuous in π is analytic. The class of continuous algebraic functions includes many cases of physical importance.

The following theorem is proved in the Appendix (Sec. A-5):

> If the functions $a_{ij}(p)$ are analytic in π, if stable equilibrium exists when
> $p < \lambda$ (where λ is some constant in π), if the equation $D(p) = 0$

---

* The terms "positive definite," "negative definite," etc., are defined in the Appendix (Sec. A-1).

possesses a minimum root p_0 in π, and if D changes its sign from positive to negative as p passes through p_0, then p_0 is the buckling load p_{cr}.

The hypotheses of this theorem are illustrated by a graph of D versus p that crosses the p-axis from above to below at p_0.

Example. Buckling of a Linkage. A column consists of three rigid struts, as shown in Fig. 6-3a. All hinges contain springs such that the moment resisting an angular deflection θ is proportional to θ. The spring

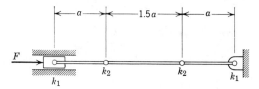

Fig. 6-3a

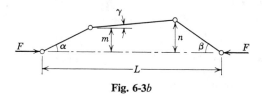

Fig. 6-3b

constants for the hinges are k_1 and k_2, as indicated in Fig. 6-3a. The hinges are frictionless and they are free to rotate either way.

The straight form of the linkage is obviously an equilibrium configuration. However, if the compression load F is large, this configuration is unstable. To determine the buckling load, we consider the system in a deflected form, as indicated by Fig. 6-3b. The angles α and β are adopted as generalized coordinates. The strain energy of the hinges is

$$U = \tfrac{1}{2}k_1\alpha^2 + \tfrac{1}{2}k_1\beta^2 + \tfrac{1}{2}k_2(\alpha - \gamma)^2 + \tfrac{1}{2}k_2(\beta + \gamma)^2 \tag{a}$$

Also,

$$\sin \alpha = \frac{m}{a}, \qquad \sin \beta = \frac{n}{a}, \qquad \sin \gamma = \frac{n - m}{1.5a}$$

where m and n are the deflections of the joints (Fig. 6-3b). Consequently,

$$\sin \gamma = \tfrac{2}{3}(\sin \beta - \sin \alpha)$$

Writing γ as a power series in α and β by means of this equation, we obtain the following approximation, which is correct through the second-degree terms in α and β:

$$\gamma = \tfrac{2}{3}(\beta - \alpha) \tag{b}$$

Equations (a) and (b) yield

$$U = (\tfrac{1}{2}k_1 + \tfrac{29}{18}k_2)(\alpha^2 + \beta^2) - \tfrac{20}{9}k_2\alpha\beta \tag{c}$$

The length L (Fig. 6-3b) is

$$L = a\cos\alpha + a\cos\beta + \tfrac{3}{2}a\cos\gamma$$

Consequently, to second powers of α and β,

$$L = a(1 - \tfrac{1}{2}\alpha^2) + a(1 - \tfrac{1}{2}\beta^2) + \tfrac{3}{2}a(1 - \tfrac{1}{2}\gamma^2) \tag{d}$$

The potential energy of the external forces is

$$\Omega = -F(\tfrac{7}{2}a - L)$$

With Eqs. (b) and (d), this yields

$$\Omega = -\tfrac{5}{6}Fa(\alpha^2 + \beta^2) + \tfrac{2}{3}Fa\alpha\beta \tag{e}$$

The total potential energy of the buckled linkage is $V = U + \Omega$. Consequently, by Eqs. (c) and (e),

$$V = (\tfrac{1}{2}k_1 + \tfrac{29}{18}k_2 - \tfrac{5}{6}Fa)(\alpha^2 + \beta^2) + (\tfrac{2}{3}Fa - \tfrac{20}{9}k_2)\alpha\beta \tag{f}$$

A power-series expansion of V has the form

$$V = V_0 + \delta V + \tfrac{1}{2}\delta^2 V + \cdots \tag{g}$$

where V_0 is a constant, δV is a linear form in (α, β), $\delta^2 V$ is a quadratic form in (α, β), etc. Since Eq. (f) represents V as a quadratic form in (α, β), $V_0 = 0$ and $\delta V = 0$. Actually, V contains cubic and higher degree terms in (α, β), but these terms have not been derived, since Eqs. (c) and (e) are second-degree approximations. Accordingly, Eq. (f) is really an exact formula for $\tfrac{1}{2}\delta^2 V$. The fact that δV vanishes verifies the hypothesis that the straight form of the linkage (defined by $\alpha = \beta = 0$) is an equilibrium configuration.

Identifying (h_1, h_2) with (α, β) and noting that $\delta^2 V$ is linear in F [hence analytic; see Eq. (f)], we conclude from the preceding theorem that the buckling criterion is that the determinant of coefficients in Eq. (f) vanish; that is,

$$D = (\tfrac{1}{2}k_1 + \tfrac{29}{18}k_2 - \tfrac{5}{6}Fa)^2 - (\tfrac{1}{3}Fa - \tfrac{10}{9}k_2)^2 = 0 \tag{h}$$

The solutions of Eq. (h) are

$$aF = k_1 + k_2^* \tag{i}$$

$$aF = \tfrac{3}{7}k_1 + \tfrac{7}{3}k_2 \tag{j}$$

The smaller value of F obtained from these two equations is the buckling load F_{cr}. Hence Eq. (i) applies if $3k_1 < 7k_2$ and Eq. (j) applies if $3k_1 > 7k_2$.

To verify that Eqs. (i) and (j) provide the buckling load, we must show that $D > 0$ if $F < F_{cr}$ and $D < 0$ if $F_{cr} < F < B$, where B is some constant. This condition is easily seen to be true, since the graph of D versus aF is the parabola shown in Fig. 6-4. The parabola cuts the horizontal axis, as shown in Fig. 6-4, if $3k_1 \neq 7k_2$. However, if $3k_1 = 7k_2$, the parabola is tangent to the horizontal axis at the point $Fa = (10/7)k_1$. In this case it is easily shown that $\delta^2 V$ is positive definite if $Fa < (10/7)k_1$ and negative definite if $Fa > (10/7)k_1$. Consequently, if $3k_1 = 7k_2$, the buckling load is determined by $Fa = (10/7)k_1$.

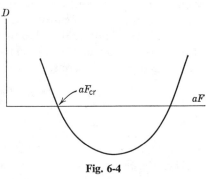

Fig. 6-4

Only Eq. (i) is obtained if, at the outset, we make the assumption $\alpha = \beta$. On the other hand, only Eq. (j) is obtained if we make the assumption $\alpha = -\beta$. Accordingly, the linkage buckles symmetrically if $3k_1 < 7k_2$, and it buckles antisymmetrically if $3k_1 > 7k_2$ (Fig. 6-5).

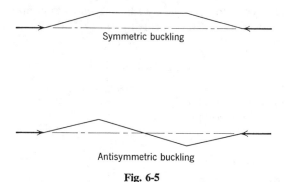

Symmetric buckling

Antisymmetric buckling

Fig. 6-5

Example. Column Supported by Elastic Foundation. As a second example, we consider a column with hinged ends that is supported laterally by a continuous distribution of springs. The spring resistance in an interval dx is $\beta y \, dx$, where β is a constant. The deflection of the column due to buckling is represented by

$$y = \sum_{n=1}^{\infty} b_n \sin \frac{n\pi x}{L} \qquad (k)$$

The strain energy of the elastic foundation is

$$U_f = \tfrac{1}{2}\beta \int_0^L y^2 \, dx = \frac{\beta L}{4} \sum_{n=1}^{\infty} b_n{}^2 \tag{l}$$

The strain energy of bending and the potential energy of the axial load are given by Eq. (2-18) and Eq. (d) of Sec. 2-2. The sum of these energy expressions is the increment of potential energy due to buckling; it does not include the strain energy due to direct compression before buckling. Also, because of quadratic approximations, the formula for ΔV is correct only to second-degree terms. Accordingly,

$$\delta^2 V = \sum_{n=1}^{\infty} \left(\frac{\pi^4 E I n^4}{2L^3} + \frac{\beta L}{2} - \frac{\pi^2 P n^2}{2L} \right) b_n{}^2 \tag{m}$$

Since δV vanishes, the straight form of the column is an equilibrium configuration, although it need not be stable.

Evidently, $\delta^2 V$ is positive definite if all coefficients in Eq. (m) are positive; it is indefinite if any coefficient is negative. Consequently, the buckling load P_{cr} is the smallest load for which a coefficient in Eq. (m) is zero. Therefore,

$$P_{cr} = \frac{\pi^2 E I n^2}{L^2} + \frac{\beta L^2}{\pi^2 n^2} \tag{n}$$

The positive integer n in Eq. (n) is to be chosen to minimize P_{cr}.

The half wavelength of the buckled form is $L/n = l$. Hence, Eq. (n) may be written

$$P_{cr} = \frac{\pi^2 E I}{l^2} + \frac{\beta l^2}{\pi^2}$$

For an infinitely long column, l is to be chosen to minimize P_{cr}. Therefore, for an infinitely long column,

$$P_{cr} = 2\sqrt{EI\beta}$$

$$l = \pi \sqrt[4]{EI/\beta}$$

6-4. GENERAL PRINCIPLES OF BUCKLING OF CONSERVATIVE SYSTEMS.

For a conservative system with finite degrees of freedom, the second variation $\delta^2 V$ of the potential energy is defined as in Sec. 1-11; for a system with infinitely many degrees of freedom, $\delta^2 V$ is interpreted as in the calculus of variations (Sec. 3-2). The second variation occupies a dominant place in the sufficiency theory of the calculus of variations (6). In fact, the theory of buckling of conservative systems lies properly in that branch of mathematics. However, the available results in the sufficiency theory of the calculus of variations are adequate for only the simplest buckling problems. Consequently, buckling problems are often treated by means of approximations that reduce continuous elastic structures to finite degrees of freedom. Although this method is effective, there is usually no feasible way to estimate the accuracy of the results.

At the buckling load, $\delta^2 V$ ordinarily changes its character from positive definiteness to negative definiteness, negative semidefiniteness, or indefiniteness. Consequently, we may anticipate that $\delta^2 V$ is positive semidefinite when $p = p_{\mathrm{cr}}$. With the hypothesis that the coefficients a_{ij} in $\delta^2 V$ are analytic functions of p, it has been proved (Appendix, Sec. A-5) that $\delta^2 V$ is necessarily positive semidefinite at the buckling load. Consequently, if $p = p_{\mathrm{cr}}$, there exist nonzero virtual displacements for which $\delta^2 V = 0$, but there is no virtual displacement for which $\delta^2 V < 0$. This conclusion may be written in a slightly different form by means of the relation $V = U + \Omega$, where U and Ω are the potential energies of internal and external forces, respectively. Also, $\Omega = -W_e$, where W_e is the work performed on the system by the conservative external forces. Consequently, if $p = p_{\mathrm{cr}}$, there exist nonzero virtual displacements such that $\delta^2 U = \delta^2 W_e$, but there is no virtual displacement such that $\delta^2 U < \delta^2 W_e$. If $p < p_{\mathrm{cr}}$, $\delta^2 U > \delta^2 W_e$ for all nonzero virtual displacements. Kelvin and Tait (42) expressed the buckling criterion in the form $\Delta U = \Delta W_e$, where ΔU and ΔW_e denote the increments of U and W_e corresponding to an infinitesimal virtual displacement. Timoshenko (84) employed the criterion $\Delta U = \Delta W_e$ to solve many practical buckling problems. However, since $\delta V = 0$ for the prebuckling configuration and second-degree approximations were used to compute ΔU and ΔW_e, the condition $\delta^2 U = \delta^2 W_e$ was actually applied. In 1891 Bryan (98) applied the criterion $\delta^2 U = \delta^2 W_e$ to investigate buckling of plates.

It is to be emphasized that if $p = p_{\mathrm{cr}}$ the relation $\delta^2 U = \delta^2 W_e$ is not true for all virtual displacements; it is valid only for special virtual displacements that coincide with the actual buckling pattern. Consequently, the criterion $\delta^2 U = \delta^2 W_e$ has often been applied in an approximate way by means of estimations of deflection functions that represent the buckled form. The relation has also been applied in an exact manner (84) by means of the condition that the load parameter p shall have the minimum value that is consistent with a nonzero solution of the equation $\delta^2 U = \delta^2 W_e$.

Criterion of Trefftz. Trefftz (146) restricted attention to the case in which the external forces are constants. Then, if no internal constraints are assumed (such as inextensionality of center lines or center planes), ΔW_e is a linear functional of the virtual increments of the displacement vector. Consequently, $\delta^2 W_e$ vanishes identically, and the stability criterion reduces to the condition that $\delta^2 U$ be positive definite. Therefore, when the buckling load is reached, $\delta^2 U$ becomes positive semi-definite; that is, the minimum value of $\delta^2 U$ is zero, and this value is attained for some nonzero virtual displacements h_i. Trefftz set $\delta^2 U = Q$, and he observed that the condition that Q possess a nontrivial minimum signifies that the variational equation $\delta Q = 0$ be satisfied identically for certain nonzero values of the functions h_i. Variations of Q are effected by adding arbitrary variations to h_i. The equation $\delta Q = 0$ yields linear homogeneous differential equations (the Euler equations of the calculus of variations). Also, the resulting natural boundary conditions are homogeneous; in other words, if certain functions satisfy the natural boundary conditions (or the differential equations), the products of those functions with arbitrary constants also satisfy the natural boundary conditions (or the differential equations). Thus the variational theory leads to the same linear eigenvalue problem as the approach by way of the differential equations of equilibrium. In the theory of buckling of shells the variational equation $\delta Q = 0$ has occasionally been used to determine the tangential components of displacement after an equation containing a few undetermined parameters has been assumed for the normal displacement.

The condition that $\delta^2 W_e = 0$ is naturally not essential in Trefftz's theory; the foregoing argument applies more generally if $\delta^2 V = Q$.

Bounds of the Buckling Load. The set of unbuckled configurations is denoted by Γ. If $p < p_{cr}$, the value of V is a relative minimum at the point X_0 in Γ that corresponds to p. Therefore, in configuration space there exists a deleted* neighborhood N of point X_0 such that $\Delta V > 0$ for all points X in N where ΔV denotes $V(X, p) - V(X_0, p)$. On the other hand, if $p_{cr} > p$, there exist points X in any deleted neighborhood of X_0 such that $\Delta V \leq 0$.

Let S denote the entire configuration space. If we introduce arbitrary assumptions about the nature of the deformation, we effectively constrain the system to a subset S' of points in space S. We consider the case for which Γ lies in S'. This is the usual type of approximation; the assumptions restrict the deformations that accompany buckling, but they do not restrict the unbuckled form.

Let N' be the points of neighborhood N that lie in S'. Let p_{cr}' be the buckling load, computed with the assumption that the system is constrained to S'. The load p_{cr}' is defined in the same way as the true buckling load p_{cr}, but the configuration space is taken to be S'. If $p_{cr} > p_{cr}'$ and p lies in the range $p_{cr}' < p < p_{cr}$, $\Delta V > 0$ for all points X in N. Hence $\Delta V > 0$ for all points X in N', since N' is a subset of N. However, this

---

* See Sec. 1-1 for definition of a deleted neighborhood.

conclusion conflicts with the definition of p_{cr}', for, when $p_{cr}' < p$, the point X_0 does not provide a relative minimum to V in S'. Consequently, the inequality $p_{cr} > p_{cr}'$ is impossible.

In other words, *if the deformations that accompany buckling are assumed to lie in a designated class the computed load for which the unbuckled configuration ceases to be a configuration of minimum potential energy is generally too high; it cannot be too low.*

Unfortunately, no equally versatile method for establishing a lower bound for the buckling load has been devised. Trefftz (147) suggested that a lower bound is obtained if the class of admissible deflection functions is enlarged (e.g. by relaxation of boundary constraints or continuity requirements), and the problem is then solved exactly. For example, the buckling load of a plate with one or more hinged edges is a lower bound for the buckling load of the same plate with all edges clamped. Another procedure that is sometimes applicable is to simplify the expression for $\delta^2 V$ while simultaneously weakening the positive definite character of $\delta^2 V$ (e.g. by discarding certain terms that are essentially positive). If the simplified problem is solved exactly, a lower bound for the buckling load is evidently obtained. This method has been used successfully in the theory of buckling of rings (93).

Shear Effect in Columns. As an illustration of the preceding theory, the reduction of the buckling load of a pin-ended column because of shear deformation is considered. The slope y' is the sum of the slope β due to shear and the supplemental slope $y' - \beta$ due to bending. In view of Eqs. (2-15) and (2-19), the second variation of the total potential energy is

$$\tfrac{1}{2}\delta^2 V = Q = \tfrac{1}{2}\int_0^L [EI(y'' - \beta')^2 + \lambda\beta^2 - P(y')^2]\, dx$$

The Euler equations for the integral Q are

$$EI(y'''' - \beta''') + Py'' = 0 \qquad (a)$$

$$EI(y''' - \beta'') + \lambda\beta = 0 \qquad (b)$$

Equation (a) yields

$$EI(y'' - \beta') + Py = ax + b$$

The end conditions $y = 0$ and $y'' - \beta' = 0$ yield $a = b = 0$. Hence

$$EI(y'' - \beta') + Py = 0 \qquad (c)$$

Equations (b) and (c) yield

$$\beta = \frac{Py'}{\lambda} \qquad (d)$$

Hence, by Eq. (c),

$$y'' + \frac{k^2 y}{1 - P/\lambda} = 0, \qquad k^2 = \frac{P}{EI} \tag{e}$$

Equation (e) yields

$$y = A \sin mx + B \cos mx, \qquad m^2 = \frac{k^2}{1 - P/\lambda} \tag{f}$$

The constant B is zero because $y = 0$ at $x = 0$. Also, since $y = 0$ at $x = L$, $\sin mL = 0$. Therefore, $m = \pi/L$. This yields

$$k^2 = \frac{\pi^2}{L^2}\left(1 - \frac{P}{\lambda}\right) = \frac{P}{EI} \tag{g}$$

Solving Eq. (g) for P, we obtain

$$P = P_{\mathrm{cr}} = \frac{P_e}{1 + P_e/\lambda}, \qquad P_e = \frac{\pi^2 EI}{L^2} \tag{h}$$

The constant P_e is the Euler critical load if shear deformation is neglected. This fact is apparent from Eq. (h), since $P_{\mathrm{cr}} = P_e$ if $\lambda = \infty$.

6-5. BUCKLING OF SIMPLY SUPPORTED COMPRESSED RECTANGULAR PLATES.
Buckling of plates was first analyzed by G. H. Bryan (98) in 1891. The following treatment is similar to his. A homogeneous isotropic elastic rectangular plate of constant thickness h has two

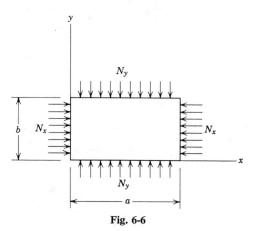

Fig. 6-6

edges coinciding with the x- and y-axes (Fig. 6-6). All edges are hinged. The plate is subjected to uniform compressions N_x and N_y. For convenience, positive N_x and N_y are considered compressive. This is opposite to the sign convention in Chap. 5.

The displacement vector of the middle plane before buckling is denoted by (u, v, w). By Eqs. (5-3) and (5-9), the strain energy of the plate corresponding to arbitrary functions (u, v, w) is

$$U = \frac{Gh}{1-v} \iint \left\{ (u_x + \tfrac{1}{2}w_x{}^2)^2 + (v_y + \tfrac{1}{2}w_y{}^2)^2 + 2v(u_x + \tfrac{1}{2}w_x{}^2)(v_y + \tfrac{1}{2}w_y{}^2) \right.$$
$$+ \tfrac{1}{2}(1 - v)(v_x + u_y + w_x w_y)^2$$
$$\left. + \frac{h^2}{12}[w_{xx}{}^2 + w_{yy}{}^2 + 2vw_{xx}w_{yy} + 2(1 - v)w_{xy}{}^2] \right\} dx\, dy \quad \text{(a)}$$

The decrease of the distance between the particles initially at points $(0, y)$ and (a, y) is

$$u(0, y) - u(a, y) = -\int_0^a u_x\, dx$$

Consequently, the potential energy of the external load N_x is $\iint N_x u_x\, dx\, dy$. Similarly, the potential energy of N_y is determined. Accordingly, the total potential energy of the external forces is

$$\Omega = \iint (u_x N_x + v_y N_y)\, dx\, dy \tag{b}$$

If (u, v, w) receive variations $\xi(x, y)$, $\eta(x, y)$, $\zeta(x, y)$, U and Ω receive increments ΔU and $\Delta \Omega$. As in Chap. 3, we write

$$\Delta U = \delta U + \tfrac{1}{2}\delta^2 U + \cdots, \qquad \Delta \Omega = \delta\Omega + \tfrac{1}{2}\delta^2\Omega + \cdots$$

Since Ω is linear in u and v, $\delta^2\Omega = 0$. For the prebuckling configuration, $u_y = 0$, $v_x = 0$, and $w = 0$. Hence by Eq. (a)

$$\delta^2 U = \frac{2Gh}{1-v} \iint \left\{ \xi_x{}^2 + u_x\zeta_x{}^2 + \eta_y{}^2 + v_y\zeta_y{}^2 + 2v\xi_x\eta_y + vu_x\zeta_y{}^2 \right.$$
$$+ vv_y\zeta_x{}^2 + \tfrac{1}{2}(1 - v)(\eta_x + \xi_y)^2$$
$$\left. + \frac{h^2}{12}[\zeta_{xx}{}^2 + \zeta_{yy}{}^2 + 2v\zeta_{xx}\zeta_{yy} + 2(1 - v)\zeta_{xy}{}^2] \right\} dx\, dy \quad \text{(c)}$$

The equilibrium relation, $\delta V = 0$, merely yields the prebuckling configuration. This may be obtained immediately, however, from Eqs. (5-17):

$$u_x = \frac{-1}{Eh}(N_x - vN_y), \qquad v_y = \frac{-1}{Eh}(N_y - vN_x) \tag{d}$$

Equilibrium is stable if $\delta^2 U$ is positive definite, since $\delta^2\Omega = 0$. Equation (c) shows that ξ and η can never contribute a negative amount to $\delta^2 U$. Consequently, to determine the range of load intensities (N_x, N_y) for which equilibrium is stable, we must set $\xi = \eta = 0$. Bryan assumed this condition from the outset. Eliminating u and v from Eq. (c) by means of Eq. (d), and dropping ξ and η, we obtain

$$\delta^2 U = \iint \{ D[\zeta_{xx}^2 + \zeta_{yy}^2 + 2\nu\zeta_{xx}\zeta_{yy} + 2(1 - \nu)\zeta_{xy}^2] $$
$$- N_x\zeta_x^2 - N_y\zeta_y^2\} \, dx \, dy \tag{e}$$

where D is the flexural rigidity defined in Eq. (5-10). Evidently, $\delta^2 U$ is positive definite if N_x and N_y are small, but it becomes indefinite if they are large.

Bryan represented ζ by a double sine series [Eq. (5-32)]. Accordingly, Eq. (e) yields

$$\delta^2 U = \frac{\pi^2}{4} \sum \sum \left[\pi^2 ab D \left(\frac{m^2}{a^2} + \frac{n^2}{b^2}\right)^2 - \frac{bN_x}{a} m^2 - \frac{aN_y}{b} n^2 \right] a_{mn}^2 \tag{f}$$

Apparently $\delta^2 U$ is positive definite if, and only if, all terms in Eq. (f) are positive. Consequently, buckling occurs when any term in Eq. (f) equals zero. Supposing that $N_y/N_x = r = $ constant, we may express the critical value of N_x in the form

$$(N_x)_{\text{cr}} = k \frac{\pi^2 D}{b^2} \tag{g}$$

Letting $a/b = c$, we obtain from Eq. (f)

$$k = \min \frac{(m^2 + n^2c^2)^2}{c^2(m^2 + n^2c^2r)} \tag{h}$$

Equation (h) means that the positive integers m and n are to be chosen to minimize k. If $N_y = 0$, $r = 0$ and $n = 1$. Then m is the nearest positive integer to c.

6-6. BUCKLING OF A UNIFORMLY COMPRESSED CIRCULAR PLATE.

Another case treated by Bryan (98) is the buckling of a circular isotropic elastic plate due to uniform radial compression. The postbuckling behavior in this case has been investigated by Friedrichs and Stoker (107).

It is assumed that the plate buckles symmetrically. The radial and lateral displacements of the middle plane are denoted by $u(r)$ and $w(r)$, where r is the radial coordinate (Fig. 6-7). The radius of the plate is a. By Eqs. (5-86),

$$U = \pi \int_0^a \left\{ \frac{Eh}{1 - \nu^2} \left[r(u_r + \tfrac{1}{2}w_r^2)^2 + \frac{u^2}{r} + 2\nu u(u_r + \tfrac{1}{2}w_r^2) \right] \right.$$
$$\left. + D(rw_{rr}^2 + \frac{1}{r} w_r^2 + 2\nu w_r w_{rr}) \right\} dr \tag{6-1}$$

As for a rectangular plate (Sec. 6-5), the tangential components of displacement that accompany buckling are zero, since they merely augment the membrane energy and increase the buckling load. Also, as for a rectangular plate, Ω is a linear functional of the tangential displacements; hence $\delta^2\Omega = 0$. Therefore, stability exists if $\delta^2 U$ is positive definite. Since, for the prebuckling configuration, $w = 0$ Eq. (6-1) yields

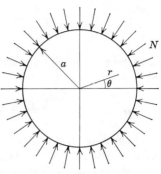

Fig. 6-7

$$\delta^2 U = 2\pi \int_0^a \left[\frac{Eh}{1 - \nu^2} (r u_r + \nu u)\zeta_r^2 \right.$$

$$\left. + D\left(r\zeta_{rr}^2 + \frac{1}{r}\zeta_r^2 + 2\nu\zeta_r\zeta_{rr} \right) \right] dr \quad \text{(a)}$$

where $\zeta(r)$ is the variation of w.

Before buckling, $N_r = N_\theta = N$ and $e_r = e_\theta = u_r$. Hence by Eqs. (5-17)

$$u = \frac{-(1 - \nu)Nr}{Eh}, \qquad w = 0 \tag{b}$$

Elimination of u from Eq. (a) by means of Eq. (b) yields

$$Q = \delta^2 U = 2\pi D \int_0^a \left(-\alpha^2 r\zeta_r^2 + r\zeta_{rr}^2 + \frac{1}{r}\zeta_r^2 + 2\nu\zeta_r\zeta_{rr} \right) dr \tag{c}$$

where, for brevity, $\alpha^2 = N/D$.

According to Trefftz's theory (Sec. 6-4), $\delta Q = 0$ at the buckling load. A symmetry condition which may be regarded as a forced boundary condition is $\zeta_r = 0$ at $r = 0$. The Euler equation for the integral Q is

$$-\frac{d}{dr}\frac{\partial F}{\partial \zeta_r} + \frac{d^2}{dr^2}\frac{\partial F}{\partial \zeta_{rr}} = 0$$

where F denotes the integrand in Eq. (c). Consequently,

$$-\frac{\partial F}{\partial \zeta_r} + \frac{d}{dr}\frac{\partial F}{\partial \zeta_{rr}} = C \tag{d}$$

where C is a constant. Equation (d) yields,

$$r\zeta_{rrr} + \zeta_{rr} + \left(\alpha^2 r - \frac{1}{r} \right)\zeta_r = C \tag{e}$$

Since ζ_r vanishes at the center of the plate, only one natural boundary condition is obtained at the center. Letting ζ take a variation $\delta\zeta$ and integrating by parts as in Sec. 3-2, we obtain the natural boundary condition,

$$\lim_{r \to 0} \left[\left(\alpha^2 r - \frac{1}{r} \right) \zeta_r + \zeta_{rr} + r\zeta_{rrr} \right] = 0$$

Accordingly, the constant C in Eq. (e) is zero. Therefore, with the substitution $\zeta_r = \phi$, we obtain

$$r^2\phi_{rr} + r\phi_r + (\alpha^2 r^2 - 1)\phi = 0 \tag{f}$$

It may be noted that ϕ is the slope of a radial section of the buckled middle surface. Equation (f) is Bessel's differential equation of the first order. The solution, which is regular at the center, is (7)

$$\phi = AJ_1(\alpha r) \tag{g}$$

where J_1 is Bessel's function of the first order and A is a constant of integration. Equation (g) satisfies the center condition $\phi = 0$.

If the plate is clamped at the edge, the forced boundary conditions at $r = a$ are $\phi = 0$ and $\zeta = 0$. If it is simply supported, the forced boundary condition at $r = a$ is $\zeta = 0$. In this case the natural boundary condition, obtained from $\delta Q = 0$, is $r\phi_r + \nu\phi = 0$ at $r = a$. The boundary condition $\zeta = 0$ is irrelevant, since it merely determines the constant of integration when the equation $\phi = \zeta_r$ is integrated.

For the clamped plate, the boundary condition $\phi = 0$ yields $J_1(\alpha a) = 0$. Choosing the first positive zero of J_1, we obtain $\alpha a = 3.832$, whence $N_{cr} = 14.68D/a^2$.

For the simply supported plate, the boundary condition $r\phi_r + \nu\phi = 0$ yields, with Eq. (g),

$$\alpha a J_0(\alpha a) = (1 - \nu)J_1(\alpha a)$$

If $\nu = 0.30$, the smallest positive root of this equation is $\alpha a = 2.049$, whence $N_{cr} = 4.20D/a^2$.

6-7. LATERAL BUCKLING OF BEAMS.

There are many problems of lateral buckling of beams, since the types of loading, the end supports, and the cross-sectional shapes are innumerable. In this article a special case is treated. An elastic beam of length L with a symmetrical cross section (Fig. 6-8) is considered bent by couples applied at the ends. A uniform axial compression may be superimposed. The ends of the beam are hinged; the hinge lines lie in the plane of symmetry.* An approximation that reduces the beam to a system with two degrees of freedom is employed.

Fig. 6-8

The following notations are used (Fig. 6-8):

h,   depth of cross section between centroids of flanges.

t,   thickness of web.

c_c, c_t,   distances from neutral axis to centroids of compression and tension flanges, respectively. Since there may be a thrust force in addition to the bending moment, the neutral axis need not coincide with the centroidal axis.

A_c, A_t,   cross-sectional areas of compression and tension flanges, respectively.

I_c, I_t,   moments of inertia of cross sections of the compression and tension flanges about the y-axis.

I_y,   moment of inertia of the entire cross section about the y-axis.

J_c, J_t,   torsion constants of the compression and tension flanges.

L,   length of the beam.

ν,   Poisson's ratio.

ϵ_c,   compression strain in the compression flange that causes buckling. The compression stress in the compression flange that causes buckling is $\sigma_c = E\epsilon_c$.

I_1, I_2,   moments of inertia of the cross section about principal axes through the centroid.

J,   torsion constant for the entire cross section.

M_{cr},   critical bending moment.

w,   lateral deflection of web (in z-direction) due to buckling.

It is assumed that

$$w = (a + by) \sin \frac{\pi x}{L} \tag{a}$$

where a and b are constants. This assumption conforms to the condition of hinged ends. Also, it conforms to the observation that the cross sections

* By the introduction of a suitable effective length, other end conditions may be admitted (84).

are rotated and translated without distortion. With Eq. (a), Eqs. (5-9) yield the following formula for the strain energy of the web:

$$\tfrac{1}{2}\delta^2 U_w = \frac{\pi^4 Dh}{4L^3}\left\{a^2 + abh + \left[\frac{h^2}{3} + \frac{2(1-\nu)L^2}{\pi^2}\right]b^2\right\} \tag{b}$$

This quantity is denoted by $\tfrac{1}{2}\delta^2 U_w$ because it is quadratic in a and b. The curvature of the compression flange due to sidewise bending is

$$\frac{1}{R_c} = -(w_{xx})\bigg|^{y=h} = \frac{\pi^2}{L^2}(a + bh)\sin\frac{\pi x}{L}$$

Hence the strain energy of the compression flange due to bending is

$$\tfrac{1}{2}EI_c \int_0^L \frac{1}{R_c{}^2}\,dx = \frac{\pi^4 EI_c}{4L^3}(a + bh)^2$$

Similarly, the strain energy of the tension flange due to bending is

$$\tfrac{1}{2}EI_t \int_0^L \frac{1}{R_t{}^2}\,dx = \frac{\pi^4 EI_t}{4L^3}a^2$$

The twist of either flange, per unit length, is

$$w_{xy} = \frac{\pi b}{L}\cos\frac{\pi x}{L}$$

Hence, by Eq. (2-17), the strain energy of the flanges due to twisting is

$$\tfrac{1}{2}G(J_c + J_t)\frac{\pi^2 b^2}{L^2}\int_0^L \cos^2\frac{\pi x}{L}\,dx = \frac{\pi^2 b^2 G(J_c + J_t)}{4L}$$

The web buckles inextensionally if the cross section is symmetrical about the y-axis, since otherwise the strain energy of buckling would be augmented by membrane energy and the buckling load would be increased.* Collecting the preceding results, we obtain the following formula for the second variation of the strain energy:

$$\tfrac{1}{2}\delta^2 U = \frac{\pi^4 Dh}{4L^3}\left\{a^2 + abh + \left[\frac{h^2}{3} + \frac{2(1-\nu)L^2}{\pi^2}\right]b^2\right\}$$
$$+ \frac{\pi^4 EI_c}{4L^3}(a + bh)^2 + \frac{\pi^4 EI_t}{4L^3}a^2 + \frac{\pi^2 b^2 G(J_c + J_t)}{4L} \tag{c}$$

---

* More general results for a beam with an off-center web were derived by the author (123). Goodier has investigated the stability of any cylindrical or prismatic elastic bar subjected to pure bending (112).

Potential Energy of External Forces. The potential energy of the external forces is

$$\Omega = -\int e\sigma_x \, dA \tag{d}$$

where σ_x is the compression stress and e is the reduction of the distance between the ends of the longitudinal fiber on which the stress σ_x acts. The incremental contraction Δe that occurs at buckling consists of two parts, $\Delta e = \Delta_1 e + \Delta_2 e$, in which $\Delta_1 e$ is the contraction caused by incremental strain of the fiber and $\Delta_2 e$ is the contraction caused by bowing of the fiber. Since $\Delta_1 e$ contributes nothing to the second variation of the potential energy, it is irrelevant. The contraction caused by bowing is

$$\Delta_2 e = \tfrac{1}{2} \int_0^L (v_x{}^2 + w_x{}^2) \, dx \tag{e}$$

in which v is the deflection in the y-direction. The y-component of deflection of a fiber of a flange results from twisting of the flange. Consequently,

$$v = -bz \sin \frac{\pi x}{L}$$

With Eqs. (a) and (e), this yields

$$\Delta_2 e = \frac{\pi^2}{4L} [b^2 z^2 + (a + by)^2] \tag{f}$$

Since the prebuckling stress σ_x varies linearly with y, we may write $\sigma_x = K(y - c_t)$, where K is a constant. Therefore, by Eqs. (d) and (f),

$$\tfrac{1}{2}\delta^2\Omega = -\frac{K\pi^2}{4L} \int [b^2 z^2 + (a + by)^2](y - c_t) \, dA$$

Here $\tfrac{1}{2}\delta^2\Omega$ is written instead of Ω, since the foregoing equation merely gives the increment of Ω due to buckling and this is of second degree in a and b. Neglecting variation of stress throughout the depth of a flange, we may write the preceding equation as follows:

$$\tfrac{1}{2}\delta^2\Omega = \frac{K\pi^2 b^2 c_t}{4L} \int_{tf} z^2 \, dA - \frac{K\pi^2 c_c}{4L} \int_{cf} [b^2 z^2 + (a + bh)^2] \, dA$$

$$- \frac{K\pi^2 t}{4L} \int_0^h (a + by)^2(y - c_t) \, dy$$

Here, the notations tf and cf on the integrals denote "tension flange" and "compression flange." Evaluating the integrals, we obtain

$$
\begin{aligned}
\tfrac{1}{2}\delta^2\Omega = -\frac{\pi^2 K}{4L} \bigg\{ &\left[\frac{ht}{2}(h - 2c_t) + c_c A_c - c_t A_t\right]a^2 \\
&+ \left[\frac{h^2 t}{3}(2h - 3c_t) + 2h c_c A_c\right]ab \\
&+ \left[\frac{h^3 t}{12}(3h - 4c_t) + h^2 c_c A_c - c_t I_t + c_c I_c\right]b^2 \bigg\}
\end{aligned} \tag{g}
$$

Denoting the compression strain in the compression flange at which buckling occurs by ϵ_c, we introduce the following dimensionless parameter:

$$
\lambda = \frac{\epsilon_c h}{c_c} \tag{h}
$$

In view of the definition of K, $\lambda = hK/E$. Since $\delta^2 V = \delta^2 U + \delta^2\Omega$, Eqs. (c) and (g) yield

$$
\delta^2 V = a_{11}a^2 + 2a_{12}ab + a_{22}b^2 \tag{i}
$$

If a_{11}, a_{12}, a_{22} are modified by the respective factors, $2L/(\pi^2 Eh^2)$, $2L/(\pi^2 Eh^3)$, and $2L/(\pi^2 Eh^4)$, the following dimensionless formulas are obtained:

$$
\begin{aligned}
a_{11} =& \frac{\pi^2(I_c + I_t)}{h^2 L^2} + \frac{\pi^2 t^3}{12(1 - \nu^2)hL^2} - \frac{\lambda t}{2h}\left(1 - 2\frac{c_t}{h}\right) \\
&+ \frac{\lambda}{h^3}(c_t A_t - c_c A_c)
\end{aligned}
$$

$$
\begin{aligned}
a_{22} =& \frac{\pi^2 I_c}{h^2 L^2} + \frac{J_c + J_t}{2(1 + \nu)h^4} + \frac{\pi^2 t^3}{12(1 - \nu^2)hL^2}\left[\frac{1}{3} + \frac{2(1 - \nu)L^2}{\pi^2 h^2}\right] \\
&- \frac{\lambda t}{12h}\left(3 - 4\frac{c_t}{h}\right) - \frac{\lambda c_c A_c}{h^3} + \frac{\lambda}{h^5}(c_t I_t - c_c I_c)
\end{aligned} \tag{j}
$$

$$
a_{12} = \frac{\pi^2 I_c}{h^2 L^2} + \frac{\pi^2 t^3}{24(1 - \nu^2)hL^2} - \frac{\lambda t}{6h}\left(2 - 3\frac{c_t}{h}\right) - \frac{\lambda c_c A_c}{h^3}
$$

The buckling criterion is that the determinant of the matrix (a_{ij}) shall vanish; that is,

$$
a_{11}a_{22} = a_{12}{}^2 \tag{k}
$$

Doubly Symmetric Beam. For a doubly symmetric cross section, $A_c = A_t$ and $I_c = I_t$. Also, for pure bending, $c_c = c_t = h/2$. Then Eq. (j) yields approximately

$$a_{11} = \frac{\pi^2 I_y}{h^2 L^2}$$

$$a_{22} = \frac{\pi^2 I_c}{h^2 L^2} + \frac{J}{2(1+\nu)h^4} - \epsilon_c\left(\frac{t}{6h} + \frac{A_c}{h^2}\right) \tag{l}$$

$$a_{12} = \frac{\pi^2 I_y}{2h^2 L^2} - \epsilon_c\left(\frac{t}{6h} + \frac{A_c}{h^2}\right)$$

Hence, by Eq. (k),

$$\frac{\pi^2 I_y}{h^2 L^2}\left[\frac{\pi^2 I_c}{h^2 L^2} + \frac{J}{2(1+\nu)h^4}\right] = \frac{\pi^4 I_y^2}{4h^4 L^4} + \epsilon_c^2\left(\frac{t}{6h} + \frac{A_c}{h^2}\right)^2 \tag{m}$$

For a beam with flanges, the moment of inertia of the cross section of the web about its vertical center line is ordinarily negligible. Consequently, $I_c \approx \frac{1}{2}I_y = \frac{1}{2}I_1$. Moreover, the moment of inertia of the cross section of the beam about its horizontal centroidal axis is

$$I_2 = \frac{A_c h^2}{2} + \frac{h^3 t}{12}$$

Hence, Eq. (m) yields

$$\epsilon_c = \frac{\pi h}{2LI_2}\left[\frac{\pi^2 h^2 I_1^2}{4L^2} + \frac{I_1 J}{2(1+\nu)}\right]^{\frac{1}{2}}$$

Accordingly, the critical bending moment is

$$M_{cr} = \frac{\pi E}{L}\left[\frac{I_1 J}{2(1+\nu)}\right]^{\frac{1}{2}}\left[1 + \frac{(1+\nu)\pi^2 h^2 I_1}{2JL^2}\right]^{\frac{1}{2}} \tag{6-2}$$

This result agrees with the classical formula for the critical bending moment of a doubly symmetric I-beam (84).

Example. Buckling of a T-Beam. Equation (6-2) applies only for beams with identical flanges. If the flanges are unequal, we use Eqs. (j) and (k). For example, a beam with the cross section shown in Fig. 6-9 is considered. The beam is subjected to pure bending, so that the flange is placed in tension. The length L is taken to be 30 in. Elementary numerical calculations yield

$$c_t = 0.167, \quad c_c = 0.833, \quad I_1 = I_y = 0.0200, \quad I_2 = 0.0075$$
$$J_c = 0, \quad J_t = 0.0000180, \quad t = 0.030$$
$$A_t = 0.0600, \quad I_t = 0.0200, \quad h = 1.000$$
$$\nu = 0.30, \quad a_{11} = 0.000219, \quad a_{22} = 0.0000104 - 0.003\epsilon_c$$
$$a_{12} = -0.0090\epsilon_c$$

By Eq. (k),

$$219(10.4 - 3000\epsilon_c) = 81 \times 10^6\epsilon_c{}^2$$

The positive root corresponds to bending for which the flange is placed in tension; its value is $\epsilon_c = 0.00262$. Consequently, $M_{cr} = 0.0000236E$. This agrees very closely with the result obtained by Goodier's theory (123).

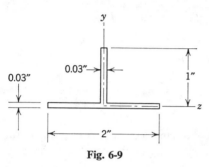

Fig. 6-9

Equation (6-2), which applies only for beams with equal flanges, gives about twice the correct result in the present example.

6-8. TORSIONAL-FLEXURAL BUCKLING OF COLUMNS.

Thin-walled columns with cross sections in the forms of the letters I, C, H, T, Z, etc., are commonly used in engineering structures. It is observed that twisting often accompanies the buckling of such a column. To explain this phenomenon, H. Wagner (149) presented a theory of torsional buckling in 1929. However, he disregarded the bending effect. In 1937 R. Kappus (117) developed a more general theory that accounts for the combined effects of bending and twisting. Subsequently, several investigators have presented theories of torsional-flexural buckling from different points of view (111, 112, 113, 116, 127). The following derivation is taken from an article by the author (122).

Beginning with the general equations of the theory of elasticity, we consider a prismatic or cylindrical bar with a simply connected cross section. The bar is referred to rectangular coordinates (x, y, z), so that the z-axis is the centroidal axis. The ends of the bar lie at $z = 0$ and $z = L$. Modes of buckling are considered for which the transverse stresses σ_x, σ_y, τ_{xy} are zero. Then $\epsilon_x = \epsilon_y = -\nu\epsilon_z$ and $\gamma_{xy} = 0$. Consequently, by Eq. (4-50),

$$U + \Delta U = \tfrac{1}{2}G\iiint [2(1 + \nu)\epsilon_z{}^2 + \gamma_{xz}{}^2 + \gamma_{yz}{}^2]\, dx\, dy\, dz \qquad (a)$$

where ΔU is the increment of strain energy due to buckling.

Initially, $u = \nu\epsilon x$, $v = \nu\epsilon y$, $w = -\epsilon z$, where ϵ is the axial compression strain before buckling. Accordingly, we set $u = \nu\epsilon x + \xi$, $v = \nu\epsilon y + \eta$, $w = -\epsilon z + \zeta$, where $\xi(x, y, z)$, $\eta(x, y, z)$, $\zeta(x, y, z)$ are variations of the displacement components. Substituting these relations into Eqs. (4-3) and neglecting ϵ in expressions of the type $1 + \epsilon$, as well as all quadratic terms in ϵ, we obtain

$$\epsilon_z = -\epsilon + \zeta_z + \tfrac{1}{2}(\xi_z^2 + \eta_z^2 + \zeta_z^2)$$

$$\gamma_{xz} = \xi_z + \zeta_x + \xi_z\xi_x + \eta_z\eta_x + \zeta_z\zeta_x$$

$$\gamma_{yz} = \zeta_y + \eta_z + \xi_y\xi_z + \eta_y\eta_z + \zeta_y\zeta_z$$

With Eq. (a), these expressions yield

$$\delta^2 U = -E\epsilon \iiint (\xi_z^2 + \eta_z^2)\, dx\, dy\, dz$$

$$+ G \iiint [2(1 + \nu)\zeta_z^2 + (\xi_z + \zeta_x)^2 + (\eta_z + \zeta_y)^2]\, dx\, dy\, dz \qquad \text{(b)}$$

Since Ω depends linearly on the displacement vector, $\delta^2\Omega = 0$. Consequently, buckling occurs when $\delta^2 U$ ceases to be positive definite.

Observations of torsional buckling suggest that the cross sections are translated, rotated, and warped when buckling occurs. The warping is represented by the function ζ. To represent translation and rotation of the cross sections in their planes, we set

$$\xi = \alpha + \gamma y, \qquad \eta = \beta - \gamma x \qquad \text{(c)}$$

where α, β, γ are functions of z alone.

Temporarily regarding ξ and η as known, we determine ζ to minimize $\delta^2 U$. The second integral in Eq. (b) is denoted by Λ. The first variation of this integral, corresponding to a variation of ζ is

$$\delta\Lambda = 2 \iiint [2(1 + \nu)\zeta_z\, \delta\zeta_z + (\xi_z + \zeta_x)\, \delta\zeta_x + (\eta_z + \zeta_y)\, \delta\zeta_y]\, dx\, dy\, dz$$

Transformation of this integral by integration-by-parts [Eq. (4-62)] yields

$$\delta\Lambda = -2 \iiint [\zeta_{xx} + \zeta_{yy} + 2(1 + \nu)\zeta_{zz}]\, \delta\zeta\, dx\, dy\, dz$$

$$+ 2 \iint [l(\xi_z + \zeta_x) + m(\eta_z + \zeta_y) + 2(1 + \nu)n\zeta_z]\, \delta\zeta\, dS \qquad \text{(d)}$$

where (l, m, n) are the direction cosines of the outward normal to the surface of the bar. The surface integral extends over the lateral surface and the ends of the bar. On the end faces $l = m = 0$, and on the lateral surface $n = 0$.

The volume integral in Eq. (d) must vanish for all variations $\delta\zeta$. Consequently, ζ is a solution of the differential equation,

$$\zeta_{xx} + \zeta_{yy} + 2(1 + \nu)\zeta_{zz} = 0 \qquad \text{(e)}$$

The lateral surface of the bar is free from constraint. By Eq. (d), the natural boundary condition for the lateral surface is

$$l(\alpha' + \gamma'y + \zeta_x) + m(\beta' - \gamma'x + \zeta_y) = 0$$

or

$$\frac{d\zeta}{dn} = -\frac{d}{ds}[\alpha'y - \beta'x + \tfrac{1}{2}\gamma'(x^2 + y^2)] \tag{f}$$

where $d\zeta/dn$ is the directional derivative along the outward normal to the lateral surface, and d/ds denotes the derivative with respect to arc length on the periphery of the cross section.

If the column is clamped at the ends, the end conditions are $\alpha = \beta = \gamma = \zeta = 0$. If it is loaded by point forces at the centroids of the end sections and rotation of the ends is prevented (pin-ended case), the quantities (α, β, γ) again vanish at the ends and Eq. (d) yields the natural boundary condition $\zeta_z = 0$ for the end sections.

Pin-Ended Column. For the pin-ended column, it is assumed that

$$\alpha = a \sin\frac{\pi z}{L}, \qquad \beta = b \sin\frac{\pi z}{L}, \qquad \gamma = c \sin\frac{\pi z}{L} \tag{g}$$

where (a, b, c) are constants. These functions satisfy the end conditions $\xi = \eta = 0$. A solution of Eqs. (e) and (f) that satisfies the natural end condition $\zeta_z = 0$ is

$$\zeta = F(x, y) \cos\frac{\pi z}{L} \tag{h}$$

where $F(x, y)$ is a solution of the differential equation

$$F_{xx} + F_{yy} - 2(1 + \nu)\frac{\pi^2}{L^2}F = 0 \tag{i}$$

with the boundary condition

$$\frac{dF}{dn} = -\frac{\pi}{L}\frac{d}{ds}[ay - bx + \tfrac{1}{2}c(x^2 + y^2)] \tag{j}$$

The function $F(x, y)$ is determined uniquely by Eqs. (i) and (j). Evidently, $F(x, y)$ is a homogeneous linear function of (a, b, c).

Equation (b) now yields the following expression for $\delta^2 U$ if an irrelevant factor $\pi^2 E/2L$ is removed:

$$\delta^2 U = -\epsilon[A(a^2 + b^2) + I_p c^2] + \iint F^2 \, dx \, dy + \frac{1}{2(1 + \nu)}$$

$$\times \iint\left\{\left[(a + cy) + \frac{LF_x}{\pi}\right]^2 + \left[(b - cx) + \frac{LF_y}{\pi}\right]^2\right\} dx \, dy \tag{k}$$

Here, A is the cross-sectional area of the column and I_p is the polar moment of inertia of the cross section about its centroid. Accordingly, $\delta^2 U$ is a homogeneous quadratic function of (a, b, c); that is,

$$\delta^2 U = a_{11}a^2 + a_{22}b^2 + a_{33}c^2 + 2a_{23}bc + 2a_{31}ca + 2a_{12}ab \tag{l}$$

Relation to the Kappus Theory. The Kappus equations are obtained if the last term in Eq. (i) is neglected. This appears to be a reasonable approximation,* since L is ordinarily large compared to the dimensions of a cross section. It is convenient to introduce the substitution,

$$F(x, y) = -\frac{\pi}{L}[ax + by + c\Psi(x, y)] \tag{m}$$

Then, if the last term in Eq. (i) is neglected,

$$\Psi_{xx} + \Psi_{yy} = 0 \tag{n}$$

and the boundary condition is

$$\frac{d\Psi}{dn} = \frac{1}{2}\frac{d}{ds}(x^2 + y^2) \tag{o}$$

In view of Eqs. (n) and (o), the function Ψ is identified as the warping function of the Saint Venant torsion theory (51). The solution of Eqs. (n) and (o) contains an arbitrary additive constant. This loss of uniqueness results from the discarding of the last term in Eq. (i). The additive constant must be chosen to minimize $\delta^2 U$. This condition is satisfied if the mean value of Ψ is zero; that is,

$$\iint \Psi \, dx \, dy = 0 \tag{p}$$

The expression for $\delta^2 U$ [Eq. (k)] now becomes

$$\delta^2 U = -\epsilon[(a^2 + b^2)A + c^2 I_p] + \iint \left\{ \frac{\pi^2}{L^2}(ax + by + c\Psi)^2 \right.$$
$$\left. + \frac{c^2}{2(1 + \nu)}[(\Psi_x - y)^2 + (\Psi_y + x)^2] \right\} dx \, dy \tag{q}$$

It is shown in the Saint Venant theory of torsion (122) that

$$\iint [(\Psi_x - y)^2 + (\Psi_y + x)^2] \, dx \, dy = J \tag{r}$$

* An example treated in (122) shows that the last term in Eq. (i) introduces only a minor effect in the case of a column of circular cross section.

where GJ is the torsional stiffness of the bar.* Three section constants of Kappus are defined as follows:

$$\Gamma = \iint \Psi^2 \, dx \, dy, \qquad R_x = \iint y\Psi \, dx \, dy$$

$$R_y = \iint x\Psi \, dx \, dy \tag{s}$$

Consequently, the coefficients in Eq. (l) are

$$a_{11} = I_{yy} - \frac{AL^2\epsilon}{\pi^2}, \qquad a_{22} = I_{xx} - \frac{AL^2\epsilon}{\pi^2}$$

$$a_{33} = \Gamma + \frac{JL^2}{2(1+\nu)\pi^2} - \frac{I_p L^2 \epsilon}{\pi^2} \tag{t}$$

$$a_{12} = a_{21} = I_{xy}, \qquad a_{13} = a_{31} = R_y, \qquad a_{23} = a_{32} = R_x$$

where I_{xx}, I_{yy}, I_{xy} are the moments and product of inertia with respect to the x- and y-axes. It is to be noted that $a_{13} = 0$ if the section is symmetrical about the x-axis and $a_{23} = 0$ if it is symmetrical about the y-axis.

The buckling criterion is

$$\begin{vmatrix} a_{11} & a_{12} & a_{13} \\ a_{21} & a_{22} & a_{23} \\ a_{31} & a_{32} & a_{33} \end{vmatrix} = 0 \tag{u}$$

With the coefficients given by Eq. (t), Eq. (u) is identical to the result of Kappus. To apply Eqs. (t) and (u), we must have a practical way to compute the warping function Ψ for a thin-walled column so that the constants Γ, R_x, R_y may be calculated. This question has been discussed by Goodier (113, 111, 122).

If the coordinate axes are oriented so that $I_{xy} = 0$ and the cross section is symmetrical about the x- and y-axes, Eq. (u) reduces to $a_{11}a_{22}a_{33} = 0$. The conditions $a_{11} = 0$ and $a_{22} = 0$ yield the Euler formulas for flexural buckling in the x- or y-directions. The condition $a_{33} = 0$ yields

$$\epsilon_{\mathrm{cr}} = \frac{1}{I_p}\left[\frac{J}{2(1+\nu)} + \frac{\pi^2\Gamma}{L^2}\right] \tag{v}$$

This is Wagner's formula for pure torsional buckling (149).

The preceding results may be applied to a column with clamped ends if the effective length is taken to be half the actual length.

---

* For a thin-walled column, J is approximated by $J = bt^3/3$, where t is the thickness of the wall and b is the developed length of the cross section.

PROBLEMS

1. The frictionless pin bears on the rigid semicircular body which is supported by perfect rollers (Fig. P6-1). If the body moves horizontally a distance s, the restoring force of the two springs is ks, where k is a constant. Compute the load P_{cr} at which the body slips from under the pin.

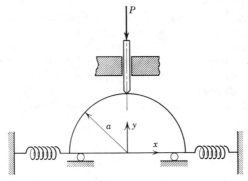

Fig. P6-1

2. Solve Prob. 1 for the case in which the profile of the body is the curve $x^4 + y^4 = a^4$ and the springs provide the nonlinear restoring force ks^3.

3. The bar is rigid and the hinge is frictionless (Fig. P6-3). The hinge contains a spring such that the moment resisting an angular deflection θ is $k\theta$. Compute the buckling load P_{cr}, supposing that the force P remains vertical when the bar rotates. Compute P_{cr}, supposing that the force P rotates only one third as much as the bar.

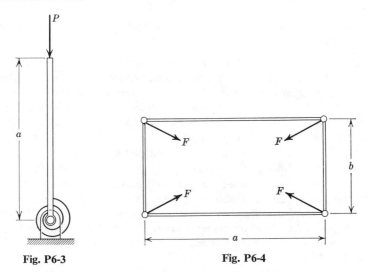

Fig. P6-3

Fig. P6-4

4. The rectangular frame consists of rigid members joined by frictionless hinges. The hinges contain springs such that the moment resisting a relative angular displacement θ of two connected bars is $k\theta$. Calculate the critical load F_{cr}, supposing that the forces remain directed along the diagonals of the parallelogram when the frame buckles. Calculate F_{cr}, supposing that the forces retain their original directions when the frame buckles and the bottom bar is fixed (Fig. P6-4).

5. The regular hexagonal frame consists of rigid bars that are hinged at the ends. The hinges contain springs, such that the moment resisting a relative angular displacement θ of two connected bars is $k\theta$. Assuming that the frame remains symmetrical about the horizontal and vertical center lines, calculate the buckling load for the following cases: (a) the forces remain directed along the diagonals; (b) the forces retain their original directions; (c) the forces remain directed along the bisectors of the angles (Fig. P6-5).

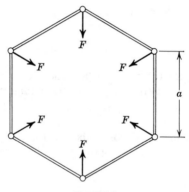

Fig. P6-5

6. A column of length L and constant flexural stiffness EI is subjected to an axial load P. The ends of the column cannot move laterally. Rotational springs at each end of the column exert restoring moments $k\theta$, where θ is the angle of rotation of the end. Approximate the deflection curve by $y = ax(L - x)$, where a is a constant. Hence, derive P_{cr}. Compare with Euler's solution in the case $k = 0$.

7. For a uniform Euler column with clamped ends, assume $y = ax^2(L - x)^2$ and determine the critical load. Compare with Euler's solution.

8. A rigid hemispherical shell with center of mass at the mid-point of the radius rests on a fixed hemispherical dome of the same radius as the shell. Non-linear springs are installed so that the moment resisting an angular displacement θ of the line connecting the centers of the hemispheres is $k\theta^3$, $k > 0$. Determine the critical weight W_{cr} of the shell at which it rolls over (see Fig. 1-7).

9. An Euler column is hinged at the end $x = 0$ and clamped at the end $x = L$. Let

$$V = \tfrac{1}{2} \int_0^L [EI(y'')^2 - P(y')^2] \, dx$$

Set $\delta y = \eta$ and $\delta^2 V = Q$. Determine $\eta(x)$ by the condition that Q satisfies Euler's equation of the calculus of variations when $P = P_{cr}$. Hence determine P_{cr}.

10. A rectangular elastic plate of length a and width b is subjected to a uniform axial compression of intensity N (see Fig. P6-10). Edges $x = 0$, $x = a$, and $y = 0$ are hinged. Edge $y = b$ is free. Assume that the deflection due to buckling is $w = Cy \sin \pi x / a$ and compute N_{cr}.

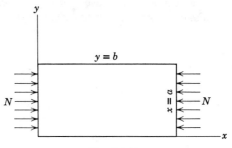

Fig. P6-10

11. The potential energy of a system with two generalized coordinates (x, y) is $V = (2 - 6p)x^2 - 4(1 - p)xy + (9 - 2p)y^2$, where p is a load parameter. Determine the value of p for which the configuration $x = y = 0$ ceases to be stable.

12. A uniform upright elastic column of length L is clamped at the bottom and free at the top. A vertical compression load F acts at the top. Assuming that the deflection y of the buckled column is given by $y = ax^2 + bx^3$, derive a potential energy expression correct to second powers of (a, b). Hence, by investigating $\delta^2 V$, calculate the buckling load F_{cr}. Compare with Euler's solution.

13. A flat uniform rectangular elastic plate is subjected to axial compression of intensity N in the x-direction. The edges $x = 0$ and $x = a$ are simply supported. The edges $y = \pm b$ are clamped. Set $w = f(y) \sin n\pi x / a$ and derive the characteristic equation that determines N_{cr} by means of von Kármán's large-deflection equation (5-14). *Hint.* $N/D > n^2\pi^2/a^2$.

14. Study postbuckling behavior of a simple column by means of the assumption that the axis of the buckled column is a circular arc. Compute F/F_{cr} for a 30° rotation of either end and compare with the result obtained in Sec. 6-2.

15. A flanged rectangular plate is simply supported on the edges $x = 0$, $x = a$, and $y = 0$. The flanged edge ($y = b$) is free. The flange is subjected to an

axial compression load P. Neglecting torsional stiffness of the flange, derive a formula for P_{cr} by means of the assumption $w = Cy \sin \pi x/a$.

16. The weight of a uniform flagpole per unit length is γ. If the flagpole buckles under its own weight, the deflection curve is approximated by $y = C(1 - \cos \pi x/2L)$, where x is the height above the base. Derive a formula for the maximum length that the flagpole can have without buckling.

17. In Eq. (e) of Sec. 6-5, set $\delta^2 U = Q$. Write the Euler equation for Q; thus derive the differential equation whose eigenvalues determine the buckling load of a compressed flat plate.

18. A deep elastic beam without flanges has a rectangular cross section of depth h and thickness t. The ends of the beam are simply supported. Derive the bending moment M_{cr} which causes lateral buckling.

19. For the linkage of Fig. 6-3a, prove that symmetrical buckling occurs if $3k_1 < 7k_2$ and antisymmetric buckling occurs if $3k_1 > 7k_2$.

7 Hamilton's principle and the equations of Lagrange and Hamilton

We must gather and group appearances, until the scientific imagination discerns their hidden laws, and unity arises from variety; and then from unity we must rededuce variety, and force the discovered law to utter its revelations of the future.

W. R. HAMILTON

7-1. KINETIC ENERGY OF A SYSTEM WITH FINITE DEGREES OF FREEDOM. Let (X, Y, Z) be rectangular coordinates attached to a Newtonian reference frame and let $(x_1, x_2, \cdots, x_n)$ be generalized coordinates of a mechanical system with respect to the same reference frame. The ideas will be clarified if we consider that the mechanical system consists of a finite number of particles—say p particles. Since no question of limits is involved, there is no logical difficulty in passing to the case of an infinite number of particles. The particles may be numbered $1, 2, \cdots, p$. The rectangular coordinates of particle ν are denoted by (X_ν, Y_ν, Z_ν). Since the variables x_i determine the locations of all the particles, relations of the following type exist:

$$X_\nu = f_\nu(x_1, x_2, \cdots, x_n), \qquad Y_\nu = g_\nu(x_1, x_2, \cdots, x_n),$$
$$Z_\nu = h_\nu(x_1, x_2, \cdots, x_n) \tag{a}$$

Such equations exist for each particle; they are mathematical expressions of the constraints.

Using dots over letters to denote derivatives with respect to time, we obtain by differentiation of Eq. (a)

$$\dot{X}_\nu = \sum_{r=1}^{n} \frac{\partial X_\nu}{\partial x_r} \dot{x}_r, \qquad \dot{Y}_\nu = \sum_{r=1}^{n} \frac{\partial Y_\nu}{\partial x_r} \dot{x}_r, \qquad \dot{Z}_\nu = \sum_{r=1}^{n} \frac{\partial Z_\nu}{\partial x_r} \dot{x}_r \qquad \text{(b)}$$

The quantities $(\dot{x}_1, \dot{x}_2, \cdots, \dot{x}_n)$ are called "components of generalized velocity."

The kinetic energy of the system is defined by

$$T = \tfrac{1}{2} \sum_{\nu=1}^{p} m_\nu (\dot{X}_\nu^2 + \dot{Y}_\nu^2 + \dot{Z}_\nu^2) \qquad \text{(c)}$$

where m_ν denotes the mass of particle ν. Substitution of Eq. (b) into Eq. (c) yields a relationship of the form

$$T = \tfrac{1}{2} \sum_{i=1}^{n} \sum_{j=1}^{n} b_{ij} \dot{x}_i \dot{x}_j \qquad \text{(7-1)}$$

The coefficients b_{ij} are functions of the generalized coordinates x_i. They are specified to be symmetrical; that is, $b_{ij} = b_{ji}$. Equation (7-1) signifies the following:

> The kinetic energy of a system with finite degrees of freedom is a quadratic form in the components of generalized velocity, provided that the generalized coordinates specify the configuration of the system with respect to the Newtonian reference frame for which the kinetic energy is computed.

In general, a homogeneous quadratic relationship in the generalized velocity is not obtained if the generalized coordinates specify the configuration of the system relative to a reference frame that itself moves in a prescribed way with respect to the Newtonian reference frame for which the kinetic energy is computed. For example, if a block of mass m slides on the floor of a moving elevator, if the displacement of the block relative to the elevator is x, and if the speed of the elevator relative to the earth is v, the kinetic energy of the block relative to the earth is $T = \tfrac{1}{2}m(v^2 + \dot{x}^2)$. Although this is a quadratic expression in the generalized velocity $\dot{x}$, it is not a homogeneous expression because of the additive term $\tfrac{1}{2}mv^2$.

It is to be emphasized that kinetic energy is a relative quantity. It is always computed with respect to some Newtonian reference frame, but its value depends on the choice of that reference frame. Likewise, work is a relative quantity (Sec. 1-3).

7-2. HAMILTON'S PRINCIPLE.

Newton's laws of motion refer only to a single mass particle. The analysis of the motion of a mechanical system by Newtonian methods is a process of synthesis, in which the motions of the individual particles of the system are subjected to Newton's laws. For example, to analyze the motion of a rigid body, one integrates the inertial effects that Newton's laws impose on the infinitesimal mass elements of the body. Lagrange performed this type of synthesis in a general case. He thus derived equations that determine the motion of any holonomic unchecked system with finite degrees of freedom. Adopting different premises than Lagrange, the Irish mathematician W. R. Hamilton (1805–1865) derived a generalization of the principle of virtual work that determines the motion of any finite unchecked system, even though the number of degrees of freedom is nonenumerable.

Hamilton formulated the general problem of dynamics in an unconventional way. In a practical dynamical problem the configuration and the generalized velocity are usually given for an initial instant t_0. Then the subsequent motion is to be determined. Instead of concentrating on this problem, which is poorly formulated for a variational treatment, Hamilton inquired, "What is the motion if the configurations at two given instants, t_0 and t_1, are known?" The significance of this problem is clarified if we solve Hamilton's problem for a simple special case—a particle that moves in a plane without the action of any force. Let the particle lie at a given point (x_0, y_0) at a given instant t_0, and let it lie at another given point (x_1, y_1) at another given instant t_1. Newton's equations reduce to $d^2x/dt^2 = 0$ and $d^2y/dt^2 = 0$. Integration of these equations yields $x = at + b$, $y = ct + d$, in which a, b, c, d are constants that are determined by the given initial and final conditions:

$$x_0 = at_0 + b, \qquad y_0 = ct_0 + d; \qquad x_1 = at_1 + b, \qquad y_1 = ct_1 + d$$

Since $t_0 \neq t_1$, these equations fix the constants. Accordingly, in this example, Hamilton's problem has a unique solution.

Hamilton adduced the principle of D'Alembert, which asserts that any law of statics becomes transformed into a law of kinetics if the driving forces are augmented by inertial forces. Using this principle, he extended the principle of virtual work to kinetics. An illustrative application of D'Alembert's principle has been presented in Sec. 3-5, where the differential equation of a freely vibrating beam is derived from the differential equation for the deflection of a statically loaded beam.

Pursuing Hamilton's general reasoning, without following the many discursive investigations in his two essays, "On a General Method in Dynamics" (115), we consider an unchecked moving mechanical system that is referred to a Newtonian reference frame. We imagine that the system receives a virtual displacement that does not necessarily coincide with the true course of the motion. To avoid confusing the virtual displacement with the actual motion, we suppose that the performance of the virtual displacement consumes no time. In other words, the time variable t is conceived to remain constant while the virtual displacement is executed. An equivalent point of view is that the real motion of the system is stopped while the virtual displacement is performed. However, we must suppose that the inertial forces corresponding to the real motion persist during the virtual displacement, since we wish to calculate the virtual work of these forces.

When the system experiences the virtual displacement, the rectangular coordinates (x, y, z) of any given particle receive infinitesimal increments (ξ, η, ζ). To first-degree terms in (ξ, η, ζ), the work that inertial forces perform when a particle of mass m is displaced from point (x, y, z) to the point $(x + \xi, y + \eta, z + \zeta)$ is $-m(\ddot{x}\xi + \ddot{y}\eta + \ddot{z}\zeta)$. Since this is a linear expression in the infinitesimal virtual displacement (ξ, η, ζ), it is the first variation of the virtual work of the inertial force of the particle. Consequently, the first variation of the virtual work of all the inertial forces of the system is

$$-\sum m(\ddot{x}\xi + \ddot{y}\eta + \ddot{z}\zeta)$$

where the sum extends over all particles. As in Sec. 1-6, the first variation of the virtual work of the noninertial forces is denoted by δW. Extending the principle of virtual work to kinetics by augmenting δW with the virtual work of the inertial forces, we obtain

$$\delta W - \sum m(\ddot{x}\xi + \ddot{y}\eta + \ddot{z}\zeta) = 0 \qquad \text{(a)}$$

This is the variational equation on which Hamilton founded the general theory of dynamics. It is valid for all virtual displacements that are consistent with the constraints. Equation (a) must be restricted to unchecked systems, since otherwise the variational form of the principle of virtual work is inapplicable (see Sec. 1-6). However, if the driving forces are discontinuous only at isolated points in configuration space, Eq. (a) is valid except at those points.

One of Hamilton's greatest achievements in dynamics was a mathematical transformation of Eq. (a) which renders the equation more useful for analytical purposes. He supposed that at the time t_0 the system has a given configuration $\mathbf{X}_0$ and that at the time t_1 (where $t_1 > t_0$) it has a given configuration $\mathbf{X}_1$. The real motion during the time interval (t_0, t_1) is denoted by $\mathbf{X} = \mathbf{F}(t)$. The symbol $\mathbf{F}$ signifies that a single configuration $\mathbf{X}$ corresponds to each value of t in the range (t_0, t_1). The set of configurations $\mathbf{X}$ corresponding to all values of t in the range (t_0, t_1) has been called the "path" that the system describes. It is represented symbolically by a curve (Fig. 7-1). The problem is to find the function $F(t)$.

Hamilton considered a varied path that lies infinitesimally close to the path $X = F(t)$. The varied path is denoted by $X^* = F^*(t)$. Since the configurations are fixed at times t_0 and t_1, the two paths have the same end points; that is

$$F(t_0) = F^*(t_0) = X_0$$

$$F(t_1) = F^*(t_1) = X_1$$

The notation $X^* - X$ denotes the displacement from X to X^* (see Sec. 1-1). It is designated by

$$S = X^* - X \qquad (b)$$

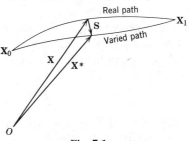

Fig. 7-1

Since X and X^* are functions of t, S is a function of t. This function vanishes if $t = t_0$ or $t = t_1$. The relations are depicted in Fig. 7-1.

The virtual displacement S imposes the displacements (ξ, η, ζ) on the particles of the system. Since S is a function of time, the variables (ξ, η, ζ) are now functions of time. This condition imposes no restrictions on the virtual displacements. Since (ξ, η, ζ) are functions of time, the following differential identities exist:

$$\ddot{x}\xi = \frac{d}{dt}(\dot{x}\xi) - \dot{x}\dot{\xi}, \text{ etc.}$$

Therefore, Eq. (a) may be written as follows:

$$\sum m(\dot{x}\dot{\xi} + \dot{y}\dot{\eta} + \dot{z}\dot{\zeta}) + \delta W - \frac{d}{dt}\sum m(\dot{x}\xi + \dot{y}\eta + \dot{z}\zeta) = 0 \qquad (c)$$

The kinetic energy of the system is defined by

$$T = \tfrac{1}{2}\sum m(\dot{x}^2 + \dot{y}^2 + \dot{z}^2)$$

Consequently, as the system describes the varied path, its kinetic energy at point X^* is

$$T + \Delta T = \tfrac{1}{2}\sum m[(\dot{x} + \dot{\xi})^2 + (\dot{y} + \dot{\eta})^2 + (\dot{z} + \dot{\zeta})^2]$$

The part of this expression that is linear in $(\dot{\xi}, \dot{\eta}, \dot{\zeta})$ is the first variation of T. Hence

$$\delta T = \sum m(\dot{x}\dot{\xi} + \dot{y}\dot{\eta} + \dot{z}\dot{\zeta})$$

Consequently, Eq. (c) may be written as follows:

$$\delta T + \delta W - \frac{d}{dt} \sum m(\dot{x}\xi + \dot{y}\eta + \dot{z}\zeta) = 0 \tag{d}$$

Hamilton disposed of the last term in Eq. (d) by integrating the equation with respect to t between the limits t_0 and t_1. Since the last term in Eq. (d) is a time derivative, it may be integrated explicitly. Since S vanishes at the end points of the path, the functions (ξ, η, ζ) vanish at the times t_0 and t_1. Consequently, integration of Eq. (d) yields

$$\int_{t_0}^{t_1} (\delta T + \delta W)\, dt = 0 \tag{7-2}$$

Equation (7-2) is the desired formulation of Hamilton's principle. It is important to note that δW is not the total virtual work; it is the virtual work of the noninertial forces. For example, if the system is a wheel that rotates freely on a fixed frictionless axle, $\delta W = 0$, since there are no noninertial forces that perform work. Then Eq. (7-2) states that the definite time integral of the kinetic energy is stationary. This integral is

$$\tfrac{1}{2}I \int_{t_0}^{t_1} \dot{\theta}^2\, dt$$

where θ is the angle through which the wheel has turned and I is the moment of inertia of the wheel about its axis. The Euler equation for this integral is $\ddot{\theta} = 0$. Consequently, $\dot{\theta} = $ constant; that is, the wheel rotates at constant angular velocity.

Derivation of Newton's Law from Hamilton's Principle. Let a moving mass particle m be located by a position vector $\mathbf{r}$. Let the particle be subjected to a variable force $\mathbf{F}$. Then, $\delta W = \mathbf{F} \cdot \delta \mathbf{r}$. Also, the kinetic energy of the particle is $T = \tfrac{1}{2}m\dot{\mathbf{r}}^2$; hence $\delta T = m\dot{\mathbf{r}} \cdot \delta\dot{\mathbf{r}}$. Therefore, by Hamilton's principle

$$\int_{t_0}^{t_1} (\delta T + \delta W)\, dt = \int_{t_0}^{t_1} (m\dot{\mathbf{r}} \cdot \delta\dot{\mathbf{r}} + \mathbf{F} \cdot \delta\mathbf{r})\, dt = 0$$

Integrating by parts and recalling that Hamilton supposed that $\delta\mathbf{r}$ vanishes at the terminal instants, t_0 and t_1, we obtain

$$\int_{t_0}^{t_1} (\mathbf{F} - m\ddot{\mathbf{r}}) \cdot \delta\mathbf{r}\, dt = 0$$

Since this relation is true for every virtual displacement $\delta\mathbf{r}$,

$$\mathbf{F} - m\ddot{\mathbf{r}} = 0$$

This is Newton's second law. Consequently, Hamilton's equation is as general as Newton's equation.

Instead of basing Hamilton's principle on Newton's law, we may logically regard Hamilton's principle as the fundamental law of dynamics and consider Newton's law to be a special consequence of it. This point of view was adopted by Hertz (33).

One important feature of Hamilton's principle is that it is formulated without reference to any particular system of coordinates. Also, it may be applied to systems with nonenumerable degrees of freedom. In this respect it is broader than the Lagrangian equations that are derived in the following article. Applications of Hamilton's principle to systems with infinitely many degrees of freedom are deferred to Chap. 8.

7-3. LAGRANGE'S EQUATIONS FOR CONSERVATIVE SYSTEMS.

Hamilton's principle is easily applied to conservative systems. In this case $\delta W = -\delta V$, where V is the potential energy of the system. Consequently, Eq. (7-2) may be written in the following form:

$$\delta A = 0, \qquad A = \int_{t_0}^{t_1} L\, dt, \qquad L = T - V \qquad (7\text{-}3)$$

The quantity L is known as the "Lagrangian function." Hamilton referred to the integral A as the "principal function of dynamics"; nowadays, it is often called "action." The variational equation, $\delta A = 0$, means the following:

> Among all motions that will carry a conservative system from a given initial configuration $\mathbf{X}_0$ to a given final configuration $\mathbf{X}_1$ in a given time interval (t_0, t_1), that which actually occurs provides a stationary value to the integral A.

If the system under consideration is nonholonomic, the variation S must be consistent with the nonholonomic constraints. Since nonholonomic constraints occur infrequently in practice, attention is restricted to holonomic systems. The variation S is then arbitrary. For a holonomic system, Hamilton's principle yields the useful conclusion that the Euler equations for the integral A are the differential equations of motion. If the system has a finite number of generalized coordinates x_i, the differential equations of motion are accordingly

$$\frac{d}{dt}\frac{\partial L}{\partial \dot{x}_i} - \frac{\partial L}{\partial x_i} = 0 \qquad (7\text{-}4)$$

Equation (7-4) is known as Lagrange's equation of motion. Lagrange was aware of the variational principle $\delta A = 0$, but he seemingly did not recognize that it is broader than Eq. (7-4).

Double Pendulum. As a simple application of Lagrange's equations, we consider the so-called double-pendulum shown in Fig. 7-2. The pendulum oscillates in the (x, y)-plane under the action of gravity. The masses are concentrated in the two bobs. The hinges are frictionless. The generalized coordinates are (θ, ϕ).

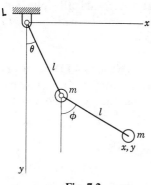

The kinetic energy of the upper bob is $\frac{1}{2}ml^2\dot{\theta}^2$. To compute the kinetic energy of the lower bob, we first write the equations for its coordinates:

$$x = l(\sin \theta + \sin \phi)$$
$$y = l(\cos \theta + \cos \phi) \qquad (a)$$

Hence

$$\dot{x} = l(\dot{\theta} \cos \theta + \dot{\phi} \cos \phi)$$
$$\dot{y} = -l(\dot{\theta} \sin \theta + \dot{\phi} \sin \phi) \qquad (b)$$

The kinetic energy of the lower bob is

$$\tfrac{1}{2}m(\dot{x}^2 + \dot{y}^2) = \tfrac{1}{2}ml^2[\dot{\theta}^2$$
$$+ \dot{\phi}^2 + 2\dot{\theta}\dot{\phi} \cos (\phi - \theta)]$$

Fig. 7-2

Consequently, the kinetic energy of the system is

$$T = ml^2[\dot{\theta}^2 + \tfrac{1}{2}\dot{\phi}^2 + \dot{\theta}\dot{\phi} \cos (\phi - \theta)] \qquad (c)$$

Incidentally, this expression is a quadratic form in the components $\dot{\theta}$, $\dot{\phi}$ of generalized velocity, as predicted in Sec. 7-1.

The potential energy of the lower bob is $-mgy$. By Eq. (a),

$$-mgy = -mgl(\cos \theta + \cos \phi)$$

The potential energy of the upper bob is $-mgl \cos \theta$. Accordingly, the potential energy of the system is

$$V = -mgl(2 \cos \theta + \cos \phi) \qquad (d)$$

With Eq. (7-3), Eqs. (c) and (d) yield

$$L = ml^2\left[\dot{\theta}^2 + \tfrac{1}{2}\dot{\phi}^2 + \dot{\theta}\dot{\phi} \cos (\phi - \theta) + \frac{g}{l}(2 \cos \theta + \cos \phi)\right] \qquad (e)$$

By Eq. (7-4), the Lagrangian equations are

$$\frac{d}{dt}\frac{\partial L}{\partial \dot\theta} - \frac{\partial L}{\partial \theta} = 0, \qquad \frac{d}{dt}\frac{\partial L}{\partial \dot\phi} - \frac{\partial L}{\partial \phi} = 0$$

Consequently, the differential equations of motion are

$$2\ddot\theta + \ddot\phi \cos(\phi - \theta) - \dot\phi^2 \sin(\phi - \theta) + \frac{2g}{l}\sin\theta = 0$$

$$\ddot\phi + \ddot\theta \cos(\phi - \theta) + \dot\theta^2 \sin(\phi - \theta) + \frac{g}{l}\sin\phi = 0$$

(f)

Equations (f) are complicated nonlinear differential relationships. In the classical theory of vibrations certain linearizing approximations are introduced. In the present case these approximations are equivalent to setting $\cos(\phi - \theta) = 1$, $\sin\theta = \theta$, $\sin\phi = \phi$, and $\dot\theta^2 \sin(\phi - \theta) = \dot\phi^2 \sin(\phi - \theta) = 0$ in Eq. (f).

7-4. TIME-DEPENDENT CONSTRAINTS. Occasionally, a mechanical system is subjected to constraints that vary with time in a prescribed way. For example, certain parts of a vibrating system may be subjected to given oscillations by means of external shaking devices. The prescribed movements of these parts are time-dependent constraints.

Hamilton's principle remains valid for systems with time-dependent constraints, provided that the terminal configurations $(\mathbf{X}_0, \mathbf{X}_1)$ and the variations **S** are restricted in accordance with these constraints, since such constraints are not excluded in the derivation of Hamilton's principle.

Systems that are subjected to time-dependent constraints are generally nonconservative in the sense that the total mechanical energy varies with time, since the constraining forces perform work on the system. However, if the instantaneous constraints are held constant, the work V that we perform against noninertial forces in giving the system any displacement may conceivably be independent of the path. Then $\delta W = -\delta V$, and V may be regarded as a time-dependent potential energy function. Since the potential energy V depends on the instantaneous constraints, it is not only a function of the coordinates of the system but also an explicit function of time. Also, the initial configuration $\mathbf{X}_0$ that serves as the zero configuration for potential energy is a function of time. Consequently, the potential energy contains an arbitrary additive function of time. This function is irrelevant, insofar as Hamilton's principle is concerned, since it does not affect variations of V. However, it does preclude the use of the formula $T + V = $ constant.

When time-dependent constraints exist, the generalized coordinates x_i may specify the configuration of the system relative to a reference frame that moves in some prescribed way with respect to a so-called "fixed reference frame." The fixed reference frame is Newtonian, and the kinetic and potential energies must be computed with respect to that reference frame.

Example. Mechanism with Time-Dependent Constraint. A rigid rod rotates clockwise in a vertical plane with constant angular velocity ω. At the time $t = 0$, the rod is horizontal. Accordingly, its angular position

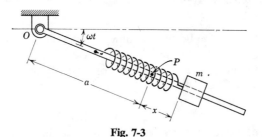

Fig. 7-3

at time t is ωt (Fig. 7-3). A body of mass m slides without friction on the bar. The body is subjected to the force of gravity and to the action of a spring which attracts the body toward a point P on the bar with a force kx, where x is the distance of the body from point P and k is a constant. The point P lies at distance a from the center of rotation O. The motion of the body is determined when x is expressed as a function of t, since the orientation of the bar is a known function of t. Accordingly, with respect to the rotating reference frame, the body has a single degree of freedom.

The requirement that the body shall rotate with the bar is a time-dependent constraint. The potential energy of the body is calculated for a fixed configuration of the bar. If the bar is fixed in its position at time t, the potential energy of the body is

$$V = -mgx \sin \omega t + \tfrac{1}{2}kx^2 \qquad (a)$$

The origin might be chosen equally well at the center of rotation. Then

$$V = -mg(x + a) \sin \omega t + \tfrac{1}{2}kx^2$$

However, the additive term $-mga \sin \omega t$ contributes nothing to the equation of motion that is derived by Lagrange's equation.

The kinetic energy of the body with respect to the fixed reference frame is

$$T = \tfrac{1}{2}m[(a + x)^2\omega^2 + \dot{x}^2] \qquad (b)$$

Consequently, the Lagrange equation (7-4) yields

$$\ddot{x} + \left(\frac{k}{m} - \omega^2\right)x = g \sin \omega t + a\omega^2 \tag{c}$$

This is an elementary type of differential equation. With the initial conditions, it determines the motion.

7-5. THE HAMILTONIAN FUNCTION. If (x_1, x_2, x_3) are rectangular coordinates of a particle that moves in a conservative force field (F_1, F_2, F_3), the Lagrangian function is

$$L = \tfrac{1}{2}m(\dot{x}_1^2 + \dot{x}_2^2 + \dot{x}_3^2) - V$$

Consequently, $\partial L/\partial \dot{x}_i = m\dot{x}_i$ and $\partial L/\partial x_i = -\partial V/\partial x_i = F_i$ [see Eq. (1-18)]. This means that $\partial L/\partial \dot{x}_i$ is the ith momentum component of the particle and $\partial L/\partial x_i$ is the ith component of force that acts on the particle. Accordingly, for a single particle, Lagrange's equation expresses Newton's law, "force equals rate of change of momentum."

The following notation is conventional:

$$\frac{\partial L}{\partial \dot{x}_i} = p_i \tag{7-5}$$

For any conservative system with enumerable degrees of freedom, the quantities p_i are called "components of generalized momentum," since they reduce to ordinary momentum components if the system is a single particle. Likewise, the quantities $\partial L/\partial x_i$ are frequently called "components of generalized force," but we avoid this terminology for generalized force has already been defined differently (Sec. 1-7). With Eq. (7-5), Lagrange's equation (7-4) becomes

$$\frac{\partial L}{\partial x_i} = \dot{p}_i \tag{7-6}$$

The Hamiltonian function is defined by

$$H = \sum_{i=1}^{n} p_i \dot{x}_i - L \tag{7-7}$$

Differentiation yields

$$\frac{dH}{dt} = \sum \dot{p}_i \dot{x}_i + \sum p_i \ddot{x}_i - \sum \frac{\partial L}{\partial x_i} \dot{x}_i - \sum \frac{\partial L}{\partial \dot{x}_i} \ddot{x}_i - \frac{\partial L}{\partial t}$$

Hence, by Eqs. (7-5) and (7-6),

$$\frac{dH}{dt} = -\frac{\partial L}{\partial t} \tag{7-8}$$

Ordinarily, L does not depend explicitly on t if there are no time-dependent constraints. Then Eq. (7-8) yields $H =$ constant.

To interpret this conclusion, we consider more specifically the form of the function L. The kinetic energy is a quadratic form in the variables $\dot{x}_i$ [see Eq. (7-1)]. Hence, since V does not depend on the quantities $\dot{x}_i$,

$$p_i = \frac{\partial L}{\partial \dot{x}_i} = \frac{\partial T}{\partial \dot{x}_i} = \sum_{j=1}^{n} b_{ij}\dot{x}_j$$

Consequently, by Eqs. (7-1) and (7-7), $H = 2T - L = T + V$. Accordingly, if L does not contain t explicitly, H is the total mechanical energy of the system (Sec. 1-9). The condition $H =$ constant then expresses the law of conservation of mechanical energy. It is to be emphasized, however, that H is not generally the total mechanical energy if the coordinates x_i specify the configuration of the system with respect to a reference frame that itself has a prescribed motion with respect to a fixed reference frame for which T and V are computed.

7-6. HAMILTON'S EQUATIONS. Let $F(u_1, u_2, \cdots, u_n; w_1, w_2, \cdots, w_m)$ be a function of $m + n$ variables and set $v_i = (\partial F/\partial u_i)_w$. The notation $(\partial F/\partial u_i)_w$ indicates that F is regarded as a function of the independent variables (u_i, w_i) and that the w's are held constant when the partial differentiation is performed. Inversion of the foregoing equations yields $u_i = u_i(v, w)$, where the symbols v and w stand collectively for all the v's and w's. Consequently, F may be regarded as a function of the v's and w's. Set

$$G(v, w) = \sum u_k v_k - F(v, w) \tag{a}$$

Then, by the Legendre transformation (Sec. 4-6),

$$\left(\frac{\partial G}{\partial v_i}\right)_w = u_i(v, w) \tag{b}$$

It is to be observed that the w's play no role in the Legendre transformation; they are merely passive parameters.

Differentiation of Eq. (a) yields

$$\left(\frac{\partial G}{\partial w_i}\right)_v = \sum_{k=1}^{n} v_k \left(\frac{\partial u_k}{\partial w_i}\right)_v - \left(\frac{\partial F}{\partial w_i}\right)_v$$

Also, if F is regarded as a function of the u's and w's,

$$\left(\frac{\partial F}{\partial w_i}\right)_v = \sum_{k=1}^{n} \left(\frac{\partial F}{\partial u_k}\right)_w \left(\frac{\partial u_k}{\partial w_i}\right)_v + \left(\frac{\partial F}{\partial w_i}\right)_u$$

Since $(\partial F/\partial u_k)_w = v_k$, substitution of the last relation into the preceding one yields

$$\left(\frac{\partial G}{\partial w_i}\right)_v = -\left(\frac{\partial F}{\partial w_i}\right)_u \tag{c}$$

In the following discussion we may drop the subscripts u, v, w, with the understanding that the Hamiltonian function H is regarded as a function of the variables (x, p, t) and the Lagrangian function L is regarded as a function of the variables $(x, \dot{x}, t)$. To derive Hamilton's equations, we identify the variables in the preceding Legendre transformation in the following way: $u_i = \dot{x}_i$; $w_i = x_i$; $F = L(x, \dot{x}, t)$. Then $v_i = \partial L/\partial \dot{x}_i = p_i$ = generalized momentum, and $G = \sum p_i \dot{x}_i - L = H(x, p, t)$ = Hamiltonian function. Consequently, by Eqs. (b) and (c),

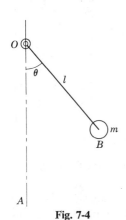

$$\frac{\partial H}{\partial p_i} = \dot{x}_i, \qquad \frac{\partial H}{\partial x_i} = -\frac{\partial L}{\partial x_i} \tag{d}$$

With Lagrange's equation (7-6), Eq. (d) yields

$$\frac{dx_i}{dt} = \frac{\partial H}{\partial p_i}, \qquad \boxed{\frac{dp_i}{dt} = -\frac{\partial H}{\partial x_i}} \tag{7-9}$$

Fig. 7-4

Equations (7-9) are Hamilton's equations. The Lagrange equations are n second-order differential equations in the n unknown functions, $x_1(t), x_2(t), \cdots, x_n(t)$. The Hamilton equations are $2n$ first-order differential equations in the $2n$ unknown functions, $x_1(t), x_2(t), \cdots, x_n(t)$; $p_1(t), p_2(t), \cdots, p_n(t)$. The first set of Hamilton equations $(dx_i/dt = \partial H/\partial p_i)$ is purely mathematical. It is the second set $(dp_i/dt = -\partial H/\partial x_i)$ that expresses a physical law, since it is this set that is derived from Lagrange's equations.

Hamilton's equations are important in statistical mechanics and in some other branches of physics. Jacobi (48, 90) developed an extensive theory for their integration. Because of their simplicity and symmetry, they are better suited for a general theory of integration than the Lagrange equations.

Example. Spherical Pendulum. Figure 7-4 represents a mass m that is suspended by an inextensible weightless thread of length l. The thread constrains the mass to a spherical surface with center O and radius l. The dihedral angle between the plane AOB and a fixed plane containing the vertical axis AO is ϕ. The generalized coordinates of the pendulum are (θ, ϕ); these are ordinary spherical coordinates. For correlation with Eq. (7-9), set $x_1 = \theta$, $x_2 = \phi$. The kinetic energy of the pendulum is

$$T = \tfrac{1}{2}ml^2(\dot{\theta}^2 + \dot{\phi}^2 \sin^2 \theta) \tag{e}$$

The potential energy is

$$V = -mgl \cos \theta \tag{f}$$

Since $L = T - V$, Eq. (7-5) yields

$$p_1 = ml^2\dot{\theta}, \qquad p_2 = ml^2\dot{\phi} \sin^2 \theta \quad \text{Natural?} \tag{g}$$

Since the reference frame is <u>Newtonian</u>, the Hamiltonian function is $H = T + V$. Consequently, if $\dot{\theta}$ and $\dot{\phi}$ are eliminated by means of Eq. (g),

$$H = \frac{p_1^2}{2ml^2} + \frac{p_2^2}{2ml^2 \sin^2 \theta} - mgl \cos \theta \tag{h}$$

Accordingly, Hamilton's equations (7-9) yield

$$\dot{\theta} = \frac{p_1}{ml^2}, \qquad \dot{\phi} = \frac{p_2}{ml^2 \sin^2 \theta}$$

$$\dot{p}_1 = \frac{p_2^2 \cos \theta}{ml^2 \sin^3 \theta} - mgl \sin \theta, \qquad \dot{p}_2 = 0 \tag{i}$$

These are four differential equations for the four unknowns, $\theta(t)$, $\phi(t)$, $p_1(t)$, $p_2(t)$.

The equation $\dot{p}_2 = 0$ yields $p_2 = $ constant. This result is easily interpreted, since Eq. (g) shows that p_2 is the moment of momentum of the pendulum about the vertical axis OA. By Eq. (g), the equation $p_2 = $ constant yields

$$\dot{\phi} \sin^2 \theta = \text{constant} \tag{j}$$

Since $p_1 = ml^2\dot{\theta}$, Eq. (i) yields

$$\ddot{\theta} = \frac{p_2^2 \cos \theta}{m^2 l^4 \sin^3 \theta} - \frac{g}{l} \sin \theta \tag{k}$$

This equation may be integrated by means of the substitution,

$$\dot{\theta} = \omega, \qquad \ddot{\theta} = \frac{d\omega}{dt} = \frac{d\omega}{d\theta}\frac{d\theta}{dt} = \omega \frac{d\omega}{d\theta}$$

Hence, since p_2 is constant, integration of Eq. (k) yields

$$\dot{\theta}^2 = \frac{-p_2^2}{m^2 l^4 \sin^2 \theta} + \frac{2g}{l} \cos \theta + \text{constant} \tag{l}$$

This differential equation is separable, but it is not integrable in terms of elementary functions. Two particular solutions are apparent. One solution is $\dot{\theta} = 0$ or $\theta = $ constant. Then, by Eq. (j), $\phi = $ constant.

Elimination of p_2 from Eq. (k) by means of Eq. (g) yields $\dot{\phi}^2 = (g/l) \sec \theta$.

Another solution is obtained if $\dot{\phi} = 0$. Then the pendulum oscillates in a fixed vertical plane. In this case $p_2 = 0$. Consequently, Eq. (l) yields $\dot{\theta}^2 = (2g/l)(\cos \theta - \cos \theta_0)$, where θ_0 is the amplitude of the swing.

7-7. LAGRANGE'S EQUATIONS FOR NONCONSERVATIVE SYSTEMS.

If the coordinates of a mechanical system receive infinitesimal virtual increments δx_i, the virtual work of all the forces that act on the system is represented by a linear differential expression [see Eq. (1-6)]:

$$\delta W = \sum_{i=1}^{n} Q_i \, \delta x_i$$

The components Q_i of generalized force may be functions of $(x_1, x_2, \cdots, x_n; \ t)$. If there are no nonholonomic constraints, the variations δx_i are arbitrary functions of t.

Since the kinetic energy T is a function of

$$(x_1, x_2, \cdots, x_n; \ \dot{x}_1, \dot{x}_2, \cdots, \dot{x}_n; \ t)$$

the theory of the first variation (Sec. 3-2) yields

$$\int_{t_0}^{t_1} \delta T \, dt = \int_{t_0}^{t_1} \sum_{i=1}^{n} \left(\frac{\partial T}{\partial x_i} - \frac{d}{dt} \frac{\partial T}{\partial \dot{x}_i} \right) \delta x_i \, dt$$

Consequently, Hamilton's principle [Eq. (7-2)] yields

$$\int_{t_0}^{t_1} \sum_{i=1}^{n} \left(\frac{\partial T}{\partial x_i} - \frac{d}{dt} \frac{\partial T}{\partial \dot{x}_i} + Q_i \right) \delta x_i \, dt = 0$$

If the system is holonomic, this relation is true for all variations δx_i that vanish at the instants t_0 and t_1. Consequently, the integrand vanishes for arbitrary variations δx_i; that is,

$$\frac{d}{dt} \frac{\partial T}{\partial \dot{x}_i} - \frac{\partial T}{\partial x_i} = Q_i \tag{7-10}$$

Equation (7-10) is the general form of Lagrange's equations.

If the system is conservative, $Q_i = -\partial V/\partial x_i$ [see Eq. (1-8)], and Eq. (7-10) reduces to Eq. (7-4). It frequently happens that some of the forces are conservative and others are not. For example, a vibrating system may contain springs and weights. The forces resulting from these parts are derivable from a potential energy function. However, frictional forces in the system are nonconservative. Consequently, it is occasionally advantageous to write Q_i in the form,

$$Q_i = R_i - \frac{\partial V}{\partial x_i} \tag{7-11}$$

where V is the potential energy function for some of the forces and R_i is the component of the remaining generalized force. Then, if $L = T - V$, Eq. (7-10) becomes

$$\frac{d}{dt}\frac{\partial L}{\partial \dot{x}_i} - \frac{\partial L}{\partial x_i} = R_i \qquad (7\text{-}12)$$

Utilizing Eq. (7-12) instead of Eq. (7-4), we may also augment Hamilton's equations (7-9) by the term R_i. Thus we obtain

$$\frac{dx_i}{dt} = \frac{\partial H}{\partial p_i}, \qquad \frac{dp_i}{dt} = -\frac{\partial H}{\partial x_i} + R_i \qquad (7\text{-}13)$$

Time-Dependent External Forces. Frequently mechanical systems are subjected to external forces that are prescribed functions of time. For example, time-dependent external forces characterize forced vibrations. If a time-dependent force $F(t)$ acts on a particle that moves through the distance y in the direction of the force, the contribution to δW from the force is $F\,\delta y$. Consequently, δW may be calculated with the understanding that the potential energy of the force is $-Fy$. Although potential energy of time-dependent forces is a useful concept, we must observe that a system that is subjected to such forces is generally nonconservative, since the total mechanical energy varies with time.

Newton's Law. If Eq. (7-10) is applied to a free-mass particle that is located by rectangular coordinates (x, y, z), $Q_1 = F_x$, $Q_2 = F_y$, $Q_3 = F_z$, where (F_x, F_y, F_z) are the components of the force that acts on the particle. Also, $T = \frac{1}{2}m(\dot{x}^2 + \dot{y}^2 + \dot{z}^2)$. Consequently, Eq. (7-10) yields Newton's equations, $F_x = m\ddot{x}$, $F_y = m\ddot{y}$, $F_z = m\ddot{z}$. Accordingly, Newton's law may be regarded as a special consequence of Lagrange's equations.

Example. Pendulum with Moving Pivot. Figure 7-5 represents a simple pendulum that is suspended from a block which moves horizontally with a prescribed motion, $z = f(t)$. The coordinates of the bob are

$$x = z - l\sin\theta, \qquad y = l\cos\theta \qquad (a)$$

Consequently,

$$\dot{x} = \dot{z} - l\dot{\theta}\cos\theta, \qquad \dot{y} = -l\dot{\theta}\sin\theta \qquad (b)$$

The kinetic energy of the pendulum is $T = \frac{1}{2}m(\dot{x}^2 + \dot{y}^2)$. Hence

$$T = \frac{1}{2}m(\dot{z}^2 - 2l\dot{\theta}\dot{z}\cos\theta + l^2\dot{\theta}^2) \qquad (c)$$

Suppose that the hinge contains a spring that exerts a restoring moment

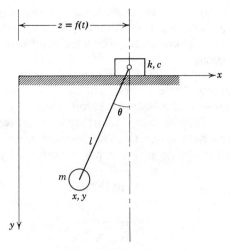

Fig. 7-5

$k\theta$ and a damper that exerts a resisting moment $c\dot{\theta}$. The potential energy due to the spring and the weight is

$$V = -mgl \cos \theta + \tfrac{1}{2}k\theta^2 \tag{d}$$

If θ receives an infinitesimal increment $\delta\theta$, the work of the damping moment is $R \, \delta\theta = -c\dot{\theta} \, \delta\theta$. Consequently, $R = -c\dot{\theta}$. Therefore, Lagrange's equation (7-12) is

$$\frac{d}{dt}\frac{\partial L}{\partial \dot{\theta}} - \frac{\partial L}{\partial \theta} = -c\dot{\theta}$$

Setting $L = T - V$, we obtain

$$\ddot{\theta} + \frac{c}{ml^2}\dot{\theta} + \frac{k}{ml^2}\theta + \frac{g}{l}\sin \theta = \frac{\ddot{z}}{l}\cos \theta \tag{e}$$

If the pivot moves with constant velocity ($\ddot{z} = 0$), Eq. (e) is the same as though the pivot were stationary. This conclusion exemplifies the fact that a uniform velocity of translation does not alter the behavior of a mechanical system. If $\ddot{z} = $ constant, Eq. (e) evidently admits a solution for which θ is constant.

7-8. KINEMATICS OF A RIGID BODY. As a preliminary to the theory of dynamics of a rigid body, a few kinematical relations are derived. The motion of a rigid body may be resolved into a translation that imparts the proper motion to a specified particle of the body and a rotation about an axis through that particle. The theory of translation is treated in elementary texts on mechanics. Consequently, attention will be confined to the rotary part of the motion. Accordingly, a rigid body with one fixed point is considered. The fixed point may be regarded as a ball-and-socket joint.

Let (x, y, z) be right-handed rectangular coordinates attached to a Newtonian reference frame. These will be called "fixed coordinates." Let the fixed point of the body lie at the origin O of the coordinates (x, y, z). Let (ξ, η, ζ) be right-handed orthogonal axes scribed in the body and let them also have origin O. The direction cosines of the (ξ, η, ζ) axes with respect to the (x, y, z) axes are represented by the rows in the following table:

TABLE 7-1

	x	y	z
ξ	l_1	m_1	n_1
η	l_2	m_2	n_2
ζ	l_3	m_3	n_3

Each entry in this table is the cosine of the angle between the axes designated at the top of its column and the left end of its row.

Let $(\mathbf{i}, \mathbf{j}, \mathbf{k})$ and $(\mathbf{a}, \mathbf{b}, \mathbf{c})$ be unit vectors along the (x, y, z) and (ξ, η, ζ) axes, respectively. Since the projections of a unit vector on rectangular coordinate axes are the direction cosines of the vector, $\mathbf{a} = \mathbf{i}l_1 + \mathbf{j}m_1 + \mathbf{k}n_1$, $\mathbf{i} = \mathbf{a}l_1 + \mathbf{b}l_2 + \mathbf{c}l_3$, etc. Since $\mathbf{a} \cdot \mathbf{a} = 1, \mathbf{i} \cdot \mathbf{i} = 1$, etc., these equations yield $l_1^2 + m_1^2 + n_1^2 = 1$, $l_1^2 + l_2^2 + l_3^2 = 1$, etc. Accordingly, *the sum of the squares of the terms in any row or any column of Table* 7-1 *is* 1. Also, since $\mathbf{a} \cdot \mathbf{b} = 0$, $\mathbf{i} \cdot \mathbf{j} = 0$; etc., $l_1 l_2 + m_1 m_2 + n_1 n_2 = 0$, $l_1 m_1 + l_2 m_2 + l_3 m_3 = 0$, etc. Accordingly, *the sum of the products of corresponding terms in any two rows or any two columns of Table* 7-1 *is* 0.

Since $\mathbf{b} \times \mathbf{c} = \mathbf{a}$, $\mathbf{c} \times \mathbf{a} = \mathbf{b}$, and $\mathbf{a} \times \mathbf{b} = \mathbf{c}$, we obtain

$$l_1 = m_2 n_3 - m_3 n_2, \quad m_1 = n_2 l_3 - n_3 l_2, \quad n_1 = l_2 m_3 - l_3 m_2$$
$$l_2 = m_3 n_1 - m_1 n_3, \quad m_2 = n_3 l_1 - n_1 l_3, \quad n_2 = l_3 m_1 - l_1 m_3 \quad (7\text{-}14)$$
$$l_3 = m_1 n_2 - m_2 n_1, \quad m_3 = n_1 l_2 - n_2 l_1, \quad n_3 = l_1 m_2 - l_2 m_1$$

Since $\mathbf{b} \times \mathbf{c} = \mathbf{a}$ and $\mathbf{a} \cdot \mathbf{a} = 1$, we obtain $\mathbf{a} \cdot \mathbf{b} \times \mathbf{c} = 1$. Accordingly, *the determinant of* Table 7-1 *is* 1.

The foregoing relationships, which may be augmented by others, show that the nine quantities in Table 7-1 are not all independent. In fact, these quantities may all be expressed as functions of three independent parameters (θ, ϕ, ψ), known as "Euler's angles." These angles are illustrated by Fig. 7-6. The angle between the z and ζ axes is denoted by θ. The dihedral angle between the planes xOz and $zO\zeta$ is denoted by ϕ. The dihedral angle between the planes $zO\zeta$ and $\xi O\zeta$ is denoted by ψ. In Fig. 7-6 these angles are represented with the aid of arcs on a sphere with center at the origin. Arc A is the intersection of plane $\xi O\zeta$ with the sphere.

Since θ, ϕ are ordinary spherical coordinates of the ζ-axis,

$$l_3 = \sin\theta \cos\phi, \qquad m_3 = \sin\theta \sin\phi, \qquad n_3 = \cos\theta \qquad \text{(a)}$$

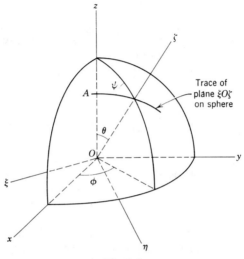

Trace of plane $\xi O\zeta$ on sphere

Fig. 7-6

Since the dihedral angle between the planes $zO\zeta$ and $\xi O\zeta$ is the same as the angle between the normals to these planes, ψ is the angle between the η-axis and the normal to plane $zO\zeta$. The direction cosines of the normal-to-plane $zO\zeta$ are evidently $\sin\phi$, $-\cos\phi$, 0. Hence, by the scalar-product relation for vectors,

$$\cos\psi = l_2 \sin\phi - m_2 \cos\phi \qquad \text{(b)}$$

Also,

$$l_2 l_3 + m_2 m_3 + n_2 n_3 = 0 \qquad \text{(c)}$$

and

$$l_2^2 + m_2^2 + n_2^2 = 1 \qquad \text{(d)}$$

The quantities (l_3, m_3, n_3) may be eliminated from Eq. (c) by means of Eq. (a). Then Eqs. (b), (c), and (d) may be solved for (l_2, m_2, n_2). Thus the direction cosines (l_2, m_2, n_2) are expressed in terms of (θ, ϕ, ψ). Then (l_1, m_1, n_1) are expressed in terms of (θ, ϕ, ψ) by Eq. (7-14). The results are assembled in Table 7-2.

TABLE 7-2

	x	y	z
ξ	$\sin \phi \sin \psi$ $- \cos \theta \cos \phi \cos \psi$	$- \cos \phi \sin \psi$ $- \cos \theta \sin \phi \cos \psi$	$\sin \theta \cos \psi$
η	$\sin \phi \cos \psi$ $+ \cos \theta \cos \phi \sin \psi$	$- \cos \phi \cos \psi$ $+ \cos \theta \sin \phi \sin \psi$	$- \sin \theta \sin \psi$
ζ	$\sin \theta \cos \phi$	$\sin \theta \sin \phi$	$\cos \theta$

It may be verified that the sum of the squares of the terms in any row or any column of Table 7-2 is 1 and that the sum of the products of corresponding terms in any two rows or any two columns is 0. Also, the vector product of any two rows yields the other row, and the determinant of the matrix is 1.

If $\theta = 0$, Table 7-2 reduces to

TABLE 7-3

	x	y	z
ξ	$- \cos (\phi + \psi)$	$- \sin (\phi + \psi)$	0
η	$\sin (\phi + \psi)$	$- \cos (\phi + \psi)$	0
ζ	0	0	1

Table 7-3 shows that if $\theta = 0$ the coordinates ϕ and ψ are not determined uniquely by the direction cosines given in Table 7-1. Also, Fig. 7-6 shows that, if $\theta = 0$, angle ϕ is indeterminate. Hence, if the axes (ξ, η, ζ) are scribed in a rigid body that rotates about the origin, the Euler angles are not suitable coordinates if θ takes the value zero. From a computational standpoint, equations involving the Euler angles are poorly conditioned if θ is small. Therefore, the Euler angles are unsuitable for studying small oscillations in the neighborhood of the value $\theta = 0$. According to the definition given in Sec. 1-2, the Euler angles cease to be regular coordinates when $\theta = 0$, since there is no longer a one-to-one correspondence between values of (θ, ϕ, ψ) and configurations of the body.

Relation of Euler's Angles to Angular Velocity. It is known that the motion of a rigid body with one fixed point is a vectorial angular velocity $\boldsymbol{\omega}$ about an axis through the fixed point. The vector $\boldsymbol{\omega}$ is conveniently represented by its projections (ω_1, ω_2, ω_3) on the (ξ, η, ζ) axes. It is desirable to express (ω_1, ω_2, ω_3) in terms of the time derivatives ($\dot{\theta}$, $\dot{\phi}$, $\dot{\psi}$).

Let ($\mathbf{a}$, $\mathbf{b}$, $\mathbf{c}$) be unit vectors along the (ξ, η, ζ) axes. It follows from the definition of the vector angular velocity (49) that

$$\dot{\mathbf{a}} = \boldsymbol{\omega} \times \mathbf{a}, \qquad \dot{\mathbf{b}} = \boldsymbol{\omega} \times \mathbf{b}, \qquad \dot{\mathbf{c}} = \boldsymbol{\omega} \times \mathbf{c}$$

Consequently, since $\boldsymbol{\omega} = \mathbf{a}\omega_1 + \mathbf{b}\omega_2 + \mathbf{c}\omega_3$,

$$\dot{\mathbf{a}} = \mathbf{b}\omega_3 - \mathbf{c}\omega_2, \qquad \dot{\mathbf{b}} = \mathbf{c}\omega_1 - \mathbf{a}\omega_3, \qquad \dot{\mathbf{c}} = \mathbf{a}\omega_2 - \mathbf{b}\omega_1 \qquad \text{(e)}$$

Since the (x, y, z) components of the vectors $(\mathbf{a}, \mathbf{b}, \mathbf{c})$ are the respective rows in Table 7-1, we obtain from Eq. (e)

$$\dot{l}_1 = \omega_3 l_2 - \omega_2 l_3, \qquad \dot{l}_2 = \omega_1 l_3 - \omega_3 l_1, \qquad \dot{l}_3 = \omega_2 l_1 - \omega_1 l_2$$
$$\dot{m}_1 = \omega_3 m_2 - \omega_2 m_3, \qquad \dot{m}_2 = \omega_1 m_3 - \omega_3 m_1, \qquad \dot{m}_3 = \omega_2 m_1 - \omega_1 m_2 \quad \text{(f)}$$
$$\dot{n}_1 = \omega_3 n_2 - \omega_2 n_3, \qquad \dot{n}_2 = \omega_1 n_3 - \omega_3 n_1, \qquad \dot{n}_3 = \omega_2 n_1 - \omega_1 n_2$$

Also, comparing Tables 7-1 and 7-2, we obtain by differentiation

$$\dot{l}_1 = \dot{\theta} n_1 \cos \phi - m_1 \dot{\phi} + l_2 \dot{\psi}, \qquad \dot{m}_1 = \dot{\theta} n_1 \sin \phi + l_1 \dot{\phi} + m_2 \dot{\psi}$$
$$\dot{n}_1 = \dot{\theta} \cos \theta \cos \psi + n_2 \dot{\psi}, \qquad \dot{l}_2 = \dot{\theta} n_2 \cos \phi - m_2 \dot{\phi} - l_1 \dot{\psi} \qquad \text{(g)}$$
$$\dot{m}_2 = \dot{\theta} n_2 \sin \phi + l_2 \dot{\phi} - m_1 \dot{\psi}, \qquad \dot{n}_2 = -\dot{\theta} \cos \theta \sin \psi - \dot{\psi} \sin \theta \cos \psi$$
$$\dot{l}_3 = \dot{\theta} \cos \theta \cos \phi - \dot{\phi} \sin \theta \sin \phi, \qquad \dot{m}_3 = \dot{\theta} \cos \theta \sin \phi + \dot{\phi} \sin \theta \cos \phi$$
$$\dot{n}_3 = -\dot{\theta} \sin \theta$$

The quantities $\dot{l}_1$, $\dot{m}_1$, etc., may be eliminated from Eqs. (g) by means of Eqs. (f). Also, the quantities l_1, m_1, etc., may be eliminated by means of Table 7-2. Thus, nine equations are obtained that relate ω_1, ω_2, ω_3 to $\dot{\theta}$, $\dot{\phi}$, $\dot{\psi}$. Naturally, some of them are redundant. Solving three of them for ω_1, ω_2, ω_3, we obtain

$$\omega_1 = -\dot{\theta} \sin \psi + \dot{\phi} \sin \theta \cos \psi$$
$$\omega_2 = -\dot{\theta} \cos \psi - \dot{\phi} \sin \theta \sin \psi \qquad \text{(7-15)}$$
$$\omega_3 = \dot{\phi} \cos \theta + \dot{\psi}$$

By substituting Eqs. (7-15) into Eqs. (f) and expressing the direction cosines by Table 7-2, we may verify that Eqs. (g) are all satisfied. A different derivation of Eqs. (7-15), based on elementary geometrical arguments, is also available (62). Equations (7-15) are known as "Euler's kinematical equations."

7-9. EULER'S DYNAMICAL EQUATIONS.

Let the axes (ξ, η, ζ) be the principal axes of inertia of a rigid body that has a fixed point at the origin of these axes. Let the body be subjected to a variable external couple $\mathbf{M}$ with components (M_1, M_2, M_3) on the (ξ, η, ζ) axes, respectively. If the Euler angles are given infinitesimal virtual increments $(\delta\theta, \delta\phi, \delta\psi)$, the virtual work of the couple $\mathbf{M}$ is

$$\delta W = Q_1 \, \delta\theta + Q_2 \, \delta\phi + Q_3 \, \delta\psi \qquad \text{(a)}$$

The coefficients (Q_1, Q_2, Q_3) are the components of generalized force (Sec. 1-7).

Any displacement of a rigid body with one fixed point can be effected by a rotation about an axis (90). Furthermore (67), an infinitesimal angular displacement is a vector $\delta\boldsymbol{\beta}$. The components of this vector on the axes (ξ, η, ζ) are denoted by $\delta\beta_1, \delta\beta_2, \delta\beta_3$. If the angular displacement occurs in a time interval δt, $\delta\boldsymbol{\beta} = \boldsymbol{\omega}\delta t$, or $\delta\beta_1 = \omega_1\,\delta t$, $\delta\beta_2 = \omega_2\delta t$, etc., where $\boldsymbol{\omega}$ is the angular velocity. Consequently, by Eqs. (7-15),

$$\delta\beta_1 = -\sin\psi\,\delta\theta + \sin\theta\cos\psi\,\delta\phi$$
$$\delta\beta_2 = -\cos\psi\,\delta\theta - \sin\theta\sin\psi\,\delta\phi \tag{b}$$
$$\delta\beta_3 = \cos\theta\,\delta\phi + \delta\psi$$

Since the virtual work of the couple $\mathbf{M}$ may be expressed in the form, $\delta W = \mathbf{M}\cdot\delta\boldsymbol{\beta} = M_1\,\delta\beta_1 + M_2\,\delta\beta_2 + M_3\,\delta\beta_3$, we obtain from Eqs. (a) and (b)

$$Q_1 = -M_1\sin\psi - M_2\cos\psi$$
$$Q_2 = M_1\sin\theta\cos\psi - M_2\sin\theta\sin\psi + M_3\cos\theta \tag{c}$$
$$Q_3 = M_3$$

The principal moments of inertia of the body are denoted by (A, B, C). These are the respective moments of inertia about the (ξ, η, ζ) axes. By the theory of dynamics of rigid bodies (49), the kinetic energy of the body is

$$T = \tfrac{1}{2}(A\omega_1{}^2 + B\omega_2{}^2 + C\omega_3{}^2) \tag{7-16}$$

By Eq. (7-10), the Lagrange equation corresponding to ψ is

$$\frac{d}{dt}\frac{\partial T}{\partial\dot\psi} - \frac{\partial T}{\partial\psi} = Q_3 \tag{d}$$

Equations (7-15) and (7-16) yield $T/\partial\dot\psi = C\omega_3$ and $\partial T/\partial\psi = A\omega_1\omega_2 - B\omega_2\omega_1$. Hence Eqs. (c) and (d) yield

$$C\dot\omega_3 - (A - B)\omega_1\omega_2 = M_3 \tag{e}$$

Since the labeling of the principal axes of inertia is arbitrary, analogous relations hold for ω_1 and ω_2. The complete set of equations is

$$A\dot\omega_1 - (B - C)\omega_2\omega_3 = M_1$$
$$B\dot\omega_2 - (C - A)\omega_3\omega_1 = M_2 \tag{7-17}$$
$$C\dot\omega_3 - (A - B)\omega_1\omega_2 = M_3$$

Equations (7-17) are known as "Euler's dynamical equations." Instead of inferring two of these equations from symmetry conditions, we may derive them from the other two Lagrange equations, just as Eq. (e) was derived. However, the algebra is more complicated than in the derivation of Eq. (e). Elimination of $\dot\omega_3$ must be accomplished with the aid of Eq. (e).

For a stationary body, $\omega_1 = \omega_2 = \omega_3 = 0$. Accordingly, Eqs. (7-17) yield the equilibrium criterion $\mathbf{M} = 0$ for a stationary body.

7-10. MOTION OF AN IDEAL TOP. The motion of a top on a horizontal surface may be treated conveniently with the aid of Hamilton's equations. The tip of the top is considered to remain at a fixed point. This condition requires the existence of a horizontal reactive force by the surface on which the top spins. The axes (ξ, η, ζ) are fixed in the top with the origin at the tip O; the ζ-axis is the axis of symmetry (Fig. 7-7). The moments of inertia of the top about the (ξ, η, ζ) axes are (A, B, C), respectively. Because of symmetry, $A = B$. The components of the angular velocity $\boldsymbol{\omega}$ on the (ξ, η, ζ) axes are $(\omega_1, \omega_2, \omega_3)$. Since $A = B$, Eq. (7-16) yields

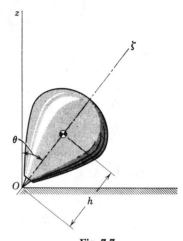

Fig. 7-7

$$T = \tfrac{1}{2}(A\omega_1{}^2 + A\omega_2{}^2 + C\omega_3{}^2) \quad \text{(a)}$$

The Euler angles (θ, ϕ, ψ) are adopted as generalized coordinates. The angle of the axis of the top (Fig. 7-7) with the vertical is θ. The angle of plane $zO\zeta$ with a fixed vertical plane is ϕ. The angle of rotation of the top about the ζ-axis is ψ (Fig. 7-6). Eliminating $\omega_1, \omega_2, \omega_3$ from Eq. (a) by means of Eqs. (7-15), we obtain

$$T = \tfrac{1}{2}A(\dot\theta^2 + \dot\phi^2 \sin^2\theta) + \tfrac{1}{2}C(\dot\psi + \dot\phi \cos\theta)^2 \quad \text{(b)}$$

The potential energy of the top is

$$V = mgh \cos\theta \quad \text{(c)}$$

where h is the distance from the origin O to the center of mass (Fig. 7-7).

The Lagrangian function is $L = T - V$. Since V does not depend on the time derivatives $(\dot\theta, \dot\phi, \dot\psi)$, $\partial L/\partial\dot\theta = \partial T/\partial\dot\theta$, etc. Consequently, by Eq. (7-5), the components of generalized momentum are $p_1 = \partial T/\partial\dot\theta$, $p_2 = \partial T/\partial\dot\phi$, $p_3 = \partial T/\partial\dot\psi$. Hence

$$\begin{aligned} p_1 &= A\dot\theta \\ p_2 &= (A \sin^2\theta + C \cos^2\theta)\dot\phi + C\dot\psi \cos\theta \\ p_3 &= C(\dot\psi + \dot\phi \cos\theta) \end{aligned} \quad \text{(d)}$$

Since the reference frame is Newtonian, the Hamiltonian function is $H = T + V$. Consequently, if $(\dot{\theta}, \dot{\phi}, \dot{\psi})$ are eliminated from Eq. (b) by means of Eqs. (d),

$$H = \frac{p_1^2}{2A} + \frac{(p_2 - p_3 \cos \theta)^2}{2A \sin^2 \theta} + \frac{p_3^2}{2C} + mgh \cos \theta \tag{e}$$

Hamilton's equations (7-9) are

$$\dot{\theta} = \frac{\partial H}{\partial p_1}, \qquad \dot{\phi} = \frac{\partial H}{\partial p_2}, \qquad \dot{\psi} = \frac{\partial H}{\partial p_3}$$

$$\dot{p}_1 = -\frac{\partial H}{\partial \theta}, \qquad \dot{p}_2 = -\frac{\partial H}{\partial \phi}, \qquad \dot{p}_3 = -\frac{\partial H}{\partial \psi}$$

The first three of these equations give relations that are equivalent to Eqs. (d). The last three yield

$$\dot{p}_1 = \frac{-(p_2 - p_3 \cos \theta)(p_3 - p_2 \cos \theta)}{A \sin^3 \theta} + mgh \sin \theta \tag{f}$$

$$\dot{p}_2 = 0, \qquad \dot{p}_3 = 0 \tag{g}$$

Equations (g) yield $p_2 = AM = $ constant and $p_3 = CK = $ constant. Hence by Eq. (d)

$$\dot{\psi} + \dot{\phi} \cos \theta = K$$

$$\dot{\phi} \sin^2 \theta + bK \cos \theta = M, \qquad b = \frac{C}{A} \tag{h}$$

By Eqs. (d), $\dot{p}_1 = A\ddot{\theta}$. To integrate Eq. (f), set

$$\dot{\theta} = v, \qquad \ddot{\theta} = \frac{dv}{dt} = \frac{dv}{d\theta}\frac{d\theta}{dt} = v\frac{dv}{d\theta}$$

Introducing these relations into Eq. (f) and noting that $p_2 = AM$ and $p_3 = CK$, we obtain by integration

$$\dot{\theta}^2 = \frac{-(M^2 + b^2K^2 - 2bKM \cos \theta)}{\sin^2 \theta} - a \cos \theta + \text{constant} \tag{i}$$

where $a = 2mgh/A$. Writing the additive constant in Eq. (i) in the form $N + b^2K^2$ and introducing the substitution $u = \cos \theta$, we obtain from Eq. (i)

$$\dot{u}^2 = (1 - u^2)(N - au) - (M - bKu)^2 \tag{j}$$

Also, Eqs. (h) yield

$$\dot{\psi} + u\dot{\phi} = K \tag{k}$$

$$\dot{\phi}(1 - u^2) + bKu = M \tag{l}$$

Equations (j), (k), and (l) are three first-order differential equations that determine θ, ϕ, and ψ as functions of t. Equation (k) merely asserts that the component of angular velocity on the axis of the top is constant, since, by Eqs. (7-15) and (k), $\omega_3 = K$.

Equation (j) is separable, but the integral cannot be evaluated in terms of elementary functions. However, the general nature of the solution is known (62). It may be shown that u is a periodic function of t; hence $\theta(t)$ is a periodic function. After $u(t)$ is determined, the function $\phi(t)$ is determined by Eq. (l).

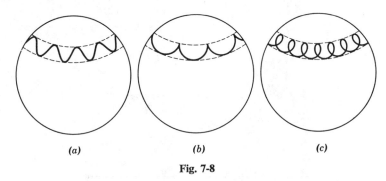

(a)   (b)   (c)

Fig. 7-8

The motion may be visualized by means of the curve that the axis of the top traces on a fixed spherical surface with center at the tip of the top. The relations $\theta = \theta(t)$, $\phi = \phi(t)$ are parametric equations of this curve. There are three possible types of curves, depending on the initial conditions (Fig. 7-8).

The motion may be studied with the aid of certain approximations if the fluctuations of θ are small. It is then convenient to set $\theta = \theta_0 + \alpha$, where θ_0 is the minimum value of θ. Accordingly, α is non-negative. Attention is confined to the case represented by Fig. (7-8b), in which a row of cusps appears. In this case $\dot{\theta} = \dot{\phi} = 0$ when $\theta = \theta_0$. Then, by Eqs. (h), $bK \cos \theta_0 = M$. Hence Eqs. (h) yield

$$\dot{\phi} \sin^2 \theta = bK(\cos \theta_0 - \cos \theta) \tag{m}$$

Likewise, Eq. (i) reduces to

$$\dot{\theta}^2 = (\cos \theta_0 - \cos \theta)\left[\frac{-b^2 K^2(\cos \theta_0 - \cos \theta)}{\sin^2 \theta} + a\right] \tag{n}$$

Because of the initial condition, $\dot{\theta} = 0$ when $\theta = \theta_0$, there is no additive constant of integration in this equation.

Since $\theta \geq \theta_0$, the bracketed expression in Eq. (n) must be positive. Consequently, if K^2 is large compared to a (i.e., if ω_3 is large, as usually happens), $\cos \theta_0 - \cos \theta$ is small. Consequently, α is small. Therefore, since

$$\cos \theta_0 - \cos \theta = 2 \sin \tfrac{1}{2}(\theta + \theta_0) \sin \tfrac{1}{2}(\theta - \theta_0)$$

we obtain the approximation

$$\cos \theta_0 - \cos \theta = \alpha \sin \theta_0 \qquad\qquad (o)$$

Then, since $\dot{\alpha} = \dot{\theta}$, Eq. (n) yields, when the defining equations for a and b are introduced,

$$\frac{d\alpha}{dt} = \frac{C}{A}\left[\alpha\left(\frac{2mAgh \sin \theta_0}{C^2} - K^2\alpha\right)\right]^{\frac{1}{2}}$$

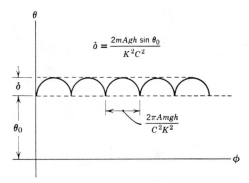

Fig. 7-9

This differential equation is separable and integrable. Integration yields

$$\alpha = \frac{2mAgh \sin \theta_0}{K^2C^2} \sin^2 \frac{CKt}{2A} \qquad\qquad (p)$$

The constant of integration has been eliminated by the condition $\alpha = 0$ when $t = 0$.

Equation (m) now yields

$$\dot{\phi} = \frac{2mgh}{KC} \sin^2 \frac{CKt}{2A} \qquad\qquad (q)$$

Integration of Eq. (q) yields

$$\phi = \frac{mgh}{KC}\left(t - \frac{A}{CK} \sin \frac{CKt}{A}\right) \qquad\qquad (r)$$

In the (ϕ, θ) plane, Eqs. (p) and (r) are parametric equations of the curve shown in Fig. 7-9. This curve agrees with Fig. 7-8b; it is a modified cycloid.

The angle ϕ is called the angle of precession, and $\dot{\phi}$ is called the speed of precession. Equation (q) shows that $\dot{\phi}$ fluctuates between 0 and $2mgh/KC$.

The mean speed of precession is $\Omega = \lim \phi/t$ as t becomes infinite Consequently, by Eq. (r),

$$\Omega = \frac{mgh}{KC} \tag{s}$$

This equation shows that the mean speed of precession is inversely proportional to the rate of spin K.

7-11. THE GYROSCOPE.   A rigid body with one fixed point is called a "gyro" (18). The top discussed in Sec. 7-10 is an example of a gyro. In devices such as the gyrocompass and gyrostabilizer the gyro is usually mounted on a shaft with bearings on the innermost of two hinged rings. called "gimbals" (Fig. 7-10). Thus the center point of the shaft of the gyro is fixed. The ring that supports the shaft of the gyro is called the "inner gimbal," and the other ring is called the "outer gimbal." The

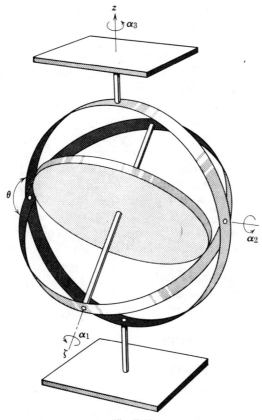

Fig. 7-10

entire apparatus is commonly called a "gyroscope" or "gyrostat." The gimbals are merely supports; their inertia and the friction of their bearings usually exert undesirable perturbations on the motion of the gyro.

The configuration of a gyroscope is conveniently specified by the angular displacement α_1 of the gyro relative to the inner gimbal, the angular displacement α_2 of the inner gimbal relative to the outer gimbal, and the angular displacement α_3 of the outer gimbal relative to the fixed reference frame.* These angles are identical to the Euler angles (Sec. 7-8), in fact, $\alpha_1 = \psi$, $\alpha_2 = \theta$, and $\alpha_3 = \phi$. These relations are apparent from Fig. 7-10. The angle θ has been defined as the angle between the ζ- and z-axes; hence it is identical to the dihedral angle between the planes of the inner and outer gimbals. Accordingly, $\theta = \alpha_2$, provided that the zero reference line of α_2 is chosen properly.

The plane of the ζ- and z-axes will be called plane P. The angle ϕ has been defined as the dihedral angle between plane P and a fixed plane that contains the z-axis. Since plane P is perpendicular to the plane of the outer gimbal, ϕ is accordingly equal to the angle α_3, through which the outer gimbal rotates about the z-axis.

The plane of the inner gimbal is perpendicular to plane P. The angle ψ has been defined to be the dihedral angle between plane P and a rotating meridional plane of the gyro. Consequently, ψ is likewise the dihedral angle between a rotating meridional plane of the gyro and the plane of the inner gimbal; that is, $\psi = \alpha_1$.

In an electrically operated gyroscope a motor mounted on the inner gimbal rotates the gyro at constant angular velocity relative to the inner gimbal; that is, $\dot{\psi} = $ constant. It should be recalled, however, that $\dot{\psi}$ is different from the component ω_3 of the angular velocity on the ζ-axis. By Eqs. (7-15), $\omega_3 = \dot{\phi} \cos \theta + \dot{\psi}$. According to the theory of the frictionless top (Sec. 7-10), $\omega_3 = $ constant. Consequently, for a frictionless gyroscope with massless gimbals, $\omega_3 = $ constant. The condition, $\dot{\psi} = $ constant, which is imposed by an electric motor, consequently does not conform to the motion of an ideal free gyro.

The kinetic energy of any gyro is given by Eqs. (7-15) and (7-16). The kinetic energy of the outer gimbal is

$$T_o = \tfrac{1}{2} I \dot{\phi}^2 \qquad\qquad\qquad (a)$$

where I is the moment of inertia of the outer gimbal about the axis of its fixed hinges.

If the ξ-axis is the hinge line of the inner gimbal and the ζ-axis is the axis of symmetry of the gyro, the Euler angles of the inner gimbal are

---

* These variables were used by Poritsky, who studied effects of torques applied to the gyro and the gimbals and effects of inertia of the gimbals (132).

θ, ϕ, $\pi/2$ (see Figs. 7-6 and 7-10). Consequently, by Eqs. (7-15), the components of angular velocity of the inner gimbal on the (ξ, η, ζ) axes of that gimbal are $\omega_1 = -\dot{\theta}$, $\omega_2 = -\dot{\phi} \sin \theta$, $\omega_3 = \dot{\phi} \cos \theta$. Ordinarily, the principal axes of inertia of the inner gimbal are the hinge line of the inner gimbal (ξ-axis), the axis of symmetry of the gyro (ζ-axis), and a line perpendicular to the plane of the inner gimbal (η-axis). Letting the moments of inertia of the inner gimbal with respect to these (ξ, η, ζ) axes be (A', B', C'), respectively, we have the following formula for the kinetic energy of the inner gimbal:

$$T_i = \tfrac{1}{2}(A'\omega_1^2 + B'\omega_2^2 + C'\omega_3^2)$$

With the preceding expressions for ω_1, ω_2, ω_3, this yields

$$T_i = \tfrac{1}{2}(A'\dot{\theta}^2 + B'\dot{\phi}^2 \sin^2 \theta + C'\dot{\phi}^2 \cos^2 \theta) \qquad (b)$$

The gimbals are ordinarily balanced so that their centers of mass lie at the fixed point on the axis of the gyro. Then the potential energy of the gimbals is zero. Consequently, if the center of mass of the gyro lies at distance h from the fixed point on the axis of the gyro and the mass of the gyro is m, the potential energy of the entire apparatus due to the action of gravity is

$$V = mgh \cos \theta \qquad (c)$$

Accordingly, in view of Eq. (b) of Sec. 7-10, the Lagrangian function is

$$L = \tfrac{1}{2}A(\dot{\theta}^2 + \dot{\phi}^2 \sin^2 \theta) + \tfrac{1}{2}C(\dot{\psi} + \dot{\phi} \cos \theta)^2 + \tfrac{1}{2}I\dot{\phi}^2$$
$$+ \tfrac{1}{2}(A'\dot{\theta}^2 + B'\dot{\phi}^2 \sin^2 \theta + C'\dot{\phi}^2 \cos^2 \theta) - mgh \cos \theta \qquad (7\text{-}18)$$

where (A, A, C) are the principal moments of inertia of the gyro.

Frequently torques are applied to the gimbals at the hinges because of friction or extraneous effects. If these torques are R_θ and R_ϕ, the virtual work that they perform when the coordinates (θ, ϕ) receive variations ($\delta\theta$, $\delta\phi$) is $R_\theta\, \delta\theta + R_\phi\, \delta\theta$. Consequently, ($R_\theta$, R_ϕ) are the components of generalized force that appear in Lagrange's equation (7-12) and Hamilton's equations (7-13).

With the foregoing equations, it is easy to write the dynamical equations of the gyroscope by means of Lagrange's equation or Hamilton's equations. If the inertia of the gimbals and the torques (R_θ, R_ϕ) are neglected, the theory is identical to that of the top (Sec. 7-10).

PROBLEMS

1. The lowest corner of the homogeneous cubical block moves along the x-axis (Fig. P7-1). Adopt x and θ as generalized coordinates. Verify that the kinetic energy is a homogeneous quadratic function of the components of generalized velocity.

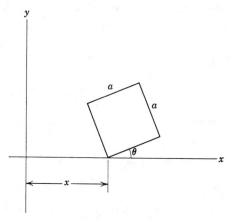

Fig. P7-1

2. A homogeneous solid ball of mass m and radius r rolls outward on an oscillating seesaw that forms an angle θ with the horizontal. The distance from the hinge of the seesaw to the point where the ball touches the board is x. The moment of inertia of the board about its hinge is I. Determine the kinetic energy of the system and show that it is a quadratic form in $\dot{x}$ and θ.

3. Show that, for a statical system, Hamilton's principle reduces to the principle of virtual work.

4. A simple pendulum has a damper in the hinge that causes a resisting moment proportional to the cube of the angular velocity. Derive the differential equation of free oscillations by means of Hamilton's principle.

5. Neglecting friction and taking the moment of inertia of the pulley to be I, determine $x(t)$ for Atwood's machine by means of Lagrange's equation (see Fig. P7-5).

6. For the compound Atwood machine, the two pulleys have the same moment of inertia I. Masses of the parts are indicated in Fig. P7-6. Let the generalized coordinates be the angles θ and ϕ through which the pulleys rotate. Neglecting friction, derive the differential equations for θ and ϕ by means of Lagrange's equations.

7. A particle of mass m that moves in space is located by spherical coordinates (r, θ, ϕ), where θ is colatitude and ϕ is longitude. The particle is attracted toward the origin by a force F that is a function of r alone. Write Lagrange's equations for r, θ, ϕ. Let $\phi = 0$ for $t = 0$ and simplify the differential equations accordingly. Show that the particle describes a plane curve. Show that a line drawn from the origin to the particle sweeps out equal areas in equal times.

8. A horizontal platform of mass M is suspended by strings of length l (see Fig. P7-8). As the platform swings, a homogeneous ball of mass m and radius r rolls on it. Using x and θ as generalized coordinates, derive the

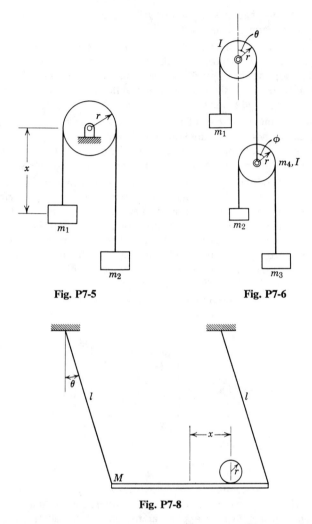

Fig. P7-5   Fig. P7-6

Fig. P7-8

differential equations for large oscillations by means of Lagrange's equations. Reduce the equation corresponding to x to the first order. Linearize the differential equations by supposing that θ is small.

9. The string of a simple pendulum is elastic. The spring constant for the string is k, and the mass of the bob is m. The free length of the string is l. Adopt polar coordinates (r, θ) with origin at the point of suspension as generalized coordinates for the bob. Neglecting the mass of the string, write the differential equations of motion by means of Lagrange's equations.

10. Write Lagrange's equation for an arbitrary conservative system with one degree of freedom that is located with respect to a Newtonian reference

frame. Show that this equation yields the conclusion that the total mechanical energy is constant.

11. A simple pendulum of length l and mass m is suspended from the rim of a wheel of radius r which rotates with constant angular velocity ω in a vertical plane. Adopting the angle θ that the pendulum forms with the vertical as the generalized coordinate, derive the differential equation for $\theta(t)$ by means of Lagrange's equation.

12. A homogeneous solid ball of radius r oscillates on a horizontal platform that executes horizontal simple harmonic motion defined by the equation $s = s_0 \sin \omega t$. Let x be the displacement of the center of the ball relative to the platform. Derive the function $x(t)$ by means of Lagrange's equation.

13. A small body of mass m falls to earth from a great height while remaining at constant latitude α. The attraction of gravity is $F = km/r^2$, where r is the distance from the center of the earth. Let the distance r and the longitude ϕ relative to the earth be generalized coordinates. Taking into account the angular velocity ω of the earth, write the differential equations of motion by means of Lagrange's equations. Eliminate ϕ. Transform the resulting differential equation by regarding r as a function of ϕ, and integrate the equation with the aid of the substitution, $r = 1/u$.

14. Solve Probs. 5, 6, 7, 8, and 9 by Hamilton's equations.

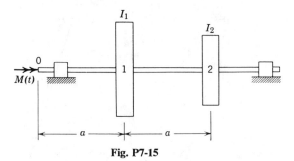

Fig. P7-15

15. Figure P7-15 represents two flywheels that are keyed to a uniform elastic shaft. The moments of inertia of the flywheels are I_1 and I_2. A time-dependent torque $M(t)$ is applied at the end of the shaft. The angular displacements at sections 0, 1, and 2 are $\theta_0(t)$, $\theta_1(t)$, and $\theta_2(t)$. The torsional stiffness of the shaft is GJ. Regard θ_0, θ_1, θ_2 as generalized coordinates and set up the differential equations of motion by means of Lagrange's equations. Eliminate θ_0 from the equations. Hence show that the equations yield the conclusion that M is equal to the rate of change of moment of momentum of the system. Neglect inertia of the shaft.

16. A particle of mass m moves in the (x, y) plane under the action of a force with components (F_x, F_y). The coordinate system (x, y) rotates in its plane with angular velocity ω relative to a Newtonian reference frame. Adopting (x, y) as generalized coordinates, derive the differential equations of motion

by means of Lagrange's equations. Write the equations of motion in the form $P_x = m\ddot{x}$, $P_y = m\ddot{y}$ and interpret the several components of the apparent force $(P_{q'}, P_y)$.

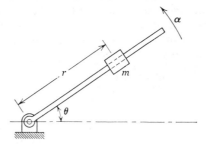

Fig. P7-17

17. The mass m slides on the rigid rod which rotates in a horizontal plane with constant angular acceleration α (Fig. P7-17). There is a constant frictional force F. By means of Lagrange's equation, derive the differential equation for r, letting $\dot{\theta} = 0$ for $t = 0$.

18. Let $(\mathbf{a}, \mathbf{b}, \mathbf{c})$ be unit vectors along the (ξ, η, ζ) axes (see Table 7-1, Sec. 7-8). Prove that $(\mathbf{a} - \mathbf{i}) \cdot (\mathbf{b} - \mathbf{j}) \times (\mathbf{c} - \mathbf{k}) = 0$. Hence prove that the direction cosines in Table 7-1 satisfy the relation,

$$\begin{vmatrix} l_1 - 1 & m_1 & n_1 \\ l_2 & m_2 - 1 & n_2 \\ l_3 & m_3 & n_3 - 1 \end{vmatrix} = 0$$

19. Show that all two-rowed minors of the determinant in Prob. 18 cannot be zero unless all nine terms in the determinant are zero. Hence, by using the result of Prob. 18, show that there is a unique line L determined by $x = \xi$, $y = \eta$ $z = \zeta$. Show that a suitable rotation of the body about line L carries the axes (ξ, η, ζ) into the respective axes (x, y, z).

20. A rigid body has one particle fixed at the origin. Initially, the Euler angles are $\theta = 30°$ $\phi = 0$, $\psi = 0$. After a displacement of the body, $\theta = 90°$, $\phi = 60°$, $\psi = 120°$. Using the result of Prob. 19, determine the fixed axis L about which the body could be rotated to effect this displacement.

21. Determine the Euler angles corresponding to the following table of direction cosines:

	x	y	z
ξ	$\frac{2}{3}$	$\frac{2}{3}$	$\frac{1}{3}$
η	0	$1/\sqrt{5}$	$-2/\sqrt{5}$
ζ	$-\sqrt{5}/3$	$4\sqrt{5}/15$	$2\sqrt{5}/15$

22. Derive Eqs. (g) and (7-15) of Sec. 7-8.

23. Derive the first two of Eqs. (7-17) by means of Lagrange's equations.

24. A body is mounted on a rigid shaft with fixed bearings. The principal axes of inertia of the body through a point O on the axis of the shaft form angles (α, β, γ) with the axis of the shaft. The angular speed is ω. By Euler's dynamical equations, determine the components (M_1, M_2, M_3) of the moment M that the shaft exerts on the body.

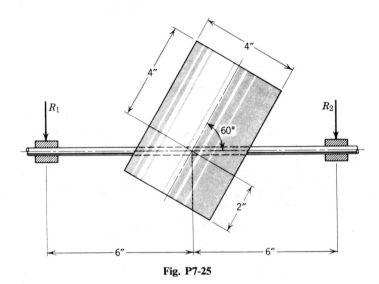

Fig. P7-25

25. Figure P7-25 represents a homogeneous cylindrical body that is mounted obliquely on a shaft that rotates with angular velocity 40 radians/sec. The body weighs 12 lb, and $g = 386$ in./sec². Neglecting the dead load on the bearings, compute R_1 and R_2 with the aid of Euler's dynamical equations.

26. Suppose that the body described in Prob. 25 is motionless and that a torque of 100 lb-in. is applied suddenly to the shaft. Compute the immediate reaction of either bearing, neglecting dead load on the bearings.

27. Derive Eqs. (h) of Sec. 7-10 by means of Lagrange's equations.

28. Suppose that a motor is installed in a gyroscope to maintain $\dot{\psi}$ constant. Neglecting friction and inertia of the gimbals, set $\dot{\psi} = \omega$ and derive the differential equations for θ and ϕ by means of Lagrange's equations. Determine $\dot{\phi}$ for $\theta = 90°$. Use notations of Sec. 7-10.

29. The center of mass of a gyro lies at the fixed point on the axis of the shaft. Suppose that the bearings are frictionless but that there are linear springs with constant k in the hinges of the gimbals that tend to restore them to the configuration for which $\theta = \pi/2$ and $\phi = 0$. Taking the inertia of the gimbals into account, derive the differential equations of motion by means of

Lagrange's equations. Show that $\omega_3 = $ constant. Linearize the equations by supposing that ϕ does not differ much from 0 and θ does not differ much from $\pi/2$.

30. A sphere of radius a rolls on a plane. Let (x, y) be rectangular coordinates of the point of contact between the sphere and the plane and let (θ, ϕ, ψ) be Euler angles of the sphere. Derive the differential relations that express the condition that any infinitesimal displacement $(dx, dy, d\theta, d\phi, d\psi)$ is performed without slipping.

31. If the lower joint of the double pendulum discussed in Sec. 7-3 is locked, $\theta = \phi$. Equations (f) are not consistent with this condition. Explain why.

8 Theory of vibrations

Pythagoras discovered the importance of dealing with abstractions; and in particular directed attention to number as characterizing the periodicities of notes of music.

A. N. WHITEHEAD

8-1. SYSTEMS WITH TWO DEGREES OF FREEDOM. Although vibrating systems with single degrees of freedom are practically important, the application of energy principles to such systems requires no special consideration. Consequently, attention is directed to systems with several degrees of freedom. The main features of small free vibrations of elastic systems may be illustrated by a simple system with two degrees of freedom.

Consider a thin light cantilever beam with an offset rigid body of mass m attached to its end (Fig. 8-1). The body is fastened to the end of the beam by a rigid vertical bar. The body is considered to possess rotary inertia about its center of mass. Without adding complications to the problem, we may let the beam be tapered. The mass m is assumed to exceed the mass of the beam to such an extent that the beam mass may be neglected. The beam is considered to have much more flexibility for sidewise bending than for vertical bending, so that the vertical component of the motion is negligible. Then the horizontal displacement x_1 of the end of the beam and the rotation x_2 of the vertical bar serve as generalized coordinates (Fig. 8-1). The system bears a crude likeness to the rear part of an airplane. The beam may be considered as the part of the fuselage behind the wing, and the rigid body may be considered as the tail assembly.

The strain energy of the beam due to sidewise bending is $\frac{1}{2}ax_1^2$, where a is a constant. The strain energy due to twisting is $\frac{1}{2}bx_2^2$, where b is a constant. Consequently, if effects of gravity are neglected, the potential energy of the system is

$$V = \tfrac{1}{2}ax_1{}^2 + \tfrac{1}{2}bx_2{}^2 \tag{a}$$

The length r (Fig. 8-1) locates the center of mass of the rigid body. The velocity of the center of mass of the body is $\dot{x}_1 + r\dot{x}_2$. Consequently, the kinetic energy of translation of the body is $\tfrac{1}{2}m(\dot{x} + r\dot{x}_2)^2$. The kinetic

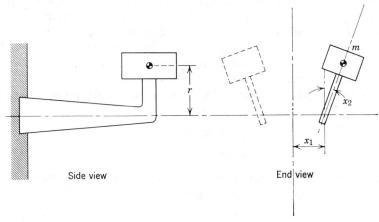

Side view   End view

Fig. 8-1

energy of rotation is $\tfrac{1}{2}m\rho^2\dot{x}_2{}^2$, where ρ is the radius of gyration of the body with respect to its center of mass. Consequently, the kinetic energy of the vibrating system is

$$T = \tfrac{1}{2}m(\dot{x}_1 + r\dot{x}_2)^2 + \tfrac{1}{2}m\rho^2\dot{x}_2{}^2 \tag{b}$$

Accordingly, the Lagrangian function is

$$L = \tfrac{1}{2}m(\dot{x}_1 + r\dot{x}_2)^2 + \tfrac{1}{2}m\rho^2\dot{x}_2{}^2 - \tfrac{1}{2}ax_1{}^2 - \tfrac{1}{2}bx_2{}^2 \tag{c}$$

The Lagrangian equations of motion (7-4) now yield

$$\ddot{x}_1 + r\ddot{x}_2 + Ax_1 = 0$$
$$r\ddot{x}_1 + (r^2 + \rho^2)\ddot{x}_2 + Bx_2 = 0 \tag{d}$$

where $A = a/m$ and $B = b/m$.

Equations (d) are homogeneous linear differential equations with constant coefficients. Their general solution may be obtained readily by well-known methods. In the theory of vibrations especial importance attaches to particular solutions of the form

$$x_1 = z_1 \sin(\omega t - \gamma), \qquad x_2 = z_2 \sin(\omega t - \gamma) \tag{e}$$

in which z_1, z_2, ω, and γ are constants. Such a solution is known as a *natural mode* of vibration (also called an "eigenvibration"). When the system vibrates in a natural mode, all particles perform simple harmonic oscillations that are in phase with each other and that have the same period. Accordingly, all particles attain their maximum velocities simultaneously as they pass through their neutral positions, and they all reach their extreme displacements simultaneously. The constants z_1 and z_2 are known as the *amplitudes* of the motion; the constant ω is called the *angular frequency* (or "circular frequency") of the motion, and the variable $\omega t - \gamma$ is called the *phase* of the motion. Consequently, the initial phase is $-\gamma$. The number of oscillations per second (called the "frequency") is $\omega/2\pi$. Accordingly, the time interval in which an oscillation is performed (called the "period") is $2\pi/\omega$.

To determine the constants z_1, z_2, ω, we substitute Eq. (e) into Eqs. (d). Thus the following equations are obtained:

$$(A - \omega^2)z_1 - r\omega^2 z_2 = 0$$
$$r\omega^2 z_1 + [(r^2 + \rho^2)\omega^2 - B]z_2 = 0 \qquad \text{(f)}$$

Equations (f) yield nonzero values of z_1 and z_2 if, and only if, the constant ω is a root of the determinantal equation,

$$\begin{vmatrix} A - \omega^2 & -r\omega^2 \\ r\omega^2 & (r^2 + \rho^2)\omega^2 - B \end{vmatrix} = 0$$

or

$$\rho^2\omega^4 - [A(r^2 + \rho^2) + B]\omega^2 + AB = 0 \qquad \text{(g)}$$

Setting $r/\rho = x$ and $B/(A\rho^2) = y^2$, we may express the solution of Eq. (g) in the following dimensionless form:

$$\frac{2\omega^2}{A} = 1 + x^2 + y^2 \pm \sqrt{[x^2 + (y - 1)^2][x^2 + (y + 1)^2]} \qquad \text{(h)}$$

Equation (h) provides two real positive values of ω^2; it is known as the "frequency equation."

If ω is determined by Eq. (h), one of the equations in (f) is redundant. Consequently, only the ratio z_2/z_1 is determinate. The first of equations in (f) yields

$$\frac{rz_2}{z_1} = \frac{A}{\omega^2} - 1 \qquad \text{(i)}$$

It should be noted that γ is an arbitrary constant. Also, the values of ω and z_2/z_1 are independent of γ.

Let ω' and ω'' be the positive roots of Eq. (h). Let z_1', z_1'', γ', γ'' be any real constants. Let z_2' be determined by Eq. (i), with $z_1 = z_1'$ and $\omega = \omega'$.

Let z_2'' be determined by Eq. (i), with $z_1 = z_1''$ and $\omega = \omega''$. Then, since the sum of any two solutions of Eq. (d) is again a solution, we obtain

$$x_1 = z_1' \sin (\omega't - \gamma') + z_1'' \sin (\omega''t - \gamma'')$$
$$x_2 = z_2' \sin (\omega't - \gamma') + z_2'' \sin (\omega''t - \gamma'') \tag{j}$$

Since Eqs. (j) contain the four arbitrary constants, z_1', z_1'', γ', γ'', they are the general solution of Eqs. (d). In other words, the general solution of Eqs. (d) is a linear combination of two natural modes.

Pendulum with a Flexible Suspension. A second example that exhibits features different from the preceding problem is illustrated by Fig. 8-2. The hinge of the simple pendulum may move up and down between the frictionless guides. The coordinate x is measured from the position of the block in which the system may remain motionless. The constant of the linear spring is k. The bob of the pendulum and the guided block each have mass m.

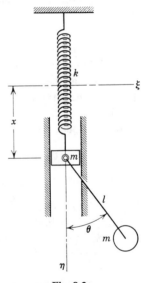

The kinetic energy of the block is $\frac{1}{2}m\dot{x}^2$. To compute the kinetic energy of the bob, we observe that the coordinates of the bob are $\xi = l \sin \theta$, $\eta = x + l \cos \theta$. The kinetic energy of the bob is $\frac{1}{2}m(\dot{\xi}^2 + \dot{\eta}^2)$. Hence the kinetic energy of the system is

$$T = m\dot{x}^2 + \tfrac{1}{2}ml^2\dot{\theta}^2 - ml\dot{\theta}\dot{x} \sin \theta$$

In the theory of small vibrations the coefficients of the generalized velocity components in the expression for the kinetic energy are approximated by their values in the statical equilibrium state. In the present case the

Fig. 8-2

equilibrium state is represented by $\theta = 0$ and $x = 0$, so T is approximated by

$$T = m\dot{x}^2 + \tfrac{1}{2}ml^2\dot{\theta}^2$$

The potential energy is

$$V = -mgx - mg\eta + \tfrac{1}{2}k(x + h)^2$$

where h is the extension of the spring for which the system can remain motionless. Since $kh = 2mg$, this reduces to

$$V = \tfrac{1}{2}kx^2 - mgl \cos \theta + \text{constant}$$

The expression for V is reduced to a quadratic form in (x, θ) by means of the approximation, $\cos \theta = 1 - \frac{1}{2}\theta^2$. Accordingly, if an irrelevant additive constant is discarded,

$$V = \tfrac{1}{2}kx^2 + \tfrac{1}{2}mgl\theta^2$$

Since $L = T - V$, the Lagrange equations yield

$$\ddot{x} + \frac{k}{2m}\,x = 0, \qquad \ddot{\theta} + \frac{g}{l}\,\theta = 0 \qquad\qquad (k)$$

The general solution of these equations is

$$x = A\cos\left(\sqrt{k/2m}\,t - a\right), \qquad \theta = B\cos\left(\sqrt{g/l}\,t - b\right)$$

The corresponding angular frequencies for x and θ are

$$\omega_x = \sqrt{k/2m}, \qquad \omega_\theta = \sqrt{g/l}$$

For a natural mode, $\omega_x = \omega_\theta = \omega$. Evidently this condition is impossible, unless $k/2m = g/l$. The special feature of this system that excludes natural modes is that the coordinates are decoupled; that is, the oscillation of x does not affect the oscillation of θ.

If the exact expressions for T and V are retained in this problem, the Lagrange equations are

$$2\ddot{x} - l\ddot{\theta}\sin\theta - l\dot{\theta}^2\cos\theta + \frac{k}{m}\,x = 0$$

$$l\ddot{\theta} + (g - \ddot{x})\sin\theta = 0 \qquad\qquad (l)$$

Equations (k) are obtained again if $\ddot{x}$ is considered negligible in comparison to g, and all nonlinear terms are discarded from Eq. (l). Equation (l) shows that the system is not completely decoupled. The decoupling results from the linearizing approximations.

8-2. VIBRATIONS OF UNDAMPED SYSTEMS WITH FINITE DEGREES OF FREEDOM.

The phrase "linear vibrations" is used to signify vibrations whose governing differential equations are linear. The examples in Sec. 8-1 illustrate linear vibrations. Linear vibrations scarcely ever occur in nature, but they are approximated closely by many systems if the amplitudes are sufficiently small. The theory of linear vibrations of systems with finite degrees of freedom was developed in a general form by Lagrange. This theory provides one of the most important engineering applications of Lagrange's equations.

Vibrations of a system are said to be free if no energy enters or leaves the system. Vibrations are said to be damped if friction causes dissipation of energy. The terms "damping" and "friction" are used synonymously in vibration theory. The present discussion is restricted to undamped systems. Accordingly, a potential energy function V exists.

A conservative unchecked mechanical system that is referred to a Newtonian reference frame may execute small oscillations about any stable equilibrium configuration $\mathbf{X_0}$. For simplicity, we choose generalized coordinates $(x_1, x_2, \cdots, x_n)$ such that the origin represents the configuration $\mathbf{X_0}$. Also, we choose the zero level of potential energy so that $V = 0$ at the origin. Then, if ΔV represents the increment of V corresponding to a displacement from the origin, $\Delta V = V$. Furthermore, by the principle of virtual work, $\delta V = 0$, since the origin is an equilibrium configuration. Consequently, $V = \frac{1}{2}\delta^2 V + \cdots$. If the oscillations are sufficiently small, only the first term in this series is important. Therefore, as a second-degree approximation [see Eq. (1-25)],

$$V = \frac{1}{2}\sum_{i=1}^{n}\sum_{j=1}^{n}a_{ij}x_i x_j \tag{8-1}$$

in which the coefficients (a_{ij}) are a symmetric matrix of constants. Since the origin is a configuration of stable equilibrium, this quadratic form is positive definite (Appendix, Sec. A-1).

The kinetic energy of the system is a quadratic form in the generalized velocity components [Eq. (7-1)]. In general, the coefficients b_{ij} are functions of the coordinates x_i, but, if the oscillations are small, these functions may be approximated by their values at the origin. In the theory of small oscillations the coefficients b_{ij} are considered to be constants. This condition, and Eq. (8-1) characterize the linear theory of vibrations.

Since the Lagrangian function is $L = T - V$, we obtain, with Eqs. (7-1) and (8-1),

$$L = \frac{1}{2}\sum_{i=1}^{n}\sum_{j=1}^{n}(b_{ij}\dot{x}_i\dot{x}_j - a_{ij}x_i x_j) \tag{8-2}$$

Hence the Lagrangian equations of motion (7-12) are

$$\sum_{j=1}^{n}(b_{ij}\ddot{x}_j + a_{ij}x_j) = R_i; \qquad i = 1, 2, \cdots, n \tag{8-3}$$

The quantities R_i may be specified periodic functions of time. They are then known as the generalized components of the exciting force. If $R_i = 0$, the vibration is free.

It is known from the theory of linear differential equations that the general solution of Eqs. (8-3) is the sum of a particular solution and the complementary solution which is the general solution of the homogeneous differential equations that remain when R_i is set equal to zero. We first consider the complementary solution, or the free vibration of the system.

As in the example in Sec. 8-1, we determine the natural modes by the equation

$$x_j = z_j \sin(\omega t - \gamma) \qquad (8\text{-}4)$$

in which z_j, ω, and γ are constants. Equations (8-3) and (8-4) yield

$$\sum_{j=1}^{n} (a_{ij} - \omega^2 b_{ij}) z_j = 0 \qquad (8\text{-}5)$$

Equation (8-5) determines nonzero values of the z's if, and only if, the determinant of the equations is zero; that is,

$$\det(a_{ij} - \omega^2 b_{ij}) = 0 \qquad (8\text{-}6)$$

where the symbol "det" stands for "determinant." Equation (8-6) is of nth degree in ω^2; it consequently possesses n roots. The symmetry conditions, $a_{ij} = a_{ji}$ and $b_{ij} = b_{ji}$, combined with the conditions that the quadratic forms V and T are positive definite, lead to the conclusion that the roots of Eq. (8-6) are positive (Appendix, Sec. A-4). Consequently, Eq. (8-6) determines n positive values of ω, denoted by $\omega_1, \omega_2, \cdots, \omega_n$. These values are called the "natural frequencies" or the "eigenfrequencies" of the system. The natural frequencies are not necessarily distinct, but the case in which two roots ω_i are equal is exceptional. If this condition occurs in a physical system, it can be eliminated by slight changes of the mass distribution or the spring constants, so that the constants b_{ij} or a_{ij} are altered slightly. Consequently, attention is confined to the case in which no two of the natural frequencies are equal.

Corresponding to any natural frequency, there are certain amplitudes z_i, determined by Eq. (8-5). These amplitudes contain an arbitrary constant factor. Consequently, only the ratios z_i/z_j are determinate. To each natural frequency, there corresponds a natural mode, determined by Eq. (8-4). Arbitrary phase angles γ_i may be assigned to the respective modes. An arbitrary linear combination of the n natural modes contains $2n$ arbitrary constants; therefore, it is the general solution of Eq. (8-3) if $R_i = 0$.

An exceptional case arises if the rank of the matrix $(a_{ij} - \omega_k^2 b_{ij})$ is less than $n - 1$ for one or more of the natural frequencies ω_k (see Appendix, Sec. A-1 for definition of rank of a matrix). If the rank is r, values of $n - r$ of the z's may be assigned arbitrarily, and Eq. (8-5) may be solved for the remaining z's (5). If $r < n - 1$, there is a certain amount of decoupling; that is, some of the x's may oscillate independently of each other. Occasionally, all coordinates are decoupled. This case is illustrated by the second example in Sec. 8-1.

The main practical difficulty that is encountered when the foregoing theory is applied to a system with more than three degrees of freedom is the solution of Eq. (8-6). In matrix notation this equation may be written as $\det (A - \omega^2 B) = 0$, where A and B denote the matrices (a_{ij}) and (b_{ij}). Since $\det B \neq 0$, the matrix B possesses an inverse B^{-1}; that is, $BB^{-1} = B^{-1}B = I$, where I is the unit matrix. Accordingly, the roots ω_i are identical to the roots of the equation $\det (B^{-1}A - \omega^2 I) = 0$. In this determinant the unknowns ω_i occur only on the principal diagonal. Practical numerical methods for finding the roots ω_i of a determinantal equation of this type are discussed in the book by Faddeeva (23).

Normal Coordinates. There is another way of attacking Eqs. (8-3) that is important in physics. Identifying $\dot{x}_i$ as y_i, we may employ the dual-transformation theory of two quadratic forms (Appendix, Sec. A-4). Let (t_{ij}) be the matrix that reduces the two quadratic forms V and T to the canonical form. Then

$$x_i = \sum_{j=1}^{n} t_{ij} u_j, \qquad \dot{x}_i = \sum_{j=1}^{n} t_{ij} v_j, \qquad \det (t_{ij}) \neq 0$$

where u_i and v_i are the new variables. Hence $v_j = \dot{u}_j$. The variables u_i are called "normal coordinates" for the system. According to the theorem in Appendix A-4, the kinetic and potential energies assume the following forms when normal coordinates are used:

$$T = \tfrac{1}{2}(\dot{u}_1{}^2 + \dot{u}_2{}^2 + \cdots + \dot{u}_n{}^2)$$
$$V = \tfrac{1}{2}(\omega_1{}^2 u_1{}^2 + \omega_2{}^2 u_2{}^2 + \cdots + \omega_n{}^2 u_n{}^2)$$

(8-7)

Hence Lagrange's equations (7-12) yield

$$\ddot{u}_i + \omega_i{}^2 u_i = R_i \qquad (8\text{-}8)$$

If R_i is a known function of t, the variables u_i are separated in Eq. (8-8). Integration yields

$$u_i = A_i \sin (\omega_i t - \gamma_i) + f_i(t) \qquad (8\text{-}9)$$

where A_i and γ_i are arbitrary constants and $f_i(t)$ is a particular solution of Eq. (8-8). For free vibrations, $R_i = 0$ and $f_i(t) = 0$.

Resonance. Any periodic shaking force R_i may be resolved by Fourier analysis into harmonic components. Consequently, a shaking force of the following type plays a fundamental role:

$$R_i = C_i \sin \Omega_i t \qquad (8\text{-}10)$$

If Ω_i does not coincide with one of the natural frequencies of the system, a particular solution of Eq. (8-8) is obtained with $u_i = H_i \sin \Omega_i t$. Substituting this relation into Eq. (8-8), we obtain

$$H_i = \frac{C_i}{\omega_i^2 - \Omega_i^2} \tag{8-11}$$

If Ω_i happens to lie near the natural frequency ω_i, H_i is very large; in fact, H_i approaches infinity if Ω_i approaches ω_i. This condition is known as "resonance." In practice, resonance is manifested by violent shaking. Damping prevents the infinite amplitude that is predicted by the preceding theory.

Rayleigh's Principle. When a conservative system vibrates freely, the total mechanical energy is constant; hence $T_0 + V_0 = T_1 + V_1$, where subscript 0 denotes the initial state and subscript 1 denotes any other configuration that the system attains. If the system vibrates in a natural mode, we may let state 0 be the equilibrium configuration about which the system vibrates; then, $V_0 = 0$. Also, we may let state 1 be the configuration of maximum displacement and zero velocity; then, $T_1 = 0$, hence $T_0 = V_1$.

Since the particles execute simple harmonic motions, the mean values of the kinetic and potential energies (i.e., the averages over a long period of time) are $\bar{T} = \frac{1}{2}T_0$ and $\bar{V} = \frac{1}{2}V_1$. Consequently, the equation $T_0 = V_1$ may be written alternatively as $\bar{T} = \bar{V}$. This conclusion is not necessarily restricted to vibrations in natural modes, for, if $f_i(t) = 0$, Eqs. (8-7) and (8-9) yield quite generally

$$\bar{T} = \bar{V} = \tfrac{1}{4}(A_1^2\omega_1^2 + A_2^2\omega_2^2 + \cdots + A_n^2\omega_n^2)$$

since the mean values of $\sin^2(\omega_i t - \gamma_i)$ and $\cos^2(\omega_i t - \gamma_i)$ are $\frac{1}{2}$.

For a conservative system with one degree of freedom, the equation $\bar{T} = \bar{V}$ or $T_0 = V_1$ determines the natural frequency. A system with more than one degree of freedom may be reduced to one degree of freedom by means of an assumption about the mode form; that is, an assumption about the ratios of amplitudes of the various parts. Then the equation $\bar{T} = \bar{V}$ provides an approximation to the frequency of the mode under consideration. For example, it might be assumed that the vibration of a uniform cantilever beam is represented by $y = A(1 - \cos \pi x/2L) \sin \omega t$. Then, by Eq. (2-8), $\bar{V} = \pi^4 EIA^2/128L^3$. Also, $\bar{T} = \frac{1}{4}\rho A^2\omega^2 L(\frac{3}{2} - 4/\pi)$, where ρ is the mass per unit length. Accordingly, the equation $\bar{T} = \bar{V}$ yields $\omega = 3.66\sqrt{EI/\rho L^4}$. This is 4.2 per cent higher than the correct value.

Rayleigh observed that an assumption concerning the mode form is equivalent to the introduction of additional springs in the system. Consequently, it may be expected to raise the frequency of the system. More precisely, the frequency computed with an assumed mode form by means of the equation $\bar{T} = \bar{V}$ is generally too high; it cannot be less than the lowest natural frequency of the system. In essence, this is Rayleigh's principle. Instead of arbitrarily reducing the system to one degree of freedom, we may assume a form of the fundamental mode that contains several parameters. The parameters are then to be chosen to minimize the frequency determined by the equation $\bar{T} = \bar{V}$. A rigorous and general proof of Rayleigh's principle is quite difficult. The matter has been treated thoroughly by Temple and Bickley (82).

8-3. SYSTEMS WITH VISCOUS DAMPING. It is often assumed in the theory of vibrations that the particles experience resistances that are proportional to their velocities. By analogy to the internal frictional forces in fluids, which are proportional to velocity gradients, the damping forces are then said to be of the "viscous type." If $(X_\alpha, Y_\alpha, Z_\alpha)$ are the rectangular coordinates of particle α, the components of the drag force on this particle are $-\mu \dot{X}_\alpha, -\mu \dot{Y}_\alpha, -\mu \dot{Z}_\alpha$, where μ is a constant friction coefficient. If the system is given an infinitesimal virtual displacement $(\delta X_\alpha, \delta Y_\alpha, \delta Z_\alpha)$, the virtual work of all the drag forces is

$$\delta W = -\mu \sum_{\alpha=1}^{p} (\dot{X}_\alpha \, \delta X_\alpha + \dot{Y}_\alpha \, \delta Y_\alpha + \dot{Z}_\alpha \, \delta Z_\alpha)$$

where the sum extends over the p particles of the system. The quantities $(\delta X_\alpha, \delta Y_\alpha, \delta Z_\alpha)$ may be expressed in terms of increments of the generalized coordinates $(x_1, x_2, \cdots, x_n)$ by differentiation of Eqs. (a) of Sec. 7-1. Also, the quantities $\dot{X}_\alpha, \dot{Y}_\alpha, \dot{Z}_\alpha$ are expressed in terms of the generalized velocity by Eqs. (b) of Sec. 7-1. Introducing these relations into the preceding expression for δW, we obtain a relationship of the form

$$\delta W = -\sum_{i=1}^{n} \sum_{j=1}^{n} c_{ij} \dot{x}_i \, \delta x_j \tag{a}$$

where (c_{ij}) is a symmetric square matrix. The coefficients c_{ij} are real functions of $(x_1, x_2, \cdots, x_n)$. Also,

$$\delta W = \sum_{j=1}^{n} R_j' \, \delta x_j \tag{b}$$

where R_i' is the part of the generalized force that represents viscous friction.

Since δx_i is arbitrary for a holonomic system, Eqs. (a) and (b) yield

$$R_i' = -\sum_{j=1}^{n} c_{ij}\dot{x}_j \qquad (8\text{-}12)$$

Equation (8-12) represents the generalized force that arises from viscous friction. Force components due to other causes may be superimposed to provide the total generalized force R_i that appears in the Lagrange equations (7-12). Conservative forces are accounted for by the potential energy function V.

By a slight extension of the theory, it may be shown that Eq. (8-12) remains valid if the frictional forces are proportional to the relative velocities of the particles rather than to their absolute velocities, provided that the coefficients c_{ij} are modified suitably (71). For free vibrations of a system with viscous damping, $R_i' = R_i$. Then, if the coefficients c_{ij} are constants, Eqs. (8-3) are generalized as follows by means of Lagrange's equations (7-12):

$$\sum_{j=1}^{n} (b_{ij}\ddot{x}_j + c_{ij}\dot{x}_j + a_{ij}x_j) = 0 \qquad (8\text{-}13)$$

Although the coefficients c_{ij} are not generally constants, they may be approximated with sufficient accuracy by their values at the origin $x_i = 0$ if the vibrations are of small amplitude. For the same reason, the coefficients a_{ij} were assumed to be constants.

Equation (8-13) possesses a solution of the type*

$$x_j = z_j e^{rt} \qquad (c)$$

in which z_j and r are constants. Substitution of Eq. (c) into Eq. (8-13) yields

$$\sum_{j=1}^{n} (r^2 b_{ij} + r c_{ij} + a_{ij})z_j = 0 \qquad (8\text{-}14)$$

These are homogeneous linear equations in z_j. They possess a nontrivial solution if, and only if, their determinant is zero; that is,

$$\det (r^2 b_{ij} + r c_{ij} + a_{ij}) = 0 \qquad (8\text{-}15)$$

Since the left side of Eq. (8-15) is a polynomial of degree $2n$ in r, the equation possesses $2n$ roots, although they are not necessarily all distinct. Again, we consider only the case in which the roots are distinct, since this condition may always be ensured by a minor modification of the physical constants of the system. Equation (c) shows that a real root r provides a motion that attenuates (or amplifies) without oscillation. Consequently,

---

* No generality is gained by writing the exponent in Eq. (c) in the form $rt - \gamma$, since the coefficients z_j contain a common arbitrary constant factor.

the complex roots are of primary interest in vibration theory. It has been shown in Sec. (8-2) that the particles of an undamped system execute periodic motions. A damped system approaches this condition if the damping factors approach zero. Consequently, if damping is small, a decaying oscillation occurs. In this case, Eq. (c) shows that the roots r_i are complex constants with negative real parts. Furthermore, since the coefficients of Eq. (8-15) are real, the roots of this equation occur in pairs $(r_i, \bar{r}_i)$, where $\bar{r}_i$ is the complex conjugate of r_i.

Let r and $\bar{r}$ be complex conjugate roots of Eq. (8-15). To these roots there correspond conjugate solutions z_i and $\bar{z}_i$ of Eq. (8-14), and not all the z_i are zero. The general solutions of Eq. (8-14), corresponding to r and $\bar{r}$, are αz_i and $\beta \bar{z}_i$, where α and β are arbitrary complex constants. Consequently, Eq. (c) yields the two solutions

$$x_j = \alpha z_j e^{rt}, \qquad x_j = \beta \bar{z}_j e^{\bar{r}t} \tag{d}$$

The real or imaginary part of either of these solutions represents a physically possible motion. Because of the arbitrary factors α and β, the two solutions given by (d) are equivalent. Consequently, attention will be confined to the first of these equations.

Let us set $\alpha = A e^{-i\gamma}$, $z_j = \rho_j e^{i\theta_j}$, $r = -c + i\omega$, where $i = \sqrt{-1}$ and $A, \gamma, \rho_j, \theta_j, c, \omega$ are real. Also, let $c > 0$ and $\omega > 0$. Then Eqs. (d) yield

$$x_j = A\rho_j e^{-ct}[\cos(\omega t + \theta_j - \gamma) + i \sin(\omega t + \theta_j - \gamma)] \tag{e}$$

Because of the arbitrary constant γ, the real and imaginary parts of Eq. (e) represent the same motion. Consequently, the general real solution obtained from Eq. (e) is

$$x_j = A\rho_j e^{-ct} \sin(\omega t + \theta_j - \gamma) \tag{8-16}$$

Equation (8-16) represents an attenuated oscillatory motion. If there is no damping, $c_{ij} = 0$, and Eq. (8-15) yields pure imaginary roots ($r = i\omega$), as shown in Sec. 8-2. Then, by Eq. (8-14), z_j is real and $\theta_j = 0$. The occurrence of the phase angles θ_j is accordingly a feature associated with damping. It indicates that the various parts of the vibrating system are not in phase with each other.

To each root r of Eq. (8-15) with a positive imaginary part ω there is a real solution of Eq. (8-13) of the type in Eq. (8-16). Each of these solutions contains two arbitrary constants, A and γ. Consequently, the solutions corresponding to all n roots $(r_1, r_2, \cdots, r_n)$ contain $2n$ arbitrary constants. An arbitrary linear combination of these n solutions—each given by Eq. (8-16)—is then the general real solution of Eq. (8-13).

8-4. FREE VIBRATIONS OF A BEAM WITH CLAMPED ENDS.

The preceding articles treat vibrations of systems with finite degrees of freedom. Motions of elastic beams provide one of the simplest examples of vibrations of systems with infinitely many degrees of freedom.

The strain energy of bending of an elastic beam is given by Eq. (2-8) if the deflections are small and the energy due to shear is negligible. Consequently, the action integral is

$$A = \tfrac{1}{2} \int_{t_0}^{t_1} \int_0^L (\rho y_t^2 - EI y_{xx}^2) \, dx \, dt$$

where ρ denotes the mass per unit length. If EI is a constant, the Euler equation for this double integral is

$$y_{xxxx} + \frac{\rho}{EI} y_{tt} = 0 \qquad (8\text{-}17)$$

Equation (8-17) is the fundamental differential equation in the theory of vibration of uniform beams. This equation was derived by a different method in Sec. 3-5.

The natural modes are solutions of the form $y = f(x)g(t)$, since, for such solutions, the particles are in phase with each other. By Eq. (8-17),

$$\frac{f''''}{f} + \frac{\rho}{EI} \frac{g''}{g} = 0$$

where primes denote derivatives. If ρ is constant, the first term is a function of x and the second term is a function of t. Consequently,

$$\frac{f''''}{f} = \beta^4, \qquad \frac{\rho}{EI} \frac{g''}{g} = -\beta^4 \qquad (8\text{-}18)$$

where β is a constant. The second of these equations yields

$$g = A_1 \sin \beta^2 \sqrt{EI/\rho}\, t + B_1 \cos \beta^2 \sqrt{EI/\rho}\, t \qquad (a)$$

This equation shows that the oscillation of any particle is a simple harmonic motion.

The first of the equations in (8-18) yields

$$f(x) = A \sinh \beta x + B \cosh \beta x + C \sin \beta x + D \cos \beta x \qquad (8\text{-}19)$$

where A, B, C, D are constants of integration.

Equation (8-19) is valid for all end conditions. The end conditions for a beam that is clamped at both ends are

$$f(x) = 0 \quad \text{at} \quad x = 0 \quad \text{or} \quad x = L, \qquad f'(x) = 0 \quad \text{at} \quad x = 0 \quad \text{or} \quad x = L$$

By virtue of the conditions at $x = 0$, Eq. (8-19) is reduced to

$$f(x) = A(\sinh \beta x - \sin \beta x) + B(\cosh \beta x - \cos \beta x) \qquad (8\text{-}20)$$

The conditions at $x = L$ now yield

$$A(\sinh \beta L - \sin \beta L) + B(\cosh \beta L - \cos \beta L) = 0$$
$$A(\cosh \beta L - \cos \beta L) + B(\sinh \beta L + \sin \beta L) = 0 \qquad (8\text{-}21)$$

These equations possess a nonzero solution if, and only if, their determinant is zero. This condition reduces to

$$\cosh \beta L \cos \beta L = 1 \qquad (8\text{-}22)$$

Equation (8-22) is known as the "frequency equation." It possesses an infinite number of roots, denoted by $\beta_0 L$, $\beta_1 L$, $\beta_2 L$, etc. These roots are

$$\beta_0 L = 0, \qquad\qquad \beta_1 L = 4.7300408, \qquad \beta_2 L = 7.8532046$$
$$\beta_3 L = 10.9956078, \qquad \beta_4 L = 14.1371655, \qquad \beta_5 L = 17.2787596$$

If $n > 5$, $\beta_n L$ is equal to $(n + \frac{1}{2})\pi$, with an accuracy of at least seven decimal places. It is to be noted that β_i determines the frequency of the ith mode by means of Eq. (a).

If Eq. (8-22) is satisfied, the second equation in (8-21) is proportional to the first. Then Eqs. (8-21) are satisfied by

$$B = 1, \qquad -A = \alpha_n = \frac{\cosh \beta_n L - \cos \beta_n L}{\sinh \beta_n L - \sin \beta_n L} \qquad (b)$$

The coefficients α_n have the following values:

$$\alpha_1 = 0.9825022158, \qquad \alpha_2 = 1.000777311, \qquad \alpha_3 = 0.999664501$$
$$\alpha_4 = 1.000001450, \qquad \alpha_5 = 0.9999999373$$

If $n > 5$, $\alpha_n = 1$, with an accuracy of at least seven decimal places.

Equation (8-20) now yields

$$f_n(x) = \cosh \beta_n x - \cos \beta_n x - \alpha_n (\sinh \beta_n x - \sin \beta_n x) \qquad (8\text{-}23)$$

The functions $f_n(x)$ are known as the "mode forms" of the beam. These functions, and their derivatives, have been tabulated by Young and Felgar (151). Mode forms for other end conditions are also tabulated.

The frequency equation (also called "characteristic equation") (8-22) is analogous to the frequency equation for a system with finite degrees of freedom [Eq. (8-6)]. The roots of this equation are known as the "eigenvalues" of the boundary-value problem. The mode forms, represented by Eq. (8-23), are called "eigenfunctions." All problems of free vibrations of undamped linear systems are eigenvalue problems.

8-5. ORTHOGONALITY PROPERTIES OF MODE FORMS OF A
BEAM. The functions $f_n(x)$ defined by Eq. (8-23) have the following property:

$$\int_0^L f_m(x) f_n(x)\, dx = 0, \qquad m \neq n \tag{8-24}$$

This condition is expressed by the statement that the functions $f_n(x)$ are orthogonal in the range $(0, L)$.

To verify Eq. (8-24), we observe that Eq. (8-18) yields

$$f_m'''' - \beta_m{}^4 f_m = 0, \qquad f_n'''' - \beta_n{}^4 f_n = 0, \qquad m \neq n \tag{a}$$

By multiplying the first of these equations by f_n and the second by f_m and subtracting, we obtain

$$f_m f_n'''' - f_n f_m'''' + (\beta_m{}^4 - \beta_n{}^4) f_m f_n = 0$$

This equation may be written as follows:

$$\frac{d}{dx}(f_m f_n''' - f_n f_m''' + f_n' f_m'' - f_m' f_n'') + (\beta_m{}^4 - \beta_n{}^4) f_m f_n = 0$$

Integration yields

$$(f_m f_n''' - f_n f_m''' + f_n' f_m'' - f_m' f_n'')\Big|_0^L + (\beta_m{}^4 - \beta_n{}^4)\int_0^L f_m f_n\, dx = 0$$

Since $f_m(0) = f_m(L) = f_n(0) = f_n(L) = 0$ and $f_m'(0) = f_m'(L) = f_n'(0) = f_n'(L) = 0$, the preceding equation reduces to Eq. (8-24).

The value of the integral in Eq. (8-24) for the case $m = n$ can be determined by direct integration, with the aid of Eqs. (8-22) and (8-23). The calculations are lengthy, but they reduce to the following simple result:

$$\int_0^L f_n{}^2\, dx = L \tag{8-25}$$

Not only are the functions $f_n(x)$ orthogonal in the range $(0, L)$ but their second derivatives are orthogonal in this range. This may be shown as follows:

Equation (a) yields

$$f_m f_n'''' - \beta_n{}^4 f_m f_n = 0, \qquad m \neq n$$

Integrating this equation and noting Eq. (8-24), we obtain

$$\int_0^L f_m f_n''''\, dx = 0$$

Integration by parts yields

$$\int_0^L f_m f_n'''' \, dx = f_m f_n''' \Big|_0^L - \int_0^L f_m' f_n''' \, dx = 0$$

Since $f_m(0) = f_m(L) = 0$, the terms outside the integral signs vanish. Integrating by parts again and noting that $f_m'(0) = f_m'(L) = 0$, we obtain

$$\int_0^L f_m'' f_n'' \, dx = 0, \qquad m \neq n \tag{8-26}$$

This is the orthogonality condition for the second derivatives.

The integral in Eq. (8-26) may be evaluated for the case $m = n$ by means of Eqs. (8-22) (8-23) and the defining equation for α_n. The result is

$$\int_0^L (f_n'')^2 \, dx = \beta_n^4 L \tag{8-27}$$

The following formulas are also useful:

$$\int_0^L f_n f_m'' \, dx = \frac{4\beta_m^2 \beta_n^2 (\alpha_n \beta_n - \alpha_m \beta_m)}{\beta_n^4 - \beta_m^4} [1 + (-1)^{m+n}], \qquad m \neq n$$

$$\int_0^L f_n f_n'' \, dx = \alpha_n \beta_n (2 - \alpha_n \beta_n L) \tag{8-28}$$

These and many other formulas for the mode forms of beams with various end conditions are given in a bulletin by Felgar (104).

Influence Function for a Clamped Beam. The mode forms for beams and various other simple vibrating systems are excellent functions for use with the Ritz procedure (Sec. 3-11), even though the problem under consideration is not concerned with vibrations. As a simple example, we consider the deflection of a clamped beam due to an off-center concentrated load (Fig. 8-3). The deflection may be represented by

$$y = \sum_{n=1}^{\infty} a_n f_n(x) \tag{b}$$

where $f_n(x)$ are the mode forms of the beam. The preceding series satisfies all the boundary conditions, irrespective of the values of the coefficients a_n. This fact and the orthogonality property are the main reasons why the functions $f_n(x)$ are advantageous in this problem. Equation (2-8) yields the following formula for the strain energy of the beam:

$$U = \tfrac{1}{2} EI \int_0^L \left[\sum a_n f_n''(x) \right]^2 dx$$

Since the functions f_n'' are orthogonal in the range $(0, L)$, this equation yields

$$U = \tfrac{1}{2}EI \sum a_n^2 \int_0^L (f_n'')^2 \, dx$$

Hence, by Eq. (8-27),

$$U = \tfrac{1}{2}EIL \sum_{n=1}^{\infty} \beta_n^4 a_n^2$$

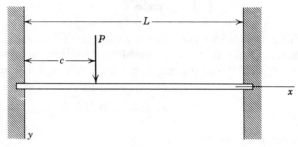

Fig. 8-3

The deflection at the point $x = c$ is

$$y_c = \sum_{n=1}^{\infty} a_n f_n(c)$$

Consequently, the potential energy of the external force is

$$\Omega = -P \sum_{n=1}^{\infty} a_n f_n(c)$$

Therefore, the total potential energy is

$$V = \tfrac{1}{2}EIL \sum_{n=1}^{\infty} \beta_n^4 a_n^2 - P \sum_{n=1}^{\infty} a_n f_n(c)$$

The condition of stationary potential energy yields

$$\frac{\partial V}{\partial a_m} = EIL\beta_m^4 a_m - Pf_m(c) = 0$$

This equation determines a_n. Hence Eq. (b) yields

$$y = \frac{PL^3}{EI} \sum_{n=1}^{\infty} \frac{f_n(x)f_n(c)}{\beta_n^4 L^4} \tag{8-29}$$

Equation (8-29) illustrates Maxwell's law of reciprocity, since the equation is not changed if x and c are permuted (Sec. 4-11).

With the tabulated values of the functions $f_n(x)$, Eq. (8-29) is easily applied. From the tables of Young and Felgar (151), $f_1(L/2) = 1.58815$, $f_2(L/2) = 0$, $f_3(L/2) = -1.40600$, $f_4(L/2) = 0$. Hence, using the values of $\beta_n L$ from Sec. 8-4, and letting $x = c = L/2$, we get $y = 0.00517 PL^3/EI$. This is an approximation for the deflection at the center due to a load at the center if only the first three terms of the series are used. The exact answer is known to be

$$y = \frac{PL^3}{192EI} = 0.005208 \frac{PL^3}{EI}$$

In his celebrated paper on variational methods of approximation (142), Ritz used products of the functions $f_n(x)$ for a clamped beam to represent deflections of a clamped plate.

8-6. VIBRATIONS OF AN AIRPLANE WING. The gust loading of an airplane wing serves as an illustration of transient motion of an elastic system under the action of external loads. If the external loads are periodic, the motion is a forced oscillation.

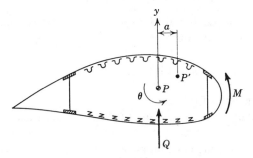

Fig. 8-4

For purposes of analysis, an airplane wing may sometimes be regarded as a beam with variable section properties and variable mass distribution. When the wing vibrates, the segment included between two neighboring cross-sectional planes is displaced in its plane as a rigid lamina. According to a theorem in kinematics (49), the displacement of the lamina may be resolved into a rotation θ about an arbitrary point P and a translation (Fig. 8-4). The rotation θ is independent of the location of point P. The displacement y of point P (Fig. 8-4) is considered to be vertical.

Besides the vertical oscillation, there is generally a fore-and-aft vibration. However, this component of the motion will be neglected, since the wing is relatively stiff when it is bent in a horizontal plane. Accordingly, the displacement y of point P (Fig. 8-4) is considered to be vertical.

Besides the distributed mass, there may be concentrated masses attached to the wing; for example, engines and landing gear. If the wing is treated as a continuous structure, these concentrated masses affect only the boundary conditions for finite segments of the wing that terminate at the concentrated masses. Consequently, concentrated masses are not considered in this discussion.

The point P (Fig. 8-4) is conveniently designated as the center of mass of the lamina. The following notations are introduced:

x a spanwise coordinate (i.e., coordinate measured axially along the wing).

y the vertical displacement of the center of mass of a lamina (i.e., the vertical displacement of point P).

θ rotation of a lamina.

$\rho\, dx$ mass of a lamina of thickness dx. ρ is a function of x.

$I_m\, dx$ mass moment of inertia of a lamina of thickness dx about point P. I_m is a function of x.

By the theory of dynamics of a rigid body the kinetic energy of a lamina of thickness dx is

$$dT = \tfrac{1}{2}(\rho y_t{}^2 + I_m\theta_t{}^2)\, dx$$

where the subscript t denotes the time derivative. Consequently, the kinetic energy of the vibrating wing is

$$T = \tfrac{1}{2}\int_0^l (\rho y_t{}^2 + I_m\theta_t{}^2)\, dx \tag{a}$$

where l is the length of the wing.

The centroid of the cross section of the structural parts of the wing (skin, stringers, and wing beams) lies at point P'. Because of nonstructural components of the wing (fuel tanks, leading edge, trailing edge, etc.), the point P' does not ordinarily coincide with point P. In Fig. 8-4 point P' is indicated to lie at a distance a forward of point P. In general, a is a function of x.

The vertical deflection of point P' is approximately $y + a\theta$. Consequently, by elementary beam theory, the strain energy of bending of the wing is

$$U_1 = \tfrac{1}{2}\int_0^l EI(y + a\theta)_{xx}{}^2\, dx \tag{b}$$

where the subscript x denotes the partial derivative with respect to x and I denotes the moment of inertia of the cross section of the structural parts about the principal axis of inertia through point P'. Also, since the twist per unit length is θ_x, the strain energy due to twisting is

$$U_2 = \tfrac{1}{2}\int_0^l GJ\theta_x{}^2\, dx \tag{c}$$

where GJ is the torsional stiffness of the wing [Eq. (2-17)]. In general, I and J are functions of x.

Suppose that the wing is subjected to a transient vertical load $Q(x, t)$ and a transient twisting moment $M(x, t)$. These functions represent time-dependent distributed loads, since they are functions of x. In problems of wing flutter Q and M are periodic functions of t, but in other cases, such as gust loading, they are nonperiodic. An inherent difficulty in the so-called aeroelastic problems of aircraft is that the aerodynamic loads on a flexible wing depend on the elastic response of the wing to these loads. In the present example this complication need not be considered.

The point at which the force $Q\,dx$ is considered to act on a lamina is arbitrary, since this point may be changed by the introduction of a compensating change in the moment function M. For convenience, we let the force $Q\,dx$ act at point P (Fig. 8-4).

The potential energy of the external forces (see Sec. 7-7) is

$$-\int_0^l (Qy + M\theta)\,dx$$

Consequently, the potential energy of all the forces is

$$V = U - \int_0^l (Qy + M\theta)\,dx \tag{d}$$

where U is the strain energy of the wing.

The action integral A is the definite time integral of $T - V$. Hence by Eqs. (a), (b), (c), and (d)

$$A = \int_{t_0}^{t_1} \int_0^l \left[\tfrac{1}{2}(\rho y_t^2 + I_m \theta_t^2) - \tfrac{1}{2}GJ\theta_x^2 \right.$$
$$\left. - \tfrac{1}{2}EI(y_{xx} + a_{xx}\theta + 2a_x\theta_x + a\theta_{xx})^2 + (Qy + M\theta) \right] dx\,dt \tag{e}$$

By Hamilton's principle the action integral is stationary. Accordingly, the integrand in Eq. (e) satisfies Euler's equations (3-20). Consequently,

$$\rho y_{tt} + (EI\phi)_{xx} = Q$$
$$I_m \theta_{tt} + EIa_{xx}\phi - (GJ\theta_x)_x - 2(EIa_x\phi)_x + (EIa\phi)_{xx} = M \tag{f}$$

where

$$\phi = y_{xx} + (a\theta)_{xx}$$

Even though the functions a, I_m, ρ, I, J are approximated by simple polynomials, Eqs. (f) are a pair of complicated differential equations. However, since they are linear, they may be solved by numerical methods with the aid of modern computing equipment.

If the points P and P' coincide ($a = 0$), the differential equations are greatly simplified. Then the differential equations for y and θ are separated. Accordingly, the torsional vibration and the bending vibration become independent of each other.

8-7. EFFECTS OF ROTARY INERTIA AND SHEAR DEFORMATION ON VIBRATIONS OF BEAMS.

Inertial effects caused by rotations of the cross sections of a vibrating beam and effects caused by shear deformation were neglected in the discussion of vibrations of beams (Sec. 8-4). These conditions have a negligible effect on the first few modes of a slender bar, but they may affect the higher modes significantly. Also, they may cause appreciable perturbations of all modes when the theory of beams is employed as a basis for a study of vibrations of complicated structures, such as wings of airplanes and hulls of ships. The effect of shear deformation is introduced by means of a modification of the strain energy formula; the effect of rotary inertia is introduced by means of an addition to the kinetic energy. The strain energy formula, including the effect of shear, is Eq. (2-15).

The rotation of a cross section equals the slope caused by bending, $y_x - \beta$. Accordingly, the angular velocity of the cross section is $y_{xt} - \beta_t$. Therefore, the rotary kinetic energy, per unit length, is $\frac{1}{2}\rho I(y_{xt} - \beta_t)^2$, where ρ is the mass per unit volume and I is the moment of inertia of the cross section about its centroidal axis. This is augmented by the kinetic energy of translation, $\frac{1}{2}\rho A y_t^2$, where A is the area of the cross section. Therefore, the total kinetic energy is

$$T = \tfrac{1}{2} \int_0^L \left[\rho A y_t^2 + \rho I(y_{xt} - \beta_t)^2 \right] dx \tag{8-30}$$

By Eqs. (2-15) and (8-30), the action integral is

$$\tfrac{1}{2} \int_{t_0}^{t_1} \int_0^L \left[\rho A y_t^2 + \rho I(y_{xt} - \beta_t)^2 - EI(y_{xx} - \beta_x)^2 - \lambda \beta^2 \right] dx \, dt$$

Because of Hamilton's principle, the Euler equations for the action integral are the differential equations of free vibrations. These equations are

$$\frac{\partial^2}{\partial x^2} \left[EI(y_{xx} - \beta_x) \right] - \frac{\partial}{\partial x} \left[\rho I(y_{xtt} - \beta_{tt}) \right] + \rho A y_{tt} = 0 \tag{8-31}$$

$$\frac{\partial}{\partial x} \left[EI(y_{xx} - \beta_x) \right] - \rho I(y_{xtt} - \beta_{tt}) + \lambda \beta = 0 \tag{8-32}$$

Differentiating Eq. (8-32) with respect to x and comparing the result with Eq. (8-31), we obtain

$$\frac{\partial}{\partial x}(\lambda \beta) = \rho A y_{tt} \tag{8-33}$$

Equations (8-31) and (8-33) are the differential equations of free vibrations. They are applicable even though A and I vary gradually with x.

If shear deformation and rotary inertia are both neglected, Eq. (8-31) reduces to

$$\frac{\partial^2}{\partial x^2}(EIy_{xx}) + \rho A y_{tt} = 0 \tag{8-34}$$

If rotary inertia is retained, but shear deformation is neglected, we set $\beta = 0$ in Eq. (8-31).

If λ and I are constants, β may be eliminated from Eq. (8-31) by means of Eq. (8-33). In this way, we obtain

$$EIy_{xxxx} - \rho I\left(1 + \frac{EA}{\lambda}\right)y_{xxtt} + \frac{\rho^2 AI}{\lambda}y_{tttt} + \rho A y_{tt} = 0 \tag{8-35}$$

This equation has been derived by Timoshenko by consideration of inertial forces (83).

8-8. FREE VIBRATIONS OF A RECTANGULAR ELASTIC PLATE.
The strain energy of bending of an isotropic elastic plate is given by Eq. (5-19). If the deflections are small, contributions to the strain energy due to stretching of the middle surface are negligible. Consequently, the action integral for a vibrating plate is

$$-\iiint \left[\tfrac{1}{2}D(w_{xx} + w_{yy})^2 - D(1-\nu)(w_{xx}w_{yy} - w_{xy}{}^2) - \tfrac{1}{2}\rho w_t{}^2\right] dx\, dy\, dt$$

where ρ is the mass per unit area of the plate. If ρ and D are constants, the Euler equation for this triple integral [Eq. (3-22)] is

$$\nabla^2\nabla^2 w + \frac{\rho}{D}w_{tt} = 0 \tag{8-36}$$

This is the fundamental differential equation in the classical theory of vibration of plates. It was derived by Lagrange.*

* In commenting on Sophie Germain's analysis of the vibrating plate, Lagrange wrote: "In adopting, like the author, $1/r + 1/r'$ for the measure of the curvature of the surface, which the elasticity tends to diminish, and to which one supposes it to be proportional, I find in the case of very small z an equation of the form

$$\frac{d^2z}{dt^2} + k^2\left(\frac{d^4z}{dx^4} + 2\frac{d^4z}{dx^2\,dy^2} + \frac{d^4z}{dy^4}\right) = 0$$

which is quite different from the preceding."

The natural modes are solutions of the type, $w = f(x, y)g(t)$. Substituting this equation into Eq. (8-36), we obtain

$$\frac{\nabla^2\nabla^2 f}{f} = \beta^4, \qquad \frac{\rho}{D}\frac{g''}{g} = -\beta^4 \qquad\qquad (a)$$

where β is a constant. The second of these equations yields

$$g = C \sin(\beta^2\sqrt{D/\rho}\, t - \gamma) \qquad\qquad (b)$$

in which C and γ are constants. Accordingly, the particles of the plate execute simple harmonic motions.

The first of the equations in (a) yields

$$\nabla^2\nabla^2 f - \beta^4 f = 0 \qquad\qquad (8\text{-}37)$$

This differential equation determines the mode forms of the plate. For a rectangular plate with simply supported edges on the lines $x = 0$ and $x = a$, the boundary conditions on these edges are $w = w_{xx} = 0$ (Sec. 5-3). Consider mode forms of the following type:

$$f(x, y) = \phi(y)\sin\frac{n\pi x}{a} \qquad\qquad (c)$$

This equation satisfies the boundary conditions at the edges $x = 0$ and $x = a$. Substitution of Eq. (c) into Eq. (8-37) yields

$$\phi'''' - 2\frac{n^2\pi^2}{a^2}\phi'' + \left(\frac{n^4\pi^4}{a^4} - \beta^4\right)\phi = 0 \qquad\qquad (d)$$

Set

$$p_n = \sqrt{\beta^2 + (n^2\pi^2/a^2)}, \qquad q_n = \sqrt{\beta^2 - (n^2\pi^2/a^2)} \qquad\qquad (e)$$

With these notations, the general solution of Eq. (d) is

$$\phi = A_n \sinh p_n y + B_n \sin q_n y + C_n \cosh p_n y + D_n \cos q_n y \qquad\qquad (f)$$

For example, let the edge $y = 0$ be simply supported and let the edge $y = b$ be free. By Sec. 5-3 and Eq. (5-28), the boundary conditions for the two edges are*

$$f = f_{yy} = 0 \quad\text{at}\quad y = 0$$
$$f_{yy} + \nu f_{xx} = 0, \qquad f_{yyy} + (2 - \nu)f_{xxy} = 0 \quad\text{at}\quad y = b$$

---

* Although these boundary conditions were derived for statical loading, we may show, by taking the first variation of the action integral, that the same boundary conditions apply for the vibrating plate.

With Eq. (c), these conditions yield

$$\phi = \phi'' = 0 \quad \text{at} \quad y = 0$$

(g)

$$\phi'' - \frac{vn^2\pi^2}{a^2}\phi = 0, \qquad \phi''' - (2-v)\frac{n^2\pi^2}{a^2}\phi' = 0 \quad \text{at} \quad y = b$$

The constants in Eq. (f) are to be determined by Eqs. (g). The boundary conditions at the edge $y = 0$ yield $C_n = D_n = 0$. The boundary conditions at the edge $y = b$ yield

$$A_n\left(p_n{}^2 - v\frac{n^2\pi^2}{a^2}\right)\sinh p_n b - B_n\left(q_n{}^2 + v\frac{n^2\pi^2}{a^2}\right)\sin q_n b = 0 \quad \text{(h)}$$

$$A_n p_n\left(q_n{}^2 + v\frac{n^2\pi^2}{a^2}\right)\cosh p_n b - B_n q_n\left(p_n{}^2 - v\frac{n^2\pi^2}{a^2}\right)\cos q_n b = 0 \quad \text{(i)}$$

Equation (i) has been simplified by means of the following relations which result from Eqs. (e):

$$p_n{}^2 - (2-v)\frac{n^2\pi^2}{a^2} = q_n{}^2 + v\frac{n^2\pi^2}{a^2}$$

(j)

$$q_n{}^2 + (2-v)\frac{n^2\pi^2}{a^2} = p_n{}^2 - v\frac{n^2\pi^2}{a^2}$$

Equations (h) and (i) determine values of A_n and B_n that are not both zero if, and only if, their determinant is zero. This condition yields

$$q_n\left(p_n{}^2 - v\frac{n^2\pi^2}{a^2}\right)^2 \tanh p_n b = p_n\left(q_n{}^2 + v\frac{n^2\pi^2}{a^2}\right)^2 \tan q_n b \quad \text{(k)}$$

For each positive integer n, Eq. (k) yields an infinite sequence of roots β_{mn}; $m = 1, 2, 3, \cdots$. These are the eigenvalues of the vibration problem. With Eq. (b), they determine the natural frequencies. Let the values of p_n and q_n, corresponding to an eigenvalue β_{mn}, be p_{mn} and q_{mn}. If $\beta = \beta_{mn}$, Eq. (i) is equivalent to Eq. (h). Let the values of A_n and B_n in Eq. (h), corresponding to β_{mn}, be A_{mn} and B_{mn}. Then, since $C_n = D_n = 0$, Eqs. (f) and (h) yield

$$\phi = \phi_{mn} = B_{mn}(\sin q_{mn}y + K_{mn}\sinh p_{mn}y) \quad \text{(l)}$$

where

$$K_{mn} = \frac{\left(q_{mn}{}^2 + v\dfrac{n^2\pi^2}{a^2}\right)\sin q_{mn}b}{\left(p_{mn}{}^2 - v\dfrac{n^2\pi^2}{a^2}\right)\sinh p_{mn}b}$$

Accordingly, by Eqs. (b) and (c), the natural mode, corresponding to the positive integers (m, n), is

$$w_{mn} = B_{mn}(\sin q_{mn}y + K_{mn} \sinh p_{mn}y) \sin \frac{n\pi x}{a} \sin (\beta_{mn}{}^2\sqrt{D/\rho}\, t - \gamma_{mn})$$

(m)

This solution contains two arbitrary constants, B_{mn} and γ_{mn}. Equation (m) yields a double infinity of natural modes. Because of the linearity and homogeneity of Eq. (8-36), the sum of any number of natural modes is a possible free vibration.

Instead of Eq. (c), we could write the more general relationship,

$$f(x, y) = \sum_{n=1}^{\infty} a_n(y) \sin \frac{n\pi x}{a}$$

Such a series can represent any mode form. Nothing new results from this approach, however. Therefore, all natural modes are represented by Eq. (m). The foregoing theory is quite parallel to the corresponding theory for buckling of a plate with the preceding boundary conditions and a uniform axial load (84, Sec. 65).

The mathematical nature of linear vibrations of undamped systems, with emphasis on the relationships to the theory of eigenvalues and eigenfunctions, is treated in the books by Courant and Hilbert (13) and Collatz (11).

8-9. SPHERICAL PRESSURE WAVES IN AN IDEAL GAS.

In the following example linearizing approximations are deferred to the end. Consequently, the problem illustrates a case of nonlinear vibration, a field that has become important in recent years.

In the theory of pressure waves in gases it is usually assumed that there is no heat transfer, since conduction of heat is a slow process compared to the rate of propagation of pressure waves. Accordingly, the adiabatic pressure-density relation is adopted:

$$\frac{p}{\rho^\gamma} = \frac{p_0}{\rho_0{}^\gamma}$$

(a)

Here, p is the pressure, ρ is the mass density, and γ is a dimensionless characteristic constant of the gas. The subscript 0 refers to the initial state. For air and other diatomic gases, γ is approximately 1.40.

The volume of a mass m of the gas is

$$v = \frac{m}{\rho}$$

(b)

If the gas expands an infinitesimal amount, the work that it performs is $p\,dv$. Consequently, the potential energy of a mass m of the gas is

$$V = -\int p\,dv \tag{c}$$

By Eq. (b), $dv = -m\,d\rho/\rho^2$. Consequently, Eq. (c) yields

$$V = \frac{m}{\gamma - 1}\frac{p_0}{\rho_0{}^\gamma}\rho^{\gamma-1} \tag{d}$$

Aside from an irrelevant additive constant of integration, V is the work that we perform in compressing the gas from density ρ_0 to density ρ. Consequently, by the first law of thermodynamics, V is identical to the internal energy of the gas.

When a spherical pressure wave passes through a gas, the particles oscillate in a direction normal to the wavefront. The radial displacement of a particle that lies initially at a distance r from the center of symmetry is denoted by $u(r, t)$. The positive sense of u is the sense of increasing r. The description of the motion by the displacement function $u(r, t)$ is a method that Lagrange applied to problems of fluid motion. For studies of oscillations, Lagrange's representation is sometimes preferable to the Eulerian description that is commonly used in fluid mechanics.

Consider the fluid that lies initially in the shell bounded by the spherical surfaces with radii r and $r + dr$. As the pressure wave passes, this shell of fluid alternately expands and contracts. At the time t, the inner and outer radii of the shell are $r + u$ and $r + u + (1 + u_r)\,dr$, respectively. Consequently, the volume of the expanded shell is $4\pi(r + u)^2(1 + u_r)\,dr$. Since the mass of fluid in the shell is the same as the initial mass, $4\pi\rho_0 r^2\,dr$, the density of fluid in the expanded shell is

$$\rho = \frac{\rho_0 r^2}{(r + u)^2(1 + u_r)}$$

Therefore, by Eq. (d), the potential energy of the gas in the expanded shell is

$$dV = \frac{4\pi p_0 r^2\,dr}{\gamma - 1}\left[\left(1 + \frac{u}{r}\right)^2(1 + u_r)\right]^{1-\gamma}$$

Consequently, if the gas is enclosed in a spherical container of radius a, its potential energy at time t is

$$V = \frac{4\pi p_0}{\gamma - 1}\int_0^a\left[\left(1 + \frac{u}{r}\right)^2(1 + u_r)\right]^{1-\gamma}r^2\,dr \tag{e}$$

The forced boundary conditions are $u = 0$ for $r = 0$ or $r = a$.

The mass of gas in the expanded shell is $4\pi\rho_0 r^2\,dr$, and its radial velocity is u_t. Consequently, the kinetic energy of the gas is

$$T = 2\pi\rho_0 \int_0^a u_t^2 r^2\,dr \tag{f}$$

Equations (e) and (f) yield the following formula for the action:

$$A = \frac{4\pi p_0}{\gamma - 1} \int_{t_0}^{t_1}\int_0^a \left[\frac{\gamma - 1}{2}\frac{\rho_0}{p_0} r^2 u_t^2 - r^2\left(1 + \frac{u}{r}\right)^{2(1-\gamma)}(1 + u_r)^{1-\gamma}\right] dr\,dt$$

The Euler equation for this double integral is

$$2r\left(1 + \frac{u}{r}\right)^{1-2\gamma}(1 + u_r)^{1-\gamma} - \frac{\partial}{\partial r}\left[r^2\left(1 + \frac{u}{r}\right)^{2(1-\gamma)}(1 + u_r)^{-\gamma}\right] = \frac{\rho_0 r^2}{p_0} u_{tt} \tag{g}$$

Since this is a complicated nonlinear partial differential equation, very few solutions of the problem of spherical blast waves are known. Departures from the perfect gas law, with consequent nonconservative forces, also complicate the problem of blast waves (12).

Equation (g) is simplified greatly if linearizing approximations are introduced. Such approximations are legitimate in acoustical problems, since u/r and u_r are small compared to unity. Linearization of Eq. (g) is accomplished by the binomial theorem. For example,

$$\left(1 + \frac{u}{r}\right)^{1-2\gamma} = 1 + (1 - 2\gamma)\frac{u}{r} + \cdots$$

Expanding the other terms in Eq. (g) similarly and discarding nonlinear terms that arise from the products, we obtain

$$u_{rr} + \frac{2}{r}u_r - \frac{2}{r^2}u = \frac{1}{c_0^2}u_{tt} \tag{h}$$

where $c_0 = \sqrt{\gamma p_0/\rho_0}$.

The general solution of Eq. (h) is

$$u = \frac{1}{r}[f'(r - c_0 t) + g'(r + c_0 t)] - \frac{1}{r^2}[f(r - c_0 t) + g(r + c_0 t)] \tag{i}$$

where f and g are arbitrary functions of one variable and primes denote derivatives. The function g is usually discarded on the grounds that an excitation can only be propagated outward from a point source. When g is discarded, Eq. (i) represents the superposition of two spherical waves that travel outward with speed c_0. The attenuation factor for the first wave is $1/r$ and that for the second wave is $1/r^2$. At a sufficiently large distance from the source, the wave with attenuation factor $1/r^2$ has almost disappeared. The remaining wave motion would be obtained from Eq. (h) if the term u/r^2 were discarded from that equation. Then Eq. (h) would be the well-known wave equation for the spherically symmetric case.

8-10. PLANE PROGRESSIVE GRAVITY WAVES IN A LIQUID.

The following example illustrates another case of nonlinear vibrations. It also illustrates the use of the Lagrange multiplier for treating an auxiliary differential equation.

Let the motion of an incompressible frictionless fluid be referred to rectangular coordinates (x, y, z), with the positive y-axis directed upward. The displacement of a particle of fluid from its initial position is represented by a vector with components (u, v, w) on the (x, y, z) axes. The motion of the fluid is defined in the Lagrangian form if the variables (u, v, w) are expressed as functions of (x, y, z, t).

Attention is restricted to plane motion, defined by the conditions $u = u(x, y, t)$, $v = v(x, y, t)$, $w = 0$. Since the fluid is considered to be incompressible, the functions (u, v) are subjected to a constraint that is expressed by the differential equation,

$$u_x + v_y + u_x v_y - u_y v_x = 0 \qquad (8\text{-}38)$$

Equation (8-38) is known as the "Lagrangian form of the continuity equation." It signifies that the volume of any part of the fluid remains constant during the motion.

Suppose that the free surface of the liquid is subjected to constant pressure. Then, if surface tension is neglected, the potential energy derives entirely from the weight of the liquid. The potential energy of a particle of mass dm is $gv\, dm$, where v is the vertical component of displacement and g is the acceleration of gravity. Since the motion is two-dimensional, we may suppose that the liquid is bounded by the planes $z = 0$ and $z = 1$. Then $dm = \rho\, dx\, dy$, where ρ is the mass density. Consequently, the potential energy of all the liquid is

$$V = \iint \rho g v\, dx\, dy \qquad (a)$$

where the double integral extends over the region initially occupied by the liquid. Likewise, the kinetic energy of the liquid is

$$T = \tfrac{1}{2} \iint \rho(u_t^2 + v_t^2)\, dx\, dy \qquad (b)$$

Consequently, the action is

$$A = \iiint [\tfrac{1}{2}\rho(u_t^2 + v_t^2) - \rho g v]\, dx\, dy\, dt \qquad (c)$$

The auxiliary differential equation (8-38) may be taken into account by the Lagrange-multiplier method (Sec. 3-8). The modified action integral is

$$\bar{A} = \iiint [\tfrac{1}{2}\rho(u_t^2 + v_t^2) - \rho g v + \rho\lambda(u_x + v_y + u_x v_y - u_y v_x)]\, dx\, dy\, dt \quad \text{(d)}$$

The Lagrange multiplier λ is a function of x, y, and t.

Suppose that the fluid is contained in a tank with end planes $x = 0$ and $x = a$ and with bottom plane $y = -b$. The free surface of the undisturbed liquid is represented by the plane $y = 0$. It is convenient to imagine a space with rectangular coordinates (x, y, t). The region of integration in Eq. (d) is a rectangular parallelepiped in this space. The limits on the time integral are (t_0, t_1).

The Euler equations for the integral $\bar{A}$ are the dynamical equations of motion. However, we must perform the variational process on the integral $\bar{A}$ to obtain the natural boundary conditions for the free surface. Let the functions (u, v) receive infinitesimal variations (ξ, η) that are functions of (x, y, t). The forced boundary conditions are $u = 0$ for $x = 0$ or $x = a$ and $v = 0$ for $y = -b$. Also, in accordance with Hamilton's viewpoint, the functions (u, v) are considered to be prescribed at the times t_0 and t_1. The first variation of the integral $\bar{A}$ is

$$\delta\bar{A} = \rho\iiint [(u_t\xi_t + v_t\eta_t) - g\eta$$
$$+ \lambda(\xi_x + \eta_y + u_x\eta_y + v_y\xi_x - u_y\eta_x - v_x\xi_y)]\, dx\, dy\, dt \quad \text{(e)}$$

This integral may be transformed by the integration-by-parts formula for multiple integrals [Eq. (4-62)]. Transforming the terms in Eq. (e) by integration by parts and observing the forced boundary conditions for ξ and η, we obtain

$$\frac{\delta\bar{A}}{\rho} = \iiint [-u_{tt} - \lambda_x - (\lambda v_y)_x + (\lambda v_x)_y]\xi\, dx\, dy\, dt$$

$$+ \iiint [-v_{tt} - g - \lambda_y - (\lambda u_x)_y + (\lambda u_y)_x]\eta\, dx\, dy\, dt$$

$$+ \iint_{y=0} [\lambda(1 + u_x)\eta - \xi\lambda v_x]\, dx\, dt + \iint_{x=0} \eta\lambda u_y\, dy\, dt$$

$$- \iint_{x=a} \eta\lambda u_y\, dy\, dt + \iint_{y=-b} \xi\lambda v_x\, dx\, dt$$

The notations $y = 0$, $x = a$, etc. beneath the double integral signs denote the planes on which the integrations are to be performed.

Since $\delta\bar{A}$ vanishes for all variations (ξ, η) that conform to the forced boundary conditions, the integrands of the triple integrals and the double integrals vanish separately. Thus the triple integrals provide the Euler equations,

$$\lambda_x(1 + v_y) - \lambda_y v_x + u_{tt} = 0 \tag{8-39}$$

$$-\lambda_x u_y + \lambda_y(1 + u_x) + g + v_{tt} = 0 \tag{8-40}$$

The double integrals provide the natural boundary conditions:

$$\lambda(1 + u_x) = 0 \quad \text{and} \quad \lambda v_x = 0 \quad \text{for} \quad y = 0 \tag{8-41}$$

$$\lambda u_y = 0 \quad \text{for} \quad x = 0 \quad \text{or} \quad x = a; \quad \lambda v_x = 0 \quad \text{for} \quad y = -b \tag{8-42}$$

Equations (8-38), (8-39), and (8-40) are the Lagrangian equations of plane flow of a frictionless liquid that moves under the action of gravity alone. Equations (8-41) are the natural boundary conditions for the free surface, and Eqs. (8-42) are the natural boundary conditions for the walls of a rectangular tank that contains the liquid. Equations (8-38) to (8-41) apply also for an infinite expanse of liquid, but the boundary conditions pertaining to the walls of the tank have no significance in this case.

There is only one known exact solution of the preceding equations that represents wave motion. It was discovered by E. Gerstner in 1802. Considering an infinite expanse of liquid and seeking a progressive wave motion for which the orbits of the particles are circular, he was led to try the relations,

$$u = r \sin 2\pi\left(\frac{t}{\tau} - \frac{x}{l}\right), \qquad v = e + r \cos 2\pi\left(\frac{t}{\tau} - \frac{x}{l}\right) \tag{8-43}$$

where e and r are functions of y alone, and τ and l are constants. These quantities may be interpreted physically; l is the wavelength, τ is the time in which the waveform advances one wavelength (called the "wave period"), r is the radius of the orbit of a particle, and e is the height of the center of an orbit above the initial level of the particle that describes that orbit.

It is necessary to adapt the parameters r, e, τ, l to the equations of motion and the natural boundary conditions. Substituting Eq. (8-43) into Eq. (8-38) and setting $\theta = 2\pi(t/\tau - x/l)$, we obtain

$$-\frac{2\pi r}{l}\cos\theta + e' + r'\cos\theta - \frac{2\pi r}{l}e'\cos\theta - \frac{2\pi r r'}{l} = 0$$

where primes denote derivatives with respect to y. Since t occurs only in θ, this equation requires that the following two equations be satisfied:

$$e' - \frac{2\pi r r'}{l} = 0 \tag{f}$$

$$\frac{2\pi r}{l}(1 + e') = r' \tag{g}$$

Equation (f) yields

$$e = \frac{\pi r^2}{l} \tag{h}$$

The additive constant of integration is zero, since r and e both approach zero as y approaches negative infinity.

Eliminating e' from Eq. (g) by means of Eq. (f) and integrating the resulting equation, we obtain

$$\log \frac{r}{a} + \frac{2\pi^2 a^2}{l^2}\left(1 - \frac{r^2}{a^2}\right) = \frac{2\pi y}{l} \tag{i}$$

The constant of integration has been eliminated by the boundary condition, $r = a$ at $y = 0$, where a is half the wave height; that is, half the height from trough to crest. Equations (h) and (i) implicitly express r and e as functions of y.

Substituting Eq. (8-43) into the dynamical equations (8-39) and (8-40), we obtain

$$\lambda_x(1 + e' + r'\cos\theta) - \frac{2\pi r}{l}\lambda_y\sin\theta - \frac{4\pi^2 r}{\tau^2}\sin\theta = 0 \tag{j}$$

$$-\lambda_x r'\sin\theta + \lambda_y\left(1 - \frac{2\pi r}{l}\cos\theta\right) + g - \frac{4\pi^2 r}{\tau^2}\cos\theta = 0 \tag{k}$$

Equations (j) and (k) are linear algebraic equations in λ_x and λ_y. In view of Eqs. (f) and (g), the determinant of these equations is 1. Hence

$$\lambda_x = \frac{2\pi r}{l}\left(\frac{2\pi l}{\tau^2} - g\right)\sin\theta$$

$$\lambda_y = r'\left(\frac{4\pi^2 r}{\tau^2} - \frac{gl}{2\pi r}\right) + r'\left(\frac{2\pi l}{\tau^2} - g\right)\cos\theta$$

Integration of these equations yields

$$\lambda = r\left(\frac{2\pi l}{\tau^2} - g\right)\cos\theta - \frac{gl}{2\pi}\log r + \frac{2\pi^2 r^2}{\tau^2} + f(t)$$

Introducing the half wave height a and absorbing a constant in $f(t)$, we may write this equation in the following form:

$$\lambda = r\left(\frac{2\pi l}{\tau^2} - g\right)\cos\theta - \frac{gl}{2\pi}\log\frac{r}{a} - \frac{2\pi^2}{\tau^2}(a^2 - r^2) + f(t) \tag{1}$$

Hence, by Eq. (i),

$$\lambda = \left(\frac{2\pi l}{\tau^2} - g\right)\left[r\cos\theta + \frac{\pi}{l}(r^2 - a^2)\right] - gy + f(t) \tag{m}$$

By Eq. (8-43), the quantities $(1 + u_x)$ and v_x do not vanish identically on the free surface. Consequently, Eqs. (8-41) require that $\lambda = 0$ for $y = 0$. Therefore, Eq. (m) yields $f(t) = 0$ and $g = 2\pi l/\tau^2$. Since the wavespeed is $c = l/\tau$, the latter relation yields

$$c = \sqrt{gl/2\pi} \tag{8-44}$$

Equation (8-44) is a well-known formula for the wavespeed in deep water.

Gerstner's wave motion accordingly satisfies all the kinematical and dynamical requirements, provided that the liquid is so deep that we may dispense with the boundary conditions at the bottom.

8-11. WAVE MOTION IN SOLIDS. The theory of wave motion in elastic solids receives a very important application in seismology (8, 22). Also, it is increasing in technological importance in investigations of elastic mechanisms, chatter of machine tools, vibrations of piezoelectric crystals, and noise output of parts with intermittent contacts such as gear teeth and tappets. For these applications, the linear theory of vibrations is usually adequate.

Let an isotropic elastic body be referred to rectangular coordinates (x, y, z). If the particles of the body perform small oscillations, the kinetic energy is

$$T = \tfrac{1}{2}\iiint \rho(u_t{}^2 + v_t{}^2 + w_t{}^2)\,dx\,dy\,dz$$

where (u, v, w) is the displacement vector and ρ is the mass density. It will be assumed that ρ is constant.

The strain energy is determined by Eqs. (4-14) and (4-50). Accordingly, when body forces and temperature effects are absent, the action is

$$A = \tfrac{1}{2}\iiiint [\rho(u_t{}^2 + v_t{}^2 + w_t{}^2) - \lambda(u_x + v_y + w_z)^2 - 2G(u_x{}^2 + v_y{}^2 + w_z{}^2)$$
$$- G(w_y + v_z)^2 - G(u_z + w_x)^2 - G(v_x + u_y)^2]\,dx\,dy\,dz\,dt$$

The Euler equations for this integral are

$$(\lambda + G)e_x + G\nabla^2 u = \rho u_{tt}, \cdots\cdots, \tag{8-45}$$

where $e = u_x + v_y + w_z$. The dots indicate that there are similar equations for v and w. In Gibbs's vector notation, Eq. (8-45) is written as follows:

$$(\lambda + G)\,\text{grad}\,e + G\nabla^2 \mathbf{q} = \rho\mathbf{q}_{tt}, \qquad e = \text{div}\,\mathbf{q} \tag{8-46}$$

Equation (8-46) is homogeneous and linear; hence a sum of solutions is again a solution. In particular, the x-, y-, and z-components of a wave motion may exist independently. There is a uniqueness theorem for the solutions of Eq. (8-46). It shows that there is not more than one vector field $\mathbf{q}(x, y, z, t)$ for which $\mathbf{q}(x, y, z, 0)$ and $\mathbf{q}_t(x, y, z, 0)$ are given functions (51).

Any continuous and differentiable vector field may be expressed as the sum of a gradient and a curl (55). Consequently, it is permissible to set

$$\mathbf{q} = \text{grad}\,\phi + \text{curl}\,\mathbf{H} \tag{8-47}$$

where $\phi(x, y, z, t)$ and $\mathbf{H}(x, y, z, t)$ are scalar and vector functions, respectively. Since $e = \text{div}\,\mathbf{q}$, Eq. (8-47) yields

$$e = \nabla^2\phi \tag{8-48}$$

If $\phi = 0$, Eq. (8-48) yields $e = 0$. Consequently, if $\phi = 0$, Eq. (4-13) shows that there is no volumetric strain (since linearized strain-displacement relations are used). The wave motion is then said to be "equivoluminal." If $\mathbf{H} = 0$, Eq. (8-47) yields curl $\mathbf{q} = 0$. It is shown in the small-deformation theory of continuous media (51) that the angular displacement of a particle is $\frac{1}{2}$curl $\mathbf{q}$. Consequently, wave motion for which $\mathbf{H} = 0$ is said to be "irrotational." Equation (8-47) shows that the most general wave motion in a solid consists of the superposition of two time-dependent displacement-vector fields, one representing equivoluminal waves and the other representing irrotational waves.

Substitution of Eqs. (8-47) and (8-48) into Eq. (8-46) yields

$$\rho\,\text{grad}\,\phi_{tt} + \rho\,\text{curl}\,\mathbf{H}_{tt} = (\lambda + 2G)\,\text{grad}\,\nabla^2\phi + G\,\text{curl}\,\nabla^2\mathbf{H}$$

This separates into the two equations,

$$\phi_{tt} - \alpha^2\nabla^2\phi = \psi, \qquad \alpha^2 = \frac{\lambda + 2G}{\rho}$$

$$\text{curl}\,(\mathbf{H}_{tt} - \beta^2\nabla^2\mathbf{H}) = -\text{grad}\,\psi, \qquad \beta^2 = \frac{G}{\rho} \tag{a}$$

where $\psi(x, y, z, t)$ is an arbitrary harmonic function.

Particular importance attaches to those solutions of (a) that represent progressive wave motion. To obtain such solutions, it is necessary to set $\psi = 0$. Then the second of the equations in (a) yields

$$\mathbf{H}_{tt} - \beta^2 \nabla^2 \mathbf{H} = \text{grad } \chi,$$

where $\chi(x, y, z, t)$ is an arbitrary function. To obtain progressive wave motion, we must set $\chi = 0$. Accordingly, wave motion in an isotropic elastic homogeneous medium is governed by the equations

$$\phi_{tt} - \alpha^2 \nabla^2 \phi = 0, \qquad \mathbf{H}_{tt} - \beta^2 \nabla^2 \mathbf{H} = 0$$

$$\alpha^2 = \frac{\lambda + 2G}{\rho}, \qquad \beta^2 = \frac{G}{\rho} \tag{8-49}$$

Equations (8-49) are known as the "wave equation." There is an extensive literature on it in treatises and articles on partial differential equations, electromagnetic theory, and acoustics. Equations (8-49) shows that irrotational waves ($\mathbf{H} = 0$) propagate with speed α and equivoluminal waves ($\phi = 0$) propagate with speed β. Evidently, irrotational waves travel faster than equivoluminal waves. This conclusion finds an application in seismology, since the difference in the times of arrival of the two types of waves indicates the distance of the receiving station from the focus of the earthquake. However, there are many complications caused by inhomogeneity and anisotropy of the rocks and by effects of the free surface of the earth (8).

Flux of Energy in Wave Motion. Let S be a closed surface within a vibrating medium, and let R be the region contained within S. By Eq. (4-16), the stress exerted by the material outside S upon the contiguous material within S is $\mathbf{p} = l\sigma_x + m\sigma_y + n\sigma_z$, where (l, m, n) is the outward-directed unit normal to S, and σ_x, σ_y, σ_z are the stress vectors on plane elements normal to the x-, y-, z-axes, respectively. If body forces are absent, the rate at which external forces perform work on the material in R is

$$W = \iint_S \mathbf{p} \cdot \mathbf{q}_t \, dS$$

where $\mathbf{q}$ is the displacement vector.

Propagation of waves is usually assumed to be an adiabatic process. Consequently, by the first law of thermodynamics, W equals the rate of increase of the sum of the kinetic energy and the internal energy of material in R. Accordingly, W is interpreted as the rate at which energy flows inward through the surface S. Therefore, if dS is any surface element with unit normal $\mathbf{n}$ and if the material on the positive side of dS (the side toward which n is directed) exerts a stress p on the material on the negative side of dS, the rate at which energy flows through dS in the sense of n is $-p \cdot q_t \, dS$. If the strain energy density is known, the vector p may be expressed in terms of the derivatives q_x, q_y, q_z by means of Eqs. (4-17) and (4-35).

Natural Modes. A natural mode of a body is a vibration in which all particles execute simple harmonic motions with the same phase and the same frequency. Consequently, a natural mode is characterized by the equation

$$q = a \cos (\omega t - \gamma) \tag{8-50}$$

in which the angular frequency ω and the initial phase $-\gamma$ are constants and $a(x, y, z)$ is a vector field that is independent of time. Substitution of Eq. (8-50) into Eqs. (8-46) yields

$$(\lambda + G) \, \text{grad div } a + G\nabla^2 a + \rho\omega^2 a = 0 \tag{8-51}$$

If the body vibrates freely, the stress p on the free surface is zero. Consequently, by Eq. (4-16), $l\sigma_x + m\sigma_y + n\sigma_z = 0$ on the free surface S of the body, where (l, m, n) is the unit normal vector to S. By means of the stress-strain relations, this boundary condition may be expressed readily in terms of the vector function $a(x, y, z)$. The boundary condition is homogeneous; that is, if vector a satisfies the boundary condition, so does ca, where c is any constant. Equation (8-51), in conjunction with the boundary condition, possesses a solution other than $a = 0$ if, and only if, ω belongs to one of an infinite sequence of values $\omega_1, \omega_2, \cdots$. These are the natural frequencies of the body. To any natural frequency ω_i, there corresponds a nonzero vector field a_i that satisfies Eq. (8-51) and the boundary conditions. This vector field contains an arbitrary constant factor; that is, ca_i is also a solution, where c is any constant. Consequently, the amplitudes of the oscillations of the particles are not determinate, but the ratio of the amplitudes for any two particles is fixed. Each particle executes a simple harmonic oscillation on the straight line in which the vector a_i lies.

Since the differential equations and the boundary conditions of the vibration problem are linear and homogeneous, a very general solution is

$$q = \sum_{i=1}^{\infty} c_i a_i \cos (\omega_i t - \gamma_i) \tag{8-52}$$

where c_i and γ_i are arbitrary constants. A general demonstration that this type of series can represent any oscillatory motion of an arbitrary elastic body is lacking, although proofs that the natural modes form a complete set of functions are available in some particular cases (51).

Plane Waves. Usually one encounters insurmountable mathematical difficulties if he attempts to determine the wave motion that results from given initial conditions and given boundary conditions Consequently, in practical investigations of wave motion much importance attaches to particular solutions of the wave equation. Plane waves are seemingly the most important special class of waves. They are characterized by the condition, $\mathbf{q} = \mathbf{q}(x, t)$. At any particular instant, the vector $\mathbf{q}$ is constant on any plane that is perpendicular to the x-axis. Consequently, the planes $x = $ constant are called "wavefronts." Lines perpendicular to the wave-fronts are called "rays." In the present case the rays are all straight lines parallel to the x-axis.

Since $\mathbf{q} = \mathbf{q}(x, t)$, $\phi = \phi(x, t)$ and $\mathbf{H} = \mathbf{H}(x, t)$. In this case the general solution of Eqs. (8-49) is (14)

$$\phi = f(x - \alpha t) + g(x + \alpha t)$$
$$\mathbf{H} = \mathbf{F}(x - \beta t) + \mathbf{G}(x + \beta t) \tag{8-53}$$

where f and g are arbitrary scalar functions and $\mathbf{F}$ and $\mathbf{G}$ are arbitrary vector functions. Consequently, by Eq. (8-47),

$$u = f'(x - \alpha t) + g'(x + \alpha t)$$
$$v = -F_z'(x - \beta t) - G_z'(x + \beta t) \tag{8-54}$$
$$w = F_y'(x - \beta t) + G_y'(x + \beta t)$$

where $(F_x, F_y, F_z) = \mathbf{F}$ and $(G_x, G_y, G_z) = \mathbf{G}$. Primes denote derivatives. In view of the arbitrariness of f, g, $\mathbf{F}$, and $\mathbf{G}$, the functions f', g', F_y', G_y', F_z', G_z' are all arbitrary. For the motion represented by Eqs. (8 54), the strain components are

$$\epsilon_x = f'' + g'', \qquad \gamma_{xy} = -F_z'' - G_z'', \qquad \gamma_{xz} = F_y'' + G_y'',$$
$$\epsilon_y = \epsilon_z = \gamma_{yz} = 0 \tag{8-55}$$

Consequently, for the irrotational plane waves, the shearing strains $\gamma_{yz}, \gamma_{zx}, \gamma_{xy}$ vanish, and for the equivoluminal plane waves the axial strains $\epsilon_x, \epsilon_y, \epsilon_z$ vanish. Therefore, irrotational plane waves are called "compression waves" and equivoluminal plane waves are called "shear waves."

Consider, for example, the solution, $u = f'(x - \alpha t)$, $v = w = 0$. For any fixed value of t, we may represent the waveform by a curve of u versus x. The curve for $t = t_1$ is the same as the curve for $t = 0$, except that it has been translated a distance αt_1 in the positive x-direction. Consequently, we may visualize the solution $u = f'(x - \alpha t)$, $v = w = 0$ as a propagation of the waveform $u = f'(x)$ without change of shape and with speed α in the positive x-direction. Waves of this type are said to be "progressive." Similarly, the solution $u = g'(x + \alpha t)$, $v = w = 0$

represents a compression wave with waveform $u = g'(x)$ that is propagated with speed α in the negative x-direction. The solution $u = w = 0$, $v = -F_z'(x - \beta t)$ likewise represents a shear wave that is propagated without change of form and with speed β in the positive x-direction. Corresponding interpretations apply to the other components of the solution given by Eqs. (8-54).

Plane compressional progressive waves defined by the following equations are especially important:

$$u = A \cos \frac{2\pi}{L} (x - \alpha t), \qquad v = w = 0 \tag{8-56}$$

This solution results from Eqs. (8-54) if $\mathbf{F} = \mathbf{G} = g = 0$ and f' is given the special form indicated by Eqs. (8-56). Equations (8-56) represent a sinusoidal waveform with wavelength L in the x-coordinate and wavelength L/α in time. The ratio L/α is evidently the time interval in which the waveform advances the distance L; it is called the "period" of the motion. The quantity $v = \alpha/L$ is known as the "frequency" of the motion. The waveform is propagated with speed α in the positive x-direction without change of shape. Equations (8-56) are also frequently written in the form

$$u = A \cos (\omega t - kx) \tag{8-57}$$

A comparison of Eqs. (8-56) and (8-57) shows that $\alpha = \omega/k$, $k = 2\pi/L$, and $v = \alpha/L = \omega/2\pi$. The constant ω is known as the "angular frequency." Equation (8-57) is also written as follows:

$$u = A e^{i(\omega t - kx)}, \qquad i = \sqrt{-1} \tag{8-58}$$

It is understood that the real and the imaginary parts of this complex function are both solutions of the wave equation. Consequently, they represent possible wave motions.

PROBLEMS

1. A mass m is suspended from a linear spring with constant k. The mass oscillates on a vertical line under the action of a time-dependent force $F(t)$. The displacement of the mass from the statical position is x. Derive the differential equation of motion by means of Lagrange's equation. Supposing that $F = F_0 \sin \omega t$, show that x becomes infinite if $\omega^2 = k/m$.

2. A uniform simple beam of length l and stiffness EI carries a mass m at the center. Neglecting gravity and the mass of the beam, derive the differential equation for free vibrations of mass m by means of Lagrange's equation.

3. A heavy mass that is suspended by a slender elastic wire executes torsional oscillations. One principal axis of inertia of the body coincides with the wire. Supposing that the frictional torque is $c\dot\theta$, where θ is the angular

displacement of the body and c is a damping constant, derive the differential equation for free vibrations by means of Lagrange's equation.

4. The two equal masses oscillate freely on a vertical line (Fig. P8-4). The constant for either spring is k. Let the generalized coordinates (x_1, x_2) be the displacements of the masses from their statical positions. Determine the natural frequencies and the corresponding ratios of amplitudes.

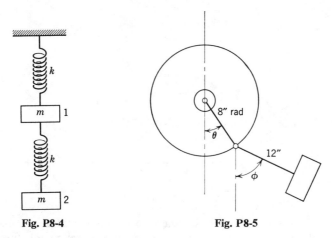

Fig. P8-4   Fig. P8-5

5. The mass of the bob of the pendulum is $\frac{1}{32}$ lb-sec²/in. (Fig. P8-5). The wheel has radius of gyration 4 in. and mass $\frac{1}{8}$ lb-sec²/in. The axle of the wheel contains a spring such that the restoring moment is 8000θ lb-in., where θ is expressed in radians. There is no damping. Adopt θ and ϕ as generalized coordinates. Compute the natural frequencies and the corresponding ratios of amplitudes by setting $\theta = u \sin \omega t$, $\phi = v \sin \omega t$.

6. The three equal masses oscillate freely on a vertical line (Fig. P8-6). The constant for each spring is k. Let the generalized coordinates (x_1, x_2, x_3) be the displacements of the masses from their statical positions. By setting $x_i = z_i \sin \omega t$, determine the natural frequencies and the corresponding ratios of amplitudes.

7. Solve Prob. 6 by introducing normal coordinates.

8. Solve the first example in Sec. 8-1 by means of normal coordinates.

9. The deflection of a uniform freely vibrating simple beam with stiffness EI and mass per unit length ρ is represented by

$$y = \sum_{n=1}^{\infty} a_n \sin \frac{n\pi x}{l}$$

where a_n is a function of t. Derive the differential equation for a_n by means of Lagrange's equations. Determine the angular frequency of the nth mode.

10. Three particles, each with mass m, move in the (x, y)-plane. There is a force of attraction between any two particles equal to kd, where d is the distance

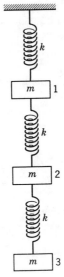

Fig. P8-6

between them, and k is a constant. The coordinates of the masses are (x_1, y_1), (x_2, y_2), (x_3, y_3). Derive the Lagrange equations of motion. Set $x_i = A_i \sin \omega t$, $y_i = B_i \sin \omega t$, and determine the positive value of ω that satisfies these equations. Show that the constants A_i and B_i are such that the center of mass remains at the origin and the moment of momentum about the origin is zero.

11. The lateral deflection of a uniform vibrating string with fixed ends is $y(x, t)$. The length of the undeflected string is l. Express the extension e of the string due to lateral deflection in terms of y, using small-deflection approximations. Hence write the expression for the strain energy, assuming that the tension N is constant. Set up the action integral (neglecting gravity) and derive the differential equation for free vibrations by means of Hamilton's principle.

12. The deflection of a uniform vibrating membrane is $w(x, y, t)$. The tractions in the membrane are $N_x = N_y = N = $ constant and $N_{xy} = 0$. The increment of area of the membrane due to deflection is

$$\tfrac{1}{2} \iint (w_x{}^2 + w_y{}^2) \, dx \, dy$$

Write the expression for the strain energy. Hence write the expression for the action integral and obtain the differential equation of free vibrations by means of Hamilton's principle.

13. Derive the differential equation for free axially symmetric small vibrations of a circular elastic plate of variable thickness $h(r)$ by means of Hamilton's

principle. Set $w = f(r) g(t)$ and derive the differential equations for $f(r)$ and $g(t)$.

14. A uniform circular elastic plate is subjected to a time-dependent axially symmetric load $p(r, t)$. Using small-deflection approximations, derive the differential equation for the deflection $w(r, t)$ by means of Hamilton's principle.

15. Supposing that the strain energy of a circular membrane of radius a is proportional to the increment of area due to deflection and using small-deflection approximations, derive the differential equation for free axially symmetric vibrations by means of Hamilton's principle. Show that the natural modes are determined by Bessel's differential equation of order zero. Determine the first few natural frequencies with the aid of a table of zeros of $J_0(x)$.

16. Derive the frequency equation for a uniform elastic beam that is hinged at the end $x = 0$ and clamped at the end $x = l$.

17. Derive the frequency equation for a uniform elastic cantilever beam. Determine the mode shapes and the lowest natural frequency.

18. Determine the natural frequencies and the mode shapes of a uniform elastic rectangular plate that is simply supported on the edges $x = 0$, $x = a$, $y = 0$, and $y = b$.

19. By Hamilton's principle, derive the differential equation for free torsional vibrations of a uniform elastic bar. Determine the natural frequencies and the mode forms for a bar that is clamped at the ends.

20. By Hamilton's principle, derive the differential equation for plane longitudinal waves in a slightly tapered elastic bar.

21. Write the action integral for a uniform elastic circular ring that vibrates freely in its plane. Obtain the differential equations for the mode forms. Neglect shear effects and rotary inertia.

22. By Hamilton's principle, derive the differential equation for plane compression waves in a gas, supposing the amplitude to be large. Show that the wave equation is obtained if nonlinear terms are neglected.

23. For ocean waves, the length is 300 ft and the vertical distance from trough to crest is 20 ft. Supposing the Gerstner theory applies, determine the wave speed. Determine the depth at which the amplitude is only 1 per cent of that at the surface.

24. By Hamilton's principle, derive the differential equations of motion for free axially symmetric vibrations of a homogeneous isotropic elastic medium with reference to cylindrical coordinates.

Appendix on
quadratic forms

A-1. TYPE OF A QUADRATIC FORM.
A quadratic form in n variables, $x_1, x_2, \cdots, x_n$, is an expression of the type,

$$Q = \sum_{i=1}^{n} \sum_{j=1}^{n} a_{ij} x_i x_j \qquad \text{(A-1)}$$

in which (a_{ij}) is a symmetric matrix of constants. The separation of the coefficient of $x_i x_j$ into the sum $a_{ij} + a_{ji}$ is arbitrary; there is consequently no loss of generality in the specification $a_{ij} = a_{ji}$.

If the coefficients a_{ij} are real and the x's are restricted to real values, Q is real. Evidently, $Q = 0$ if the x's are all zero. Let us consider the range of Q when the x's independently take all real values except the simultaneous values $0, 0, \cdots, 0$. The following conditions may occur: (a) Q may be confined to the range $Q > 0$. Then the quadratic form is said to be *positive definite*. (b) Q may be confined to the range $Q \geq 0$, there being infinitely many values of the x's for which $Q = 0$. Then the quadratic form is said to be *positive semidefinite*. (c) Q may be confined to the range $Q < 0$. Then the quadratic form is said to be *negative definite*. (d) Q may be confined to the range $Q \leq 0$, there being infinitely many values of the x's for which $Q = 0$. Then the quadratic form is said to be *negative semidefinite*. (e) Q may cover the range $-\infty < Q < \infty$. Then the quadratic form is said to be *indefinite*. Stability of a conservative mechanical system is usually characterized by positive definiteness of the second variation of the potential energy. At the critical load this quantity becomes positive semidefinite. At higher loads it is negative definite, negative semidefinite, or indefinite.

A quadratic form is said to be "definite" if it is either positive definite or negative definite. It is said to be "semidefinite" if it is either positive semidefinite or negative semidefinite. If a quadratic form vanishes at the point $(x_1, x_2, \cdots, x_n)$, it also vanishes at the point $(kx_1, kx_2, \cdots, kx_n)$, where k is any constant. This observation explains why a semidefinite quadratic form has infinitely many zeros. The following examples illustrate the five types of quadratic forms:

Positive definite,	$x_1{}^2 + x_2{}^2$
Positive semidefinite,	$x_1{}^2 + 2x_1x_2 + x_2{}^2 = (x_1 + x_2)^2$
Negative definite,	$-x_1{}^2 - x_2{}^2$
Negative semidefinite,	$-x_1{}^2 - 2x_1x_2 - x_2{}^2 = -(x_1 + x_2)^2$
Indefinite,	x_1x_2

The determination of the type of quadratic form involves the concept of *rank* of the matrix (a_{ij}). It may happen that the determinant of matrix (a_{ij}) is not zero; then the rank of the matrix is said to be n. If the determinant of matrix (a_{ij}) vanishes, we consider the determinants of submatrices, obtained by crossing out rows and columns of matrix (a_{ij}). The statement that the rank of matrix (a_{ij}) is r means that (a_{ij}) contains at least one r-rowed square submatrix with a nonzero determinant, whereas the determinants of all submatrices of (a_{ij}) with more than r rows are zero. According to a theorem on matrices (5, Sec. 20, Theorem 3), there is at least one r-rowed principal minor of matrix (a_{ij}) that is not zero [that is, a minor whose principal diagonal coincides with the principal diagonal of matrix (a_{ij})]. This conclusion follows from the fact that $a_{ij} = a_{ji}$. There are several other theorems that facilitate the calculation of the rank of a symmetrical matrix (5). It is proved in algebra (5, Sec. 47) that the quadratic form Q can be factored into a product of two linear forms if, and only if, the rank of matrix (a_{ij}) is not greater than 2. Of course, the factors may contain complex coefficients. If $r < n$, the quadratic form Q is said to be "singular."

Suppose that the x's are expressed in terms of other variables $(y_1, y_2, \cdots, y_n)$ by the linear equations

$$x_i = \sum_{j=1}^{n} t_{ij}y_j \qquad \text{(A-2)}$$

in which the coefficients t_{ij} are real constants with a nonzero determinant, det (t_{ij}). Then, conversely, the y's are determined uniquely by the x's. Equation (A-2) is said to define a nonsingular linear transformation of the variables. The term "nonsingular" signifies that det $(t_{ij}) \neq 0$.

If Eq. (A-2) is substituted into Eq. (A-1), Q becomes a real quadratic form in the y's. It is shown in algebra (5) that the coefficients t_{ij} can be chosen so that Q is reduced to the following form:

$$Q = y_1{}^2 + y_2{}^2 + \cdots + y_I{}^2 - y_{I+1}^2 - \cdots - y_r^2 \qquad \text{(A-3)}$$

where r is the rank of matrix (a_{ij}). Equation (A-3) is called the *canonical form* of the quadratic form Q. The number of positive terms in Eq. (A-3) (denoted by I) is called the *index* of the quadratic form. It is determined uniquely by the matrix (a_{ij}).

The type of a quadratic form is determined by the rank r and the index I, as follows:

(a) Q is positive definite if, and only if, $I = n$.

(b) Q is positive semidefinite if, and only if, $I = r < n$.

(c) Q is negative definite if, and only if, $I = 0$ and $r = n$.

(d) Q is negative semidefinite if, and only if, $I = 0$ and $r < n$.

(e) Q is indefinite if, and only if, $0 < I < r$.

These conditions are immediate consequences of Eq. (A-3). In the first place, Q obviously takes no negative values if all the terms in Eq. (A-3) are positive. Conversely, Q takes no positive values if all the terms in Eq. (A-3) are negative. Accordingly, in either of these cases Q is definite or semidefinite. If Eq. (A-3) contains both positive and negative terms, Q takes both positive and negative values, and therefore Q is indefinite. Accordingly, statement (e) is verified. The cases that remain to be considered are those in which all terms in Eq. (A-3) have the same sign. If all terms in Eq. (A-3) are positive (or negative) and $r = n$, Q is zero only if all the y's are zero. Then, since det $(t_{ij}) \neq 0$, all the x's are zero. Consequently, Q is positive definite (or negative definite). If all the terms in Eq. (A-3) are positive (or negative) and $r < n$, Q is zero only if $y_1 = y_2 = \cdots = y_r = 0$. However, we may assign arbitrary values to y_{r+1}, $y_{r+2}, \cdots, y_n$ without affecting the condition $Q = 0$, since $y_{r+1}, y_{r+2}, \cdots, y_n$ do not occur in Eq. (A-3). Then there are nonzero values of the x's for which $Q = 0$. Hence Q is positive semidefinite (or negative semidefinite) Thus statements (a), (b), (c), (d), (e) are verified.

Direct Determination of Index. The index I of the quadratic form Q may be derived from properties of the determinant D of matrix (a_{ij}) without explicit consideration of a transformation that reduces Q to sums and differences of squares. Consider the sequence of numbers $(1, M_1, M_2, \cdots, M_r)$ where r is the rank of matrix (a_{ij}), and the quantities M_i are any principal minors of determinant D with the following properties: (a) M_i is a first principal minor of M_{i+1}; that is, M_i is obtained by crossing out one row and the same column from M_{i+1}; $i = 1, 2, \cdots, r - 1$. (b) $M_r \neq 0$. (c) No two consecutive terms in the sequence of M's are zero.

A sequence of minors with these properties is called an *indicial sequence*. If $r = 0$, an indicial sequence degenerates to the single number 1. Supposing that $r > 0$, we may construct an indicial sequence as follows: choose a nonzero number on the principal diagonal of matrix (a_{ij}), if there is such a number. This number is M_1. If all elements on the principal diagonal of matrix (a_{ij}) are zero, set $M_1 = 0$, and let M_2 be any nonzero two-rowed principal minor of (a_{ij}). Such a minor exists, for, if all one-rowed and two-rowed principal minors of (a_{ij}) are zero, all principal minors of (a_{ij}) are zero. This conclusion follows from the following theorem (5, Sec. 20, Theorem 2):

If all the $(k + 1)$-rowed and $(k + 2)$-rowed principal minors of a symmetric matrix (b_{ij}) are zero, the rank of the matrix is k or less.

If $M_1 \neq 0$, augment M_1 by numbers from one row and the same column of matrix (a_{ij}) to form M_2. If possible, choose the augmenting row and column so that $M_2 \neq 0$. Then augment M_2 by numbers from one row and the same column of matrix (a_{ij}) to form M_3, so that $M_3 \neq 0$, if possible. If $r > 2$ and $M_2 = 0$, it is certainly possible to construct $M_3 \neq 0$. This conclusion follows from the following theorem (5, Sec. 20, Theorem 1):

If an r-rowed principal minor of a symmetric matrix (a_{ij}) is not zero, and if all the principal minors obtained by adding one row and the same column, and also all those obtained by adding two rows and the same two columns, to M_r are zero, the rank of (a_{ij}) is r.

Continuing with the preceding method, we obtain nested principal minors $(1, M_1, M_2, \cdots, M_{r-1})$, such that no two consecutive minors are zero. If $M_{r-1} = 0$, the preceding theorem shows that we can select $M_r \neq 0$. Also, if $M_{r-1} \neq 0$, we can select $M_r \neq 0$, since all $(r + 1)$-rowed minors of (a_{ij}) are zero. Hence, if there were no nonzero principal minor M_r containing M_{r-1} as a first principal minor, the rank would be $r - 1$ by the preceding theorem. Thus in all cases an indicial sequence can be constructed.

Any zero in an indicial sequence lies between adjacent terms with opposite signs. This is obviously true for M_1, since, if $M_1 = 0$, all nonzero two-rowed principal minors of (a_{ij}) that contain M_1 are negative. In general, the conclusion follows from the following theorem (37, Sec. 3, Lemma 1):

If H is a real nonzero $(k + 1)$-rowed symmetric determinant, if H'' is the $(k - 1)$-rowed determinant in the upper left corner of H, and if $H_{i,j}$ denotes the minor of H obtained by deleting the ith row and the jth column of H, then

$$HH'' = H_{k+1,k+1}H_{k,k} - H_{k+1,k}^2 \tag{a}$$

To apply this theorem, let $M_k = 0$, $k < r$. Then $M_{k-1} \neq 0$ and $M_{k+1} \neq 0$. Let the rows of M_{k+1} be permuted, if necessary, and let the columns be permuted in the same way so that the elements forming M_{k-1} will lie in the upper-left corner of the transposed determinant. This permutation does not alter the values of the principal minors. Denote the permuted determinants M_{k+1} and M_{k-1} by H and H''. Since $M_k = 0$, $H_{k,k} = 0$ or $H_{k+1,k+1} = 0$. Hence, by Eq. (a), $HH'' < 0$, or $M_{k-1}M_{k+1} < 0$. Therefore, M_{k-1} and M_{k+1} have opposite signs.

Two consecutive terms in the sequence $(1, M_1, M_2, \cdots, M_r)$ are said to have a "permanence of sign" if they have the same sign. The number of permanences of sign in the sequence is not affected by signs attributed to the zeros, since a zero always lies between adjacent terms with opposite signs. The following theorem is proved in algebra (37, Theorem 4):

> *The index I of the quadratic form Q equals the number of permanences of sign in any indicial sequence $(1, M_1, M_2, \cdots, M_r)$, wherein any zero M_i is given an arbitrary sign.*

In (37) this theorem is proved for the case in which the matrix (a_{ij}) is arranged so that an indicial sequence is formed by the leading principal minors; that is, by the principal minors in the upper-left corner. Then the matrix is said to be "regular." If the matrix is not regular, it may be permuted as explained before, so that the leading principal minors are numerically equal to the terms $(M_1, M_2, \cdots, M_r)$ in any preassigned indicial sequence; more specifically, the leading i-rowed principal minor may be obtained by subjecting M_i to a permutation of rows and the same permutation of columns. This transformation of the matrix effectively permutes the variables x_i in the quadratic form Q. Since a permutation of the variables does not change the range of values of Q, the foregoing theorem remains valid for any indicial sequence.

The following theorem is occasionally useful:

> *If the determinant D of matrix (a_{ij}) contains any principal minor that is negative, Q is negative definite, negative semidefinite, or indefinite.*

For proof, a principal minor M of determinant D is postulated to be negative. A quadratic form R is formed with the coefficients that occur in M. An indicial sequence formed from determinant M is represented by $(1, M_1, M_2, \cdots, M)$. Since $M < 0$, the preceding theorem shows that the index of the quadratic form R is less than the number of rows in determinant M. Hence R is negative definite, negative semidefinite, or indefinite. The same conclusion holds for Q, since $Q = R$ if those x's that do not occur in R are set equal to zero.

A demonstration quite similar to the preceding argument shows that Q is not positive definite if the determinant D contains any principal minor that is zero. Consequently, Q is positive definite if, and only if, all principal minors of D are positive. However, the signs of all principal minors need not be determined. It suffices to show that $r = n$ and that all terms in an indicial sequence are positive.

For example, consider the quadratic form

$$Q = 2x^2 + xy + 3xz - 2xw + yx + 5y^2 + 4yz + 3yw + 3zx$$
$$+ 4zy + 2z^2 + 7zw - 2wx + 3wy + 7wz - 10w^2$$

The coefficients possess the following sequence of subdeterminants:

$$
\begin{vmatrix} 2 & 1 & 3 & -2 \\ 1 & 5 & 4 & 3 \\ 3 & 4 & 2 & 7 \\ -2 & 3 & 7 & -10 \end{vmatrix} = 0, \qquad M_3 = \begin{vmatrix} 2 & 1 & 3 \\ 1 & 5 & 4 \\ 3 & 4 & 2 \end{vmatrix} = -35
$$

$$
M_2 = \begin{vmatrix} 2 & 1 \\ 1 & 5 \end{vmatrix} = 9, \qquad M_1 = 2
$$

Since the fourth-order determinant is zero, but there is a nonzero third-order determinant, $r = 3$. An indicial sequence is $(1, M_1, M_2, M_3) = (1, 2, 9, -35)$. There are two permanences of sign in this sequence; hence $I = 2$. Since $I = 2$ and $r = 3$, the quadratic form is indefinite. The mere fact that M_3 is negative shows that Q is negative definite, negative semidefinite, or indefinite.

A-2. PRINCIPAL-AXIS THEORY OF QUADRATIC FORMS. The class of linear transformations that has been discussed is more general than necessary for the reduction of a quadratic form to the canonical form. The reduction can be effected by a special type of linear transformation, known as an "orthogonal transformation." For an orthogonal transformation, the sum of the products of corresponding numbers in any two rows or any two columns of matrix (t_{ij}) is zero [see Eq. (A-2)]. Also, the matrix (t_{ij}) may be normalized; that is, the sum of the squares of the terms in any row or any column may be set equal to 1.

For simplicity, we first consider a quadratic form in three variables (x, y, z). We adopt a geometric approach to the theory by regarding (x, y, z) as rectangular coordinates. A quadratic form in these variables is represented by the equation,

$$
\begin{aligned}
Q = a_{11}x^2 &+ a_{12}xy + a_{13}xz \\
&+ a_{21}yx + a_{22}y^2 + a_{23}yz \\
&+ a_{31}zx + a_{32}zy + a_{33}z^2
\end{aligned}
\tag{A-4}
$$

in which $a_{ij} = a_{ji}$. The coefficients a_{ij} are real constants. To each point in space, there corresponds a value of Q; that is, Q is a point function.

Let (ξ, η, ζ) be another rectangular coordinate system with the same origin as the system (x, y, z). Let the two systems be both right-handed or both left-handed. The cosine of the angle between any two axes of the two systems may be represented by the following matrix:

TABLE A-1

	x	y	z
ξ	l_1	m_1	n_1
η	l_2	m_2	n_2
ζ	l_3	m_3	n_3

There are numerous identities among the direction cosines in this table (see Sec. 7-8). The equations of transformation of the coordinates of any point are

$$x = l_1\xi + l_2\eta + l_3\zeta$$
$$y = m_1\xi + m_2\eta + m_3\zeta \qquad \text{(A-5)}$$
$$z = n_1\xi + n_2\eta + n_3\zeta$$

If Eqs. (A-5) are substituted into Eq. (A-4), the quadratic form adopts the following form:

$$\begin{aligned} Q = & \, b_{11}\xi^2 + b_{12}\xi\eta + b_{13}\xi\zeta \\ & + b_{21}\eta\xi + b_{22}\eta^2 + b_{23}\eta\zeta \qquad \text{(A-6)} \\ & + b_{31}\zeta\xi + b_{32}\zeta\eta + b_{33}\zeta^2 \end{aligned}$$

The coefficients b_{ij} are symmetrical; that is, $b_{ij} = b_{ji}$. It is to be shown that the direction cosines in the preceding matrix may be chosen so that Eq. (A-6) reduces to the form,

$$Q = b_{11}\xi^2 + b_{22}\eta^2 + b_{33}\zeta^2 \qquad \text{(A-7)}$$

This theorem is proved most easily by indirect means. Let S denote the surface of the unit sphere, $x^2 + y^2 + z^2 = 1$. The values of Q on S attain an absolute maximum at a point P_1, since any function that is continuous in a closed finite region attains an absolute maximum in that region. This statement does not preclude the possibility that there are other points on S at which Q equals its value at P_1. In fact, since the values of Q are equal at diametrically opposite points on S, there are at least two points on S that fulfill the requirement of point P_1.

The plane through the origin O that is perpendicular to the unit vector OP_1 intersects S in a great circle C that is the equator on S with respect to the pole P_1. The values of Q on C attain an absolute maximum at a point P_2. A third point P_3 on S is defined by the condition that the three vectors OP_1, OP_2, OP_3 are mutually perpendicular. The sense of the vector OP_3 is defined as such that the three vectors OP_1, OP_2, OP_3 form a right-handed or left-handed system, according as the coordinate axes x, y, z are right-handed or left-handed.

Let the ξ-, η-, ζ-axes coincide respectively with the unit vectors OP_1, OP_2, OP_3. In other words, the vectors OP_1, OP_2, OP_3 are the unit vectors **a**, **b**, **c** of the ξ, η, ζ coordinate system. With reference to the coordinates (ξ, η, ζ), Q assumes the form of Eq. (A-6). It is to be shown that $b_{23} = b_{31} = b_{12} = 0$, so that Eq. (A-6) reduces to Eq. (A-7). The proof requires a digression for the establishment of a lemma.

The point P_1 has coordinates $\xi = 1$, $\eta = 0$, $\zeta = 0$. Consequently, by Eq. (A-6), b_{11} is the value of Q at point P_1. Therefore, by the definition of point P_1, b_{11} is the absolute maximum of Q on S. Accordingly, $Q - b_{11} \leq 0$ in the region S.

Consider the auxiliary function

$$G = Q - b_{11}(\xi^2 + \eta^2 + \zeta^2) \tag{a}$$

where Q is represented by Eq. (A-6). In view of the preceding remark, $G \leq 0$ in the region S. On any straight line through the origin Q is proportional to the square of the distance R from the origin, since the variables ξ, η, ζ are proportional to R. Consequently, since $G \leq 0$ in the region S, G is not positive anywhere. Therefore, since $G = 0$ at point P_1, G is a negative semidefinite quadratic form.

Having established this lemma, let us set $\xi = 1$, $\eta = u$, $\zeta = 0$, where u is an independent variable. By Eq. (a),

$$G(1, u, 0) = u[2b_{12} + u(b_{22} - b_{11})] \tag{b}$$

If $b_{11} = b_{22}$, set $u = b_{12}$. Then, by Eq. (b), $G(1, u, 0) = 2b_{12}{}^2$. Hence, since G is never positive, $b_{12} = 0$.

If $b_{11} \neq b_{22}$, set $u = b_{12}/|b_{22} - b_{11}|$. Then, by Eq. (b)

$$G(1, u, 0) = \frac{b_{12}(2b_{12} \pm b_{12})}{|b_{22} - b_{11}|} \tag{c}$$

Since G is never positive, Eq. (c) shows again that $b_{12} = 0$. Accordingly, in all cases, $b_{12} = 0$.

Now, to show that $b_{23} = 0$, we introduce a function H. defined by

$$H = Q - b_{22}(\xi^2 + \eta^2 + \zeta^2) \tag{d}$$

Then, reasoning as before, we see that $H \leq 0$ on the equatorial plane, $\xi = 0$. Also,

$$H(0, 1, u) = u[2b_{23} + u(b_{33} - b_{22})]$$

Since point $(0, 1, u)$ lies on the equatorial plane, we proceed as in the preceding case to show that $b_{23} = 0$.

Since $b_{23} = b_{12} = 0$, Eq. (a) yields

$$G(1, 0, u) = u[2b_{13} + (b_{33} - b_{11})u] \qquad (e)$$

If $b_{33} = b_{11}$, set $u = b_{13}$. Then, since G is never positive, Eq. (e) shows that $b_{13} = 0$. If $b_{33} \neq b_{11}$, set $u = b_{13}/|b_{33} = b_{11}|$. Then, again, Eq. (e) yields $b_{13} = 0$.

Thus the theorem is proved; that is, Q reduces to the canonical form [Eq. (A-7)] when it is expressed in terms of the coordinates (ξ, η, ζ). The axes ξ, η, ζ are called the "principal axes" of the quadratic form.

The foregoing existence proof provides a method for determining the principal axes of a quadratic form, but a more symmetric method may be used. Equation (A-7) shows that the principal axes coincide with the vector field grad Q. It also shows that if no two of the coefficients b_{11}, b_{22}, b_{33} are equal the principal axes are the only straight lines through the origin that coincide with grad Q. If two of the coefficients b_{11}, b_{22}, b_{33} are equal, Eq. (A-7) shows that the function Q is rotationally symmetric with respect to one of the principal axes. The other two principal axes may be chosen arbitrarily, provided only that the three axes are mutually perpendicular and concurrent at the origin. If all three coefficients b_{11}, b_{22}, b_{33} are equal, any three mutually perpendicular lines through the origin are principal axes.

The problem thus reduces to the determination of the straight lines through the origin that coincide with grad Q. Any such line is a principal axis and vice versa. Consequently, the principal axes are characterized by the equation grad $Q = 2\lambda \mathbf{R}$, where $\mathbf{R} = \mathbf{i}x + \mathbf{j}y + \mathbf{k}z$, and λ is a scalar. Furthermore, Eq. (A-7) shows that λ is a constant for a principal axis. Since grad $(x^2 + y^2 + z^2) = 2\mathbf{R}$, the preceding equation is equivalent to grad $Q = \lambda$ grad $(x^2 + y^2 + z^2)$. This, in turn, is equivalent to grad $H = 0$, where

$$H = Q - \lambda(x^2 + y^2 + z^2) \qquad (A-8)$$

In other words, for any principal axis, there exists a constant λ, such that the equation grad $H = 0$ is satisfied at all points on that axis. Conversely, any straight line through the origin on which grad $H = 0$ is a principal axis, since, on that line, grad $Q = \lambda$ grad $(x^2 + y^2 + z^2) = 2\lambda \mathbf{R}$.

The equation grad $H = 0$ is equivalent to the three scalar equations $\partial H/\partial x = \partial H/\partial y = \partial H/\partial z = 0$. With Eqs. (A-4) and (A-8), these equations yield

$$(a_{11} - \lambda)x + a_{12}y + a_{13}z = 0$$

$$a_{21}x + (a_{22} - \lambda)y + a_{23}z = 0 \qquad (A-9)$$

$$a_{31}x + a_{32}y + (a_{33} - \lambda)z = 0$$

By the theory of linear equations, a necessary and sufficient condition that Eqs. (A-9) possess a solution other than $x = y = z = 0$ is

$$\begin{vmatrix} a_{11} - \lambda & a_{12} & a_{13} \\ a_{21} & a_{22} - \lambda & a_{23} \\ a_{31} & a_{32} & a_{33} - \lambda \end{vmatrix} = 0 \qquad \text{(A-10)}$$

Equation (A-10) is a cubic equation in λ. Consequently, it has three roots λ_1, λ_2, λ_3, although they are not necessarily distinct. By the theory of linear equations (5), Eqs. (A-9) possess a solution $x = a$, $y = b$, $z = c$, such that $a^2 + b^2 + c^2 = 1$, provided that λ is a root of Eq. (A-10).

Supposing that λ is a root of Eq. (A-10), let us substitute the solution (a, b, c) into Eqs. (A-9). Then, if we multiply the first, second, and third of these equations by a, b, and c, respectively, and add the resulting equations, we obtain

$$Q(a, b, c) - \lambda(a^2 + b^2 + c^2) = 0$$

Since $a^2 + b^2 + c^2 = 1$, this equation yields $Q(a, b, c) = \lambda$; that is, λ is the value of Q at a point on the unit sphere. This conclusion shows that the roots λ_1, λ_2, λ_3 are all real. This property is attributable to symmetry of the matrix (a_{ij}), since a_{ij} may be any real constants, such that $a_{ij} = a_{ji}$. It may easily be shown by a specific numerical example that some of the roots of Eq. (A-10) can be complex if the condition $a_{ij} = a_{ji}$ is waived.

If any root of Eq. (A-10) is substituted into Eqs. (A-9), then the latter equations determine a principal axis, since all solutions of Eqs. (A-9) satisfy the condition grad $Q = 2\lambda\mathbf{R}$. Since λ_1, λ_2, λ_3 are the values of Q at the intercepts of the principal axes with the unit sphere, they do not depend on the choice of the coordinate system (x, y, z); in other words, they are invariants. Consequently, we may study the nature of the solution by letting (x, y, z) be principal axes. Then $a_{23} = a_{31} = a_{12} = 0$. Accordingly, we may easily verify the following conditions, which remain valid when x, y, z are arbitrary orthogonal axes: if Eq. (A-10) has no multiple roots, then the principal axes of the quadratic form Q are unique and are determined completely by Eqs. (A-9) and (A-10). If Eq. (A-10) possesses a double root, then the function Q is axially symmetrical. The root that is not doubled determines the axis of symmetry, which is a principal axis. The other two principal axes may be any two perpendicular lines that are also perpendicular to the axis of symmetry at the origin. If the three roots of Eq. (A-10) are all equal, then the function Q is spherically symmetrical. The principal axes may be designated as any three mutually perpendicular lines through the origin.

Since $(\lambda_1, \lambda_2, \lambda_3)$ are the values of Q at the intercepts of the principal axes with the unit sphere and since Q assumes the form of Eq. (A-7) with respect to these axes, we obtain

$$Q = \lambda_1 \xi^2 + \lambda_2 \eta^2 + \lambda_3 \zeta^2 \qquad \text{(A-11)}$$

Equation (A-10) may be expressed in the form

$$\lambda^3 - J_1 \lambda^2 + J_2 \lambda - J_3 = 0 \qquad \text{(A-12)}$$

where

$$J_1 = a_{11} + a_{22} + a_{33}, \qquad J_3 = \det (a_{ij})$$

$$J_2 = \begin{vmatrix} a_{22} & a_{23} \\ a_{32} & a_{33} \end{vmatrix} + \begin{vmatrix} a_{11} & a_{13} \\ a_{31} & a_{33} \end{vmatrix} + \begin{vmatrix} a_{11} & a_{12} \\ a_{21} & a_{22} \end{vmatrix} \qquad \text{(A-13)}$$

If the rank of matrix (a_{ij}) is 2, $J_3 = 0$. Then one root of Eq. (A-12) is $\lambda = 0$. If the rank of matrix (a_{ij}) is 1, $J_2 = J_3 = 0$. Then $\lambda = 0$ is a double root of Eq. (A-12). In general, if the rank of matrix (a_{ij}) is r, zero is a root of Eq. (A-12) with multiplicity $3 - r$. Hence only r nonzero terms appear in Eq. (A-11).

Extension to n Dimensions. By means of the idea of an n-dimensional euclidean space, the principal-axis theory may be extended to quadratic forms in n variables (13). The principal axes are n mutually perpendicular lines through the origin of the n-dimensional rectangular cartesian co-ordinate system $(x_1, x_2, \cdots, x_n)$. If $(y_1, y_2, \cdots, y_n)$ are rectangular coordinates that coincide with the principal axes, then the quadratic form reduces to the canonical form,

$$Q = \lambda_1 y_1^2 + \lambda_2 y_2^2 + \cdots + \lambda_r y_r^2 \qquad \text{(A-14)}$$

where r is the rank of matrix (a_{ij}). None of the coefficients in Eq. (A-14) is zero. The coefficients λ_i are the roots of the determinantal equation

$$\begin{vmatrix} a_{11} - \lambda & a_{12} & a_{13} & \cdots & a_{1n} \\ a_{21} & a_{22} - \lambda & a_{23} & \cdots & a_{2n} \\ \cdot & \cdot & \cdot & \cdot & \cdot \\ a_{n1} & a_{n2} & a_{n3} & \cdots & a_{nn} - \lambda \end{vmatrix} = 0 \qquad \text{(A-15)}$$

Only r terms appear in Eq. (A-14) because $n - r$ of the roots of Eq. (A-15) are zero. Some of the λ's in Eq. (A-14) may be positive and the others negative. The number of positive terms in Eq. (A-14) is the index of the quadratic form Q.

A-3. THEORY OF PRINCIPAL AXES OF STRAIN. The theory of principal axes of quadratic forms presented in Sec. A-2 has important applications in the theories of stress, strain, and moments of inertia. The applications in these fields are quite similar. A brief discussion of the theory of principal axes of strain will suffice as an illustration.

The strain ϵ of a line element that passes through the point (x, y, z) in the direction (l, m, n) is given by Eq. (4-2). It is convenient to consider the quantity $\phi = \epsilon + \frac{1}{2}\epsilon^2$. Equation (4-2) shows that when the point (x, y, z) is held fixed ϕ is a quadratic form in the variables l, m, n. To apply the principal-axis theory of quadratic forms, we regard l, m, n as rectangular cartesian coordinates in space. Only points on the units sphere, $l^2 + m^2 + n^2 = 1$, have physical significance, but insofar as the theory of quadratic forms is concerned (l, m, n) may take any real values. The quantity ϕ is a point function in (l, m, n) space.

Following the theory of Sec. A-2, we consider the equation

$$\begin{vmatrix} \epsilon_x - \phi & \frac{1}{2}\gamma_{xy} & \frac{1}{2}\gamma_{xz} \\ \frac{1}{2}\gamma_{yx} & \epsilon_y - \phi & \frac{1}{2}\gamma_{yz} \\ \frac{1}{2}\gamma_{zx} & \frac{1}{2}\gamma_{zy} & \epsilon_z - \phi \end{vmatrix} = 0$$

Expansion of the determinant yields

$$\phi^3 - I_1\phi^2 + I_2\phi - I_3 = 0 \tag{A-16}$$

where I_1, I_2, I_3 are defined by Eq. (4-7). The roots ϕ_1, ϕ_2, ϕ_3 of Eq. (A-16) are real. The strains corresponding to ϕ_1, ϕ_2, ϕ_3 are called the "principal strains." According to the principal-axis theory, one of these roots is the absolute maximum value of ϕ for line elements through the point (x, y, z); another is the absolute minimum value of ϕ for line elements through that point. The third root is a stationary value of ϕ on the unit sphere, $l^2 + m^2 + n^2 = 1$, but it is not necessarily a maximum or a minimum value. The stationary character is seen from the fact that grad ϕ is normal to the unit sphere in (l, m, n) space at the points corresponding to ϕ_1, ϕ_2, ϕ_3; hence the directional derivative of ϕ in any direction tangent to the sphere is zero at each of these points.

The principal axes of the quadratic form represented by Eq. (4-2) are called the "principal axes of strain" corresponding to point (x, y, z). According to Sec. A-2, these axes are determined by the equations

$$(\epsilon_x - \phi)l + \frac{1}{2}\gamma_{xy}m + \frac{1}{2}\gamma_{xz}n = 0$$
$$\frac{1}{2}\gamma_{yx}l + (\epsilon_y - \phi)m + \frac{1}{2}\gamma_{yz}n = 0 \tag{A-17}$$
$$\frac{1}{2}\gamma_{zx}l + \frac{1}{2}\gamma_{zy}m + (\epsilon_z - \phi)n = 0$$

To any root of Eq. (A-16), there corresponds a solution of Eqs. (A-17), such that $l^2 + m^2 + n^2 = 1$. Furthermore, in all cases three solutions (l_1, m_1, n_1), (l_2, m_2, n_2), (l_3, m_3, n_3), that represent mutually perpendicular directions may be determined. Hence the three principal directions at point (x, y, z) are mutually perpendicular. The line elements through point (x, y, z) that experience maximum and minimum strains coincide with two of these directions.

It is now shown that there is no shearing strain between line elements in the principal directions. The shearing strain between line elements in any two perpendicular directions (l_1, m_1, n_1) and (l_2, m_2, n_2) is given by Eq. (4-4). This equation may be written as follows:

$$\tfrac{1}{2}(1 + \epsilon_1)(1 + \epsilon_2) \cos \theta = (\epsilon_x l_1 + \tfrac{1}{2}\gamma_{xy} m_1 + \tfrac{1}{2}\gamma_{xz} n_1)l_2$$
$$+ (\tfrac{1}{2}\gamma_{yx} l_1 + \epsilon_y m_1 + \tfrac{1}{2}\gamma_{yz} n_1)m_2 + (\tfrac{1}{2}\gamma_{zx} l_1 + \tfrac{1}{2}\gamma_{zy} m_1 + \epsilon_z n_1)n_2$$

Hence, if (l_1, m_1, n_1) and (l_2, m_2, n_2) are principal directions, Eqs. (A-17) yield

$$\tfrac{1}{2}(1 + \epsilon_1)(1 + \epsilon_2) \cos \theta = \phi_1(l_1 l_2 + m_1 m_2 + n_1 n_2)$$

Since the vectors (l_1, m_1, n_1) and (l_2, m_2, n_2) are perpendicular, this yields $\cos \theta = 0$. This conclusion means that if a body (e.g., a piece of putty) is subjected to any continuous and differentiable deformation there exist three mutually perpendicular line elements through each point that remain perpendicular under the deformation. These line elements coincide with the principal axes of strain. We may speak accordingly of the principal axes of strain for the unstrained body or the strained body. For the unstrained body, these axes have the directions (l_1, m_1, n_1), (l_2, m_2, n_2), (l_3, m_3, n_3) determined by Eqs. (A-17) from the three roots ϕ_1, ϕ_2, ϕ_3 of Eq. (A-16). Line elements in these directions pass into line elements in the principal directions $(l_1{}^*, m_1{}^*, n_1{}^*)$, $(l_2{}^*, m_2{}^*, n_2{}^*)$, $(l_3{}^*, m_3{}^*, n_3{}^*)$ for the strained body. According to the preceding conclusion, the three starred unit vectors are mutually perpendicular, as are the unstarred ones. An infinitesimal rectangular parallelepiped with edges parallel to the principal directions consequently remains a rectangular parallelepiped under the deformation.

A-4. SIMULTANEOUS TRANSFORMATION OF TWO QUADRATIC FORMS TO THE CANONICAL FORM.

Let two real quadratic forms be defined* by

$$V = \tfrac{1}{2} \sum_{i=1}^{n} \sum_{j=1}^{n} a_{ij} x_i x_j, \qquad T = \tfrac{1}{2} \sum_{i=1}^{n} \sum_{j=1}^{n} b_{ij} y_i y_j$$

* The factor $\tfrac{1}{2}$ is introduced for correlation with the theory of vibrations (Sec. 8-2).

If λ is a real parameter, $F(\lambda) = \det(a_{ij} - \lambda b_{ij})$ is an nth degree polynomial in λ; that is,

$$F(\lambda) = A_0 - A_1\lambda + A_2\lambda^2 - A_3\lambda^3 + \cdots + (-1)^n A_n\lambda^n$$

It may be shown that $A_0 = \det(a_{ij})$ and $A_n = \det(b_{ij})$. More generally, A_k is the sum of all determinants that can be formed by replacing k columns of $\det(a_{ij})$ by the corresponding columns of matrix (b_{ij}).

The equation $F(\lambda) = 0$ is called the "characteristic equation" of the two quadratic forms T and V. It possesses n roots $(\lambda_1, \lambda_2, \cdots, \lambda_n)$, provided that $\det(b_{ij}) \neq 0$, although the roots need not all be distinct. It is shown in algebra (5, Sec. 59) that the roots λ_i are all real, provided that the quadratic form T is not indefinite nor singular.

If the variables x_i and y_i are subjected to identical nonsingular linear transformations,

$$x_i = \sum_{j=1}^{n} t_{ij}u_j, \qquad y_i = \sum_{j=1}^{n} t_{ij}v_j, \qquad \det(t_{ij}) \neq 0 \qquad \text{(a)}$$

the quadratic forms are transformed to

$$V = \tfrac{1}{2}\sum_{i=1}^{n}\sum_{j=1}^{n} a_{ij}'u_iu_j, \qquad T = \tfrac{1}{2}\sum_{i=1}^{n}\sum_{j=1}^{n} b_{ij}'v_iv_j$$

It is shown in algebra (5, Sec. 57) that the roots of the equation $\det(a_{ij}' - \lambda b_{ij}') = 0$ are again $(\lambda_1, \lambda_2, \cdots, \lambda_n)$. In other words, the quantities λ_i are invariant when the variables x_i and y_i are subjected to any transformation of the type of equations in (a).

The following algebraic theorem (5, Sec. 59) is important in the theory of vibrations of systems with finite degrees of freedom:

If $2T$ is a positive definite quadratic form, and if $2V$ is any real quadratic form, the two quadratic forms can be reduced by a transformation of the type of equations in (a) to the form:,

$$\begin{aligned} V &= \tfrac{1}{2}(\lambda_1 u_1^2 + \lambda_2 u_2^2 + \cdots + \lambda_n u_n^2) \\ T &= \tfrac{1}{2}(v_1^2 + v_2^2 + \cdots + v_n^2) \end{aligned} \qquad \text{(A-18)}$$

where the λ's are the zeros of $F(\lambda)$.

Insofar as this theorem is concerned, the roots λ_i need not all be distinct. The theorem shows that if V is positive definite, the roots λ_i are all positive. Consequently, in this case, we may set $\lambda_i = \omega_i^2$, where ω_i is a positive number.

In vibration theory it is desirable to find the matrix (t_{ij}) that effects the preceding transformation. Attention is restricted to the case in which the roots λ_k are distinct. In a vibration problem equal roots can be avoided by an arbitrarily small change of the spring constants or other parameters of the system.

The relation that determines the λ's is the vanishing of the determinant of the coefficients in the equation

$$\sum_{j=1}^{n}(a_{ij} - \lambda b_{ij})t_j = 0 \qquad\qquad\text{(b)}$$

If λ is any one of the zeros λ_k of $F(\lambda)$, Eq. (b) yields a solution $(t_1, t_2, \cdots, t_n)$, such that all the t's are not zero. If no two λ's are equal, the rank of the matrix $(a_{ij} - \lambda_k b_{ij})$ is $n - 1$; this is apparent from Eqs. (A-18). Consequently, the solution $(t_1, t_2, \cdots, t_n)$ corresponding to any root λ_k is unique, except for an arbitrary constant factor. This factor is determined by the supplementary condition

$$\sum_{i=1}^{n}\sum_{j=1}^{n} b_{ij}t_i t_j = 1 \qquad\qquad\text{(c)}$$

Equation (c) still permits the choice of a sign, for if $(t_1, t_2, \cdots, t_n)$ is a solution of Eqs. (b) and (c) so is $(-t_1, -t_2, \cdots, -t_n)$. However, the sign is immaterial. The solution of Eqs. (b) and (c) for $\lambda = \lambda_k$ may be written as the kth column of a matrix (t_{ij}). It is shown in the theory of quadratic forms (5) that the matrix (t_{ij}), constructed in this way, effects the reduction of T and V to the forms of Eqs. (A-18).

There is a geometric interpretation of the preceding theory based on the idea that the variables x_i or y_i are oblique cartesian coordinates in an n-dimensional euclidean space. This theory is developed clearly in the book by Lanczos (48). If n is large, the solution of the determinantal equation for the λ's presents a formidable computational problem. Numerical schemes for treating this problem are presented in the book by Faddeeva (23). (See also the last paragraph in Sec. 8-2).

Example. As a simple illustration of the foregoing theory, we consider the two quadratic forms,

$$2V = -2x_1 x_2 + x_2^2, \qquad 2T = y_1^2 - 2y_1 y_2 + 3y_2^2$$

The quadratic form $2T$ is positive definite. The matrices of the coefficients in the two quadratic forms are

$$(a_{ij}) = \begin{bmatrix} 0 & -1 \\ -1 & 1 \end{bmatrix}, \qquad (b_{ij}) = \begin{bmatrix} 1 & -1 \\ -1 & 3 \end{bmatrix}$$

The λ-equation is accordingly

$$\begin{vmatrix} -\lambda & -1+\lambda \\ -1+\lambda & 1-3\lambda \end{vmatrix} = 0$$

This reduces to $2\lambda^2 + \lambda - 1 = 0$. The roots are $\lambda_1 = \frac{1}{2}, \lambda_2 = -1$. Hence

$$\begin{bmatrix} -\lambda_1 & -1+\lambda_1 \\ -1+\lambda_1 & 1-3\lambda_1 \end{bmatrix} = \begin{bmatrix} -\frac{1}{2} & -\frac{1}{2} \\ -\frac{1}{2} & -\frac{1}{2} \end{bmatrix}$$

Therefore, by Eq. (b), the t's corresponding to λ_1 satisfy the equation, $-\frac{1}{2}t_1 - \frac{1}{2}t_2 = 0$. Also, by Eq. (c), $t_1{}^2 - 2t_1t_2 + 3t_2{}^2 = 1$. Consequently, $t_1 = 1/\sqrt{6}, t_2 = -1/\sqrt{6}$. These values form the first column of matrix (t_{ij}).

Likewise,

$$\begin{bmatrix} -\lambda_2 & -1+\lambda_2 \\ -1+\lambda_2 & 1-3\lambda_2 \end{bmatrix} = \begin{bmatrix} 1 & -2 \\ -2 & 4 \end{bmatrix}$$

Therefore, the t's corresponding to λ_2 satisfy the equation $t_1 - 2t_2 = 0$. The second equation for the t's (namely, $-2t_1 + 4t_2 = 0$) is equivalent to the equation $t_1 - 2t_2 = 0$. This fact results from the condition that the λ's are determined to ensure the vanishing of the determinant of coefficients in the equations for the t's; hence, in general, one of the equations for the t's, corresponding to any root λ_i, is redundant. In addition to the equation $t_1 - 2t_2 = 0$, Eq. (c) yields again $t_1{}^2 - 2t_1t_2 + 3t_2{}^2 = 1$. Consequently, $t_1 = 2/\sqrt{3}, t_2 = 1/\sqrt{3}$. These values are the second column in matrix (t_{ij}). Therefore,

$$(t_{ij}) = \begin{bmatrix} 1/\sqrt{6} & 2/\sqrt{3} \\ -1/\sqrt{6} & 1/\sqrt{3} \end{bmatrix}$$

By Eq. (a), the transformation that reduces T and V to the canonical form is accordingly

$$x_1 = \frac{u_1}{\sqrt{6}} + \frac{2u_2}{\sqrt{3}}, \qquad y_1 = \frac{v_1}{\sqrt{6}} + \frac{2v_2}{\sqrt{3}}$$

$$x_2 = -\frac{u_1}{\sqrt{6}} + \frac{u_2}{\sqrt{3}}, \qquad y_2 = -\frac{v_1}{\sqrt{6}} + \frac{v_2}{\sqrt{3}}$$

Substituting these relations into the formulas for T and V, we obtain

$$2V = \frac{1}{2}u_1{}^2 - u_2{}^2 = \lambda_1 u_1{}^2 + \lambda_2 u_2{}^2$$
$$2T = v_1{}^2 + v_2{}^2$$

Thus the desired transformation of V and T is accomplished.

A-5. QUADRATIC FORMS AND BUCKLING THEORY.* The

theorem stated in Sec. 6-3 is proved in this article. A conservative holonomic system with n degrees of freedom is considered. The potential energy
V is a function of the generalized coordinates $(x_1, x_2, \cdots, x_n)$ and a parameter p that may represent load, temperature, or any other physical
quantity that affects the equilibrium configuration or the stability. The
path Γ of equilibrium states in configuration space is represented by
$x_i = x_i(p)$. By the principle of stationary potential energy, this path is
determined by the equations $\partial V/\partial x_i = 0$. By Eqs. (1-27), the second
variation of potential energy is a quadratic form,

$$Q = \sum_{i=1}^{n} \sum_{j=1}^{n} a_{ij} h_i h_j, \qquad a_{ij} = \frac{\partial^2 V}{\partial x_i \, \partial x_j} \qquad \text{(A-19)}$$

Since only equilibrium states are considered, the coefficients a_{ij} are
functions of p alone. The buckling load p_{cr} is the greatest lower bound of
the values of p for which equilibrium is not stable. Equilibrium is stable
if Q is positive definite; it is unstable if Q is negative definite, negative
semidefinite, or indefinite (see Sec. 1-11). Consequently, equilibrium is
stable if all principal minors of the determinant, $D = \det(a_{ij})$, are positive;
it is unstable if any principal minor of D is negative. This type of stability
problem arises also in electrical theory and other branches of physics.

An investigation of the character of the quadratic form Q requires
consideration of the rank r of the symmetric matrix (a_{ij}). Since only points
on path Γ are considered, r is a function of p. The function $r(p)$ has the
possible set of values $(0, 1, 2, \cdots, n)$.

Cases in which the function $r(p)$ has a cluster point of discontinuities are
excluded. Continuity of the functions $a_{ij}(p)$ does not ensure this condition.
For example, if the determinant D is determined by $D = p \sin 1/p$, the
function $r(p)$ has infinitely many discontinuities in any neighborhood of
the point $p = 0$. To exclude such conditions, the functions $a_{ij}(p)$ are
postulated as analytic in an open interval π that covers the physically
significant range of p: that is, the functions $a_{ij}(p)$ shall admit power series
expansions with nonzero intervals of convergence about any point in π.
Analytic functions have the following properties: (a) They are continuous
in π. (b) The sum or product of any two is again an analytic function.
(c) An analytic function that is not identically zero vanishes at only a
finite number of points in any closed subinterval of π.

If the functions $a_{ij}(p)$ are analytic, the rank r has only a finite set of
discontinuities in any closed subinterval of π. The proof follows from the

---

* Presented at the Sixth Congress on Theoretical and Applied Mechanics, Indian
Society of Theoretical and Applied Mechanics, Delhi, December, 1960.

observation that if r has infinitely many discontinuities some minor of D has infinitely many zeros, although it is not identically zero. This condition is impossible, since the minors of D are analytic functions.

If the functions $a_{ij}(p)$ are analytic, the function $r(p)$ cannot have a step type of discontinuity, since such a discontinuity requires that some minor M of D maintain the value $M = 0$ over a finite interval of the p-axis, although M is not identically zero. This condition is impossible, since the function $M(p)$, being analytic, has only a finite number of zeros in any closed subinterval of π. Consequently, since $r(p)$ has only a finite set of possible values, $r(p) = K =$ constant in any closed subinterval of π, except for a finite set of points (possibly an empty set) on which $r(p) \neq K$. Furthermore, the inequality $r > K$ is impossible, since this would require that some minor of D with more than K rows maintain the value zero, except at isolated points. Since the minors of D are analytic, this condition cannot occur. Accordingly, the following theorem is established:

Lemma 1. If the functions $a_{ij}(p)$ are analytic in π, $r(p) = K =$ constant in any closed subinterval of π, except for a finite set of points on which $r < K$.

A conclusion that follows from Lemma 1 is the following:

Lemma 2. If $r(p)$ is continuous at a point p' in π, the index $I(p)$ is also continuous at p'.

For proof, we observe that Lemma 1 signifies that an open interval π', containing p', may be chosen so that r is constant in π'. Also, since the coefficients $a_{ij}(p)$ are analytic in π, the interval π' may be chosen so small that the principal minors of D which are not identically zero possess no zeros in π', except possibly at p'. Then a sequence S of minors of D that is an indicial sequence for $p = p'$ remains an indicial sequence for any value of p in π' (see Sec. A-1). Since the signs of the zeros in an indicial sequence are irrelevant, the sequence maintains a constant number of permanences of sign in π'. Therefore, by the theorem stated in Sec. A-1, $I(p)$ is constant in π'.

The preceding lemmas yield a proof of the following theorem, which was stated in Sec. 6-3:

> If the functions $a_{ij}(p)$ are analytic in π, if stable equilibrium exists when $p < \lambda$ (where λ is some constant in π), if the equation $D(p) = 0$ possesses a least root p_0 in π, and if $D < 0$ in an interval $p_0 < p < B$, then p_0 is the buckling load p_{cr}.

To prove this theorem, it is necessary to show first that Q is positive definite at all points in the range $p < p_0$. Since D is continuous, it maintains a constant sign in the range $p < p_0$. Therefore, $D > 0$ in the range $p < p_0$, for, if D were negative in this range, equilibrium would be unstable in the range $p < \lambda$. Since Q is positive definite in the range $p < \lambda$, $r = I = n$ in that range. Also, since $D > 0$, $r = n$ in the range $p < p_0$. Therefore, by Lemma 2, $r = I = n$ in the range $p < p_0$. Consequently, Q is positive definite in the range $p < p_0$.

On the other hand, Q is indefinite or negative definite throughout the interval $p_0 < p < B$, since D is negative in this interval. Consequently, p_0 is the greatest lower bound of the values of p for which equilibrium is not stable. In other words, the critical value of p is $p_0 = p_{cr}$.

The preceding theorem implies that the determinant D itself usually vanishes before any of its principal minors. This theorem consequently often eliminates the need for an investigation of a sequence of minors of D. By a different type of reasoning, based on the bifurcation concept, Poincaré (131) concluded that the equation $D = 0$ is a buckling criterion, but he did not derive conditions under which the equation $D = 0$ ensures buckling. An illustrative application of the theorem is given in Sec. 6-3.

In applications of the energy method to problems of buckling, the following condition is usually assumed, but the literature on buckling seemingly has not provided an algebraic proof of it:

> If the functions $a_{ij}(p)$ are analytic, Q is positive semidefinite when $p = p_{cr}$.

For proof, we note that equilibrium is stable in the interval $p < p_{cr}$. Therefore, Q is positive definite or positive semidefinite in the range $p < p_{cr}$.

Three cases must be considered. If $r < n$ when $p < p_{cr}$, Q is positive semidefinite when $p < p_{cr}$. Then buckling may occur while I and r remain constant (e.g., because of a change in the character of the fourth variation of potential energy). In this case, the theorem is obviously true.

Another conceivable condition is that I be continuous and r discontinuous at $p = p_{cr}$. However, the following argument shows that this case cannot occur. If r is discontinuous at p_{cr}, the discontinuity reduces r (by Lemma 1). Since $I = r$ for $p < p_{cr}$, this condition signifies that $I > r$ for $p = p_{cr}$. However, by definition, I can never exceed r; therefore, the case under consideration is impossible.

By Lemma 2, I cannot be discontinuous at p_{cr} unless r is also discontinuous. Therefore, the case that remains to be considered is that in which I and r are both discontinuous at p_{cr}. Let $I = r = K$ for $p < p_{cr}$, where K is a constant (see Lemma 1). Let $I = a$ and $r = b$ for $p = p_{cr}$. Then, by Lemma 1, $a \leq b < K \leq n$. Since $b < n$, Q cannot be positive definite for $p = p_{cr}$. It may happen that $b = 0$; then Q vanishes identically for $p = p_{cr}$. This case is included under the definition of positive semidefiniteness.

Since $I = r$ when $p < p_{cr}$, D contains no negative principal minors when $p < p_{cr}$. Consequently, since the principal minors of D are continuous functions of p, none can be negative when $p = p_{cr}$. Hence, when $p = p_{cr}$, no zero occurs in an indicial sequence, for a zero is enclosed between terms with opposite signs. Accordingly, when $p = p_{cr}$, all terms in an indicial sequence are positive. Therefore, $I = r$. Consequently, since $r = b < n$, Q is positive semidefinite when $p = p_{cr}$.

Applications of the preceding theorem in buckling theory have been discussed in Sec. 6-4.

PROBLEMS

1. Determine to which of the five classes the following quadratic forms belong:
   (a) $3x^2 - y^2 + 4z^2 + 6yz + 8zx + 2xy$
   (b) $5x^2 + 5y^2 + z^2 - 8xy + 2xz - 4yz$
   (c) $10x^2 + 7y^2 + 5z^2 + 12w^2 - xy - 2xz - 3zw$
   (d) $a^2 + b^2 + \frac{1}{2}c^2 + \frac{1}{3}d^2 - cd$
   (e) $a^2 + 4b^2 + 12c^2 + 4ab + 6ac - 4ad + 12bc + 6bd - 2cd$

2. Determine which of the following quadratic forms can be factored into products of two linear forms. Factor those that can be factored.
   (a) $2a^2 + b^2 - 3c^2 + 4d^2 - ab + ac + 3ad - 2bc - 5bd + 7cd$
   (b) $7a^2 + 3b^2 + 10c^2 - 4d^2 + 22ab - 37ac + 3ad - 11bc - 11bd + 18cd$
   (c) $x^2 + y^2$
   (d) $x^2 + y^2 - z^2$
   (e) $6x^2 - 5y^2 - 7xy + 14xz + 7yz$

3. $$Q = (2 - 6x)A^2 - 4(1 - x)AB + (9 - 2x)B^2.$$
   For what range of x is Q positive definite? Positive semidefinite? Negative definite? Negative semidefinite? Indefinite?

4. Show that there are infinitely many different linear transformations of the variables that will reduce the quadratic form $a_{11}x_1^2 + 2a_{12}x_1x_2 + a_{22}x_2^2$ to the canonical form if none of the a's is zero.

5. Determine the matrix of direction cosines for the principal axes (ξ, η, ζ) of the quadratic form
   $$Q = x^2 + y^2 - z^2 + 2xy - 4xz + yz$$
   Express Q in terms of (ξ, η, ζ).

6. The matrix of coefficients of a quadratic form Q is
   $$(a_{ij}) = \begin{bmatrix} 3 & 2 & 1 & 0 \\ 2 & -1 & 0 & 4 \\ 1 & 0 & 5 & 6 \\ 0 & 4 & 6 & -2 \end{bmatrix}$$
   Reduce Q to the canonical form by an orthogonal transformation without determining the directions of the principal axes.

7. $Q = x_1^2 + 4x_1x_2 - 2x_2^2$, $P = y_1^2 + y_2^2$. Determine the matric (t_{ij}) and reduce the two quadratic forms to the canonical form.

8. $Q = 3x_1^2 - 2x_1x_2 + 4x_2^2$, $P = y_1^2 + 2y_1y_2 + 2y_2^2$. Determine the matrix (t_{ij}) and reduce the two quadratic forms to the canonical form.

9.
$$Q = 3x_1^2 - 2x_2^2 - x_3^2 + 2x_1x_2 - 2x_1x_3 + 4x_2x_3$$
$$P = 2y_1^2 + 4y_2^2 + 10y_3^2 + 2y_1y_2 + 6y_1y_3 - 4y_2y_3$$

Reduce the two quadratic forms to the canonical forms without determining the matrix (t_{ij}).

10. Develop the principal-axis theory of stress with the aid of the principal-axis theory of quadratic forms.

Answers to problems

CHAPTER 1

2. $s = D[|\theta/2| + |\sin \theta/2|]$.

3. $ds/dt = 3V/2$.

6. $T/W = r/2h$.

7. $W = -7084k/15$.

11. $W/P = 2R/(R - r)$.

13. $P_n = \frac{1}{2}Lb_n$.

14. $U = [\sqrt{x^2 + y^2} + \sqrt{(L - x)^2 + y^2}]F$.

15. $(dr/dt)^2 = 4a/(mr) - 4b/(9mr^9) + \text{constant}$, $\qquad r^8 = b/a$.

16. $\frac{1}{2}m(\dot{x}^2 + \dot{y}^2) - mgy + \frac{1}{4}k(\sqrt{x^2 + y^2} - L_0)^4 = \text{constant}$.

17. $(R - r)\dot{\theta}^2 - g \cos \theta = \text{constant}$.

18. $\omega = (n^2\pi^2/L^2)\sqrt{EI/\rho}$.

19. $28°\ 19'$.

20. $V = EIL/2R^2 - ML/R$, $\qquad M = EI/R$.

21. $\sin(\theta + \alpha) = (\frac{8}{3}) \sin \alpha$.

22. $V = Cr^{-1.2} + 8\pi k r^2$, $\qquad r = (0.075C/\pi k)^{0.3125}$.

23. $x = u \pm \sqrt{u^2 - 1}$, $\qquad r = au$.

25. $F \to -2\pi\rho$.

26. $V = 2\pi k\rho a/\sqrt{x^2 + a^2}$, $\qquad F = 2\pi k\rho ax/(x^2 + a^2)^{3/2}$.

27. $V = -k\rho \log \dfrac{x + L + \sqrt{(x + L)^2 + y^2}}{x - L + \sqrt{(x - L)^2 + y^2}}$

$\qquad F_y = -2k\rho L/(y\sqrt{L^2 + y^2})$, $\qquad F_y \to -2k\rho/y$.

28. $h/r < \sqrt{3}$.

30. $26°\ 34'$, $\qquad 45°$.

CHAPTER 2

1. $k = (n - 1)EI/L$, $\qquad$ equal.

2. $V = (2a_0L - \frac{2}{3}a_1L^3)A^2 + (6a_0L^2 - 3a_1L^4)AB + (6a_0L^3 - \frac{18}{5}a_1L^5)B^2$
$\qquad - (\frac{1}{3}b_0L^3 - \frac{1}{4}b_1L^4)A - (\frac{1}{4}b_0L^4 - \frac{1}{5}b_1L^5)B$.

3. $\sigma = \dfrac{N}{A} - \dfrac{I_\xi M_\eta + I_{\xi\eta} M_\xi}{I_\xi I_\eta - I_{\xi\eta}^2}\, \xi + \dfrac{I_\eta M_\xi + I_{\xi\eta} M_\eta}{I_\xi I_\eta - I_{\xi\eta}^2}\, \eta.$

4. 4.785.

5. $y = \dfrac{2L^4 p_0}{\pi^5 EI} \displaystyle\sum_{n=1}^{\infty} \dfrac{(-1)^{n+1}}{n^5} \sin \dfrac{n\pi x}{L}.$

6. 74.26%.

7. $\dfrac{R}{F} = \dfrac{\dfrac{1}{1-r} - \dfrac{1}{9(9-r)} - \dfrac{1}{25(25-r)} + \dfrac{1}{49(49-r)} + \dfrac{1}{81(81-r)} - \cdots}{\sqrt{2}\left[\dfrac{1}{1-r} + \dfrac{1}{9(9-r)} + \dfrac{1}{25(25-r)} + \cdots\right]}.$

8. $a_n = 0$ if n is a multiple of 4. Otherwise, $a_n = FL^3/(2\pi^4 EIn^4).$

10. $N = P/\sqrt{3}, \qquad M = Pa\left[\dfrac{3}{2\pi(1+Z)} - \dfrac{1}{\sqrt{3}}\right].$

11. $u = \dfrac{Fa}{2\pi EA(1+Z)} + a_1 \cos\theta + \dfrac{Fa}{\pi EAZ} \displaystyle\sum_{n=2}^{\infty} \dfrac{\cos n\theta}{(n^2-1)^2},$

   $\bar{v} = \left(\dfrac{Fa}{\pi EA} - a_1\right) \sin\theta - \dfrac{Fa}{\pi EAZ} \displaystyle\sum_{n=2}^{\infty} \dfrac{\sin n\theta}{n(n^2-1)^2},$

   where a_1 is an arbitrary constant.

12. $Z = -1 + 2R^2/b^2 - 2(R/b)\sqrt{R^2/b^2 - 1}$, where $2b$ is the width of the cross section and R is the radius of the centroidal axis.

13. $\sigma = \dfrac{N}{A} + \dfrac{M}{Aa}\left[1 + \dfrac{z}{(a+z)Z}\right].$

14. $u = 0.09078F^2/k^2, \qquad N_0 = 0.3013F, \qquad N_1 = 0.2804F,$
    $N_2 = 0.2130F, \qquad N_3 = 0.$

15. x = vertical displacement of joint 2, y = horizontal displacement of joint 1, z = vertical displacement of joint 1. $x = (\tfrac{5}{6})(aF/EA),\ y = -(5\sqrt{3}/36) \times (aF/EA),\ z = (\tfrac{1}{12})(aF/EA).$

16. 0.2625°.

17. Left wire 237 lb, center wire 126 lb, right wire 537 lb.

18. $N_1 = 80.75$ lb, $N_2 = -28.15$ lb, $N_3 = -64.42$ lb.

19. $N_{OA} = 0.230F, \qquad N_{AC} = 0.133F, \qquad N_{AB} = -0.266F,$
    $N_{OB} = 0.771F.$

20. $N_1 = 12,500$ lb, $\qquad N_2 = 16,800$ lb, $\qquad N_3 = 19,400$ lb.

21. $N_{AD} = -90.0$ lb, $\qquad N_{BD} = 46.8$ lb, $\qquad N_{BE} = -74.9$ lb,
    $N_{CE} = 76.6$ lb, $\qquad N_{CD} = 49.3$ lb.

22. $N_6 = 1611$ lb, $\qquad N_8 = 1889$ lb, $\qquad N_{10} = 1611$ lb, $\qquad N_{12} = 889$ lb.

23. $N_{AB} = N_{AO} = \dfrac{-P}{2(1+\sqrt{2})}, \qquad N_{AC} = \dfrac{P}{2\sqrt{2}}.$

24. $N_1 = 14,150$ lb, $\qquad N_2 = 16,420$ lb, $\qquad N_3 = 17,650$ lb.

29. $40\theta_1 + 4\theta_2 + 10\theta_4 + 0.02x_1 - 0.20x_2 = 0.$
   $4\theta_1 + 18\theta_2 + 5\theta_3 + 0.20x_1 - 0.20x_2 = 0.$
   $5\theta_2 + 18\theta_3 + 4\theta_4 + 0.20x_1 - 0.20x_2 = 0.$
   $10\theta_1 + 4\theta_3 + 60\theta_4 + 10\theta_6 + 0.02x_1 - 0.20x_2 = 0.$
   $10\theta_4 + 32\theta_6 - 0.18x_1 = 0.$
   $0.02\theta_1 + 0.20\theta_2 + 0.20\theta_3 + 0.02\theta_4 - 0.18\theta_6 + 0.02413x_1 - 0.01333x_2 = 0.$
   $-0.20\theta_1 - 0.20\theta_2 - 0.20\theta_3 - 0.20\theta_4 - 0.01333x_1 + 0.01333x_2 = 100.$

30. $\theta_1 = \theta_3 = 4Fa/129K, \qquad \theta_2 = Fa/129K.$
   $x = 26Fa^2/387K.$

31. $x = Pa^3/12EI.$

32. $0.003737°, \qquad -0.0006228°.$

33. $u = \dfrac{3\sqrt{3}M}{2EA(1 + 9r^2/L^2)}, \qquad v = 0, \qquad r = \sqrt{I/A}.$

34. $\theta_0 = \theta_3 = \dfrac{(1 + 3r)Fa/K_v}{6r + \lambda + 2r\lambda}, \qquad \theta_1 = \theta_2 = \dfrac{Fa/K_v}{6r + \lambda + 2r\lambda}.$

   $x = \dfrac{(1 + 2r)Fa^2/K_v}{6r + \lambda + 2r\lambda}, \qquad r = \dfrac{K_h}{K_v}, \qquad \lambda = \dfrac{EAa^2b^2}{K_vC^3}.$

35. $S = \dfrac{-9b^2F}{32a(a + 2b)}.$

36. Let $r = K_h/K_v.$

   $2(2 + r)\theta_1 + \theta_2 + r\theta_3 - (3/a)x_2 = (\tfrac{2}{27})Fb/K_v.$
   $\theta_1 + 2(1 + r)\theta_2 + r\theta_4 + (3/a)x_1 - (3/a)x_2 = p_0b^2/30K_v.$
   $r\theta_1 + 2(2 + r)\theta_3 + \theta_4 - (3/a)x_2 = -(\tfrac{4}{27})Fb/K_v.$
   $r\theta_2 + \theta_3 + 2(1 + r)\theta_4 + (3/a)x_1 - (3/a)x_2 = -p_0b^2/20K_v.$
   $(3/a)\theta_2 + (3/a)\theta_4 + (24/a^2 + EAb^2/K_vL^3)x_1 - (12/a^2)x_2 = F_1/K_v.$
   $-(3/a)\theta_1 - (3/a)\theta_2 - (3/a)\theta_3 - (3/a)\theta_4 - (12/a^2)x_1 + (12/a^2)x_2 = F_2/K_v.$

37. x_1 and x_2 denote displacements of joints 1 and 2 normal to plane of frame; x_3 and x_4 denote rotations of joints 1 and 2 about axis 12; x_5 and x_6 denote rotations of joints 1 and 2 about axes 01 and 02. Let $K_h/K_v = r.$
   $2(1/a^2 + r/b^2)x_1 - 2(r/b^2)x_2 - (1/a)x_3 - (r/b)x_5 - (r/b)x_6 = F/3K_v.$
   $-2(r/b^2)x_1 + 2(1/a^2 + r/b^2)x_2 - (1/a)x_4 + (r/b)x_5 + (r/b)x_6 = 0.$
   $-(3/a)x_1 + (2 + GJ/bK_v)x_3 - (GJ/bK_v)x_4 = 0.$
   $-(3/a)x_2 - (GJ/bK_v)x_3 + (2 + GJ/bK_v)x_4 = 0.$
   $-(3r/b)x_1 + (3r/b)x_2 + (2r + GJ/aK_v)x_5 + rx_6 = 0.$
   $-(3r/b)x_1 + (3r/b)x_2 + rx_5 + (2r + GJ/aK_v)x_6 = 0.$

CHAPTER 3

3. $x = a \sec \tfrac{1}{2}(\theta - b).$

4. $y = px(L - x)(L^2 + Lx - x^2)/24EI.$

5. $y = p_0x^2(L - x)(2L^2 - Lx - x^2)/120EIL.$

6. $y''' + y''/a^2 + y/a^4 = p/2k.$
   At $x = L, \qquad y''' - y/a^3 = 0 \qquad$ and $\qquad y'' + y'/a + y/a^2 = 0.$

7. $u = \dfrac{Fa}{2EAZ} \theta \sin \theta, \qquad \bar{v} = \dfrac{Fa}{2EAZ} (\theta \cos \theta - \sin \theta), \qquad \sigma = \dfrac{-Fz \cos \theta}{AZ(a + z)}.$

8. $u = \dfrac{-M}{EAZ} (1 - \cos \theta), \qquad \bar{v} = \dfrac{M}{EAZ} [(1 + Z)\theta - \sin \theta],$

$\sigma = \dfrac{M}{A(a + z)} \left[1 + \left(1 + \dfrac{1}{Z} \right) \dfrac{z}{a} \right].$

9. $M = a^2 w \left[\dfrac{1 + 2Z}{(\pi^2/4)(1 + Z) - 2} \left(\cos \theta - \dfrac{\pi}{4} \right) - \dfrac{1}{2} \cos \theta + \theta \sin \theta \right],$

$N = aw \left[\dfrac{3}{2} \cos \theta - \theta \sin \theta - \dfrac{(1 + 2Z) \cos \theta}{(\pi^2/4)(1 + Z) - 2} \right],$

where w is weight per unit length.

10. $f(x) = A [(\sin kL - \sinh kL)(\sin kx - \sinh kx)$
$\qquad\qquad + (\cos kL + \cosh kL)(\cos kx - \cosh kx)].$

11. $y = f(x) \sin nt, \qquad f(x) = A \sin (\nu \pi x / L), \qquad \nu = \text{integer}.$
$n = (\nu^2 \pi^2 / L^2) \sqrt{EI/\rho}.$

12. $K = 0.617, \qquad \lambda = 0.617, \qquad C = 0.$

14. $\dfrac{\partial F}{\partial u} - \dfrac{\partial}{\partial x} \dfrac{\partial F}{\partial u_x} - \dfrac{\partial}{\partial y} \dfrac{\partial F}{\partial u_y} + \dfrac{\partial^2}{\partial x^2} \dfrac{\partial F}{\partial u_{xx}} + \dfrac{\partial^2}{\partial x \partial y} \dfrac{\partial F}{\partial u_{xy}} + \dfrac{\partial^2}{\partial y^2} \dfrac{\partial F}{\partial u_{yy}}$

$- \dfrac{\partial^3}{\partial x^3} \dfrac{\partial F}{\partial u_{xxx}} - \dfrac{\partial^3}{\partial x^2 \partial y} \dfrac{\partial F}{\partial u_{xxy}} - \dfrac{\partial^3}{\partial x \partial y^2} \dfrac{\partial F}{\partial u_{xyy}} - \dfrac{\partial^3}{\partial y^3} \dfrac{\partial F}{\partial u_{yyy}} = 0.$

16. $w_{xxxx} + w_{xxyy} + w_{yyyy} = p/2K.$

$w_{xxxx} + w_{xxyy} + w_{yyyy} + \dfrac{\rho}{2K} w_{tt} = 0.$

17. $\displaystyle\int_0^a xy \, dx = \text{constant}, \qquad 2\rho g x y + \lambda x - \dfrac{d}{dx} \dfrac{Cxy'}{\sqrt{1 + (y')^2}} = 0.$

18. $y = \dfrac{4pL^4}{\pi^5 EI} [\sin \pi x / L + \tfrac{1}{243} \sin 3\pi x / L + \cdots].$

19. $a = 32PL^3 / \pi^4 EI.$

20. $w_0 = 0.0481 pa^4 / D, \qquad w_0 = 0.0629 pa^4 / D, \qquad \nu = 0.30.$

21. $w_0 = 0.0306 Pa^2 / D, \qquad w_0 = 0.0433 Pa^2 / D, \qquad \nu = 0.30.$

22. $u = -(Pa/8EAZ)[2 \sin \theta + (3\pi + 2\pi Z - 4 - 4/\pi) \cos \theta$
$\qquad + (4/\pi)\theta \sin \theta - 2\theta \cos \theta - 4].$
$\bar{v} = (Pa/8EAZ)[-4(1 + Z)\theta - 4 \cos \theta + (3\pi + 2\pi Z - 4) \sin \theta$
$\qquad - (4/\pi)\theta \cos \theta - 2\theta \sin \theta + 4], \qquad \theta > 0.$

23. $w_0 = \dfrac{16 pa^4 b^4}{\pi^6 D(a^2 + b^2)^2}, \qquad w_{xxxx} + 2w_{xxyy} + w_{yyyy} = \dfrac{p}{D}.$

24. $w = (q/2N)(Lx - x^2), \qquad N = \sqrt[3]{q^2 L^2 EA/24}.$

CHAPTER 4

1. $\epsilon_1 = \epsilon_2 = 2,$ $\quad \epsilon_3 = -1 + \sqrt{17},$ $\quad e = 35.11.$ $\quad l_3 = \frac{1}{2},$
   $m_3 = -\frac{1}{2},$ $\quad n_3 = 1/\sqrt{2}.$

   The other two principal directions are any two that are perpendicular to each other and to the direction (l_3, m_3, n_3).

2.

	x	y	z	ϵ_i	
1	$1/\sqrt{2}$	$1/\sqrt{2}$	0	$-1 + 3\sqrt{2},$	
2	$-1/\sqrt{2}$	$1/\sqrt{2}$	0	$-1 + 2\sqrt{2},$	$e = 70.79.$
3	0	0	1	5	

3. $\sigma_\xi = \sigma_x \cos^2 \theta + \sigma_y \sin^2 \theta + 2\tau_{xy} \sin \theta \cos \theta.$
   $\sigma_\eta = \sigma_x \sin^2 \theta + \sigma_y \cos^2 \theta - 2\tau_{xy} \sin \theta \cos \theta.$
   $\tau_{\xi\eta} = (\sigma_y - \sigma_x) \sin \theta \cos \theta + (\cos^2 \theta - \sin^2 \theta)\tau_{xy}.$

4.

	x	y	z	σ_i
1	0.235	−0.639	0.733	−9.28
2	0.281	0.766	0.578	8.13
3	−0.931	0.070	0.359	2.16

5.

	x	y	z	σ_i
1	$-1/\sqrt{6}$	$1/\sqrt{6}$	$2/\sqrt{6}$	−15,000
2	$1/\sqrt{2}$	$1/\sqrt{2}$	0	5000
3	$-1/\sqrt{3}$	$1/\sqrt{3}$	$-1/\sqrt{3}$	0

$$\tau_{\max} = 10,000.$$

6. $p_x = 3333,$ $\quad p_y = -2333,$ $\quad p_z = 5667,$ $\quad p_n = 2555,$ $\quad \tau = 6491.$

8. $(1 - 2\nu)\nabla^2 u + (\partial e/\partial x) + [(1 - 2\nu)/G]\rho F_x - 2(1 + \nu)k\theta_x = 0, \cdots$

10. 0.634 in.

11. Right-hand joint, 3.70 in. Center joint, 11.66 in.

14. Top members, 125 and −216.5. Horizontal members, −49.6 and 58.7. Vertical members, 106.5 and −143.5. Diagonal members, −73.3 and −73.3.

15. $q = PL^3/48EI + 0.30PL/GA.$

16. $q = 343P^3L^5/2160b^3K^3h^7.$

17. $u = (Pa^3/2EI)(\pi - 1).$

18. $F = \dfrac{\frac{5}{8} + 3\kappa(1 + \nu)r^2/L^2}{1 + 6\kappa(1 + \nu)r^2/L^2} W,$

   where L is the length of a span, W is the total load, and r is the radius of gyration of the cross section.

19. 0.0354 in.,   0.124 in.,   1.425°.

20. 5.406 lb.

21. $\delta = PbL^3/3E + PaL^4/12E$.

22. $M_0 = -3Pa/4\pi$, $N_0 = 3P/4\pi$.

23. P/π.

24. $\delta = \dfrac{Pa^3}{2EI} + \dfrac{Pa^3}{2GJ}(\pi - 1)$.

25. $N_0 = \dfrac{P(\bar{x}\bar{y}L - I_{xy})}{2(\bar{y}^2 L - I_{xx})}$.

   $M_0 = \dfrac{P[\bar{x}I_{xx} - hL\bar{x}\bar{y} + I_{xy}(h - \bar{y})]}{2(\bar{y}^2 L - I_{xx})}$.

26. $\alpha = \tfrac{1}{2}\pi a T_0 \left(\dfrac{1}{EI} + \dfrac{1}{GJ}\right)$, $\delta = \tfrac{1}{2}\pi a^2 T_0 \left(\dfrac{1}{EI} + \dfrac{1}{GJ}\right)$.

27. $M_0 = Pab^2/L^2$, $M_1 = -Pa^2 b/L^2$.
   $R_0 = Pb^2(3a + b)/L^3$, $R_1 = Pa^2(3b + a)/L^3$.

28. $q = \dfrac{1}{EAa^2}\displaystyle\int \left[(m + an)(M + aN) + \dfrac{mM}{Z}\right] ds$.

29. $\sigma_1 = 2\tau$, $\sigma_2 = \sigma_3 = -\tau$, $\tau_{max} = 1.5\tau$, $l = m = n = 1/\sqrt{3}$.

CHAPTER 5

1. $\bar{u}_{xx} + \tfrac{1}{2}(1 + v)\bar{v}_{xy} + \tfrac{1}{2}(1 - v)\bar{u}_{yy} = (1 + v)k\theta_x$.
   $\tfrac{1}{2}(1 - v)\bar{v}_{xx} + \bar{v}_{yy} + \tfrac{1}{2}(1 + v)\bar{u}_{xy} = (1 + v)k\theta_y$.
   On edge $x = a$, $\bar{u}_x + v\bar{v}_y = (1 + v)k\theta$ and $\bar{v}_x + \bar{u}_y = 0$.

2. $f(y) = A \cosh y + By \sinh y + p_0/D$.

   $A = \dfrac{vP_0[(1 + v)/(1 - v) \sinh b - b \cosh b]}{D[(3 + v) \sinh b \cosh b - (1 - v)b]}$.

   $B = \dfrac{vP_0 \sinh b}{D[(3 + v) \sinh b \cosh b - (1 - v)b]}$.

3. $w_{yyy} + (2 - v)w_{xxy} - (EI/D)w_{xxxx} + q/D = 0$.
   $w_{yy} + vw_{xx} - (GJ/D)w_{xxy} = 0$.

4. $w = \displaystyle\sum\sum a_{mn} \sin \dfrac{m\pi x}{a} \sin \dfrac{n\pi y}{b}$.

   $a_{mn} = \dfrac{16p \sin (m\pi/2) \sin (n\pi/2) \sin (m\pi c/a) \sin (n\pi c/b)}{\pi^6 Dmn(m^2/a^2 + n^2/b^2)^2}$.

5. $\bar{u}_{rr} + (1/r)\bar{u}_r - (\bar{u}/r^2) = k(1 + v)\theta_r$.
   At $r = a$, $\bar{u}_r + (v/a)\bar{u} = k(1 + v)\theta$.

6. $\bar{u}_{rr} + (1/r)\bar{u}_r - (\bar{u}/r^2) + w_r w_{rr} + [(1 - v)/2r]w_r^2 = 0$.
   $(1 + v)\bar{u}_r w_r + \tfrac{1}{2}w_r^3 + r\bar{u}_r w_{rr} + \tfrac{3}{2}rw_r^2 w_{rr} + r\bar{u}_{rr}w_r + v\bar{u}w_{rr}$
   $+ [(1 - v^2)/Eh]pr = 0$.

7. $U = \pi D \sin a \int_{x_1}^{x_2} \left\{ x w_{xx}^2 + (1/x) w_x^2 + 2\nu w_x w_{xx} \right.$

$$+ 12/h^2 \left[x \bar{u}_x^2 + \frac{1}{x} (\bar{u} + w \cot a)^2 + 2\nu \bar{u}_x (\bar{u} + w \cot a) \right] \left. \right\} dx.$$

8. $r w_{rr} + \nu w_r = 0,$ $r w_{rrr} + w_{rr} - (1/r) w_r = 0$ at $r = a.$

9. $w = \sum\sum a_{mn} \sin \dfrac{m\pi x}{a} \sin \dfrac{n\pi y}{b}.$

$a_{mn} = 0$ if m or n is even.

$a_{mn} = \dfrac{16k\beta(1 + \nu)}{\pi^4 mn(m^2/a^2 + n^2/b^2)}$ if m and n are odd.

12. $A = 1,$ $B = x \sin a,$ **n** outward, $\dfrac{1}{r_1} = 0.$

$r_2 = x \tan a,$ $\alpha = 1,$ $\beta = x \sin a + z \cos a.$

13. $A = a,$ $B = a \sin \phi,$ **n** outward, $r_1 = r_2 = a,$
$\alpha = a + z,$ $\beta = (a + z) \sin \phi.$

14. $A = 1,$ $B = \sqrt{a^2 \sin^2 \theta + b^2 \cos^2 \theta},$ $1/r_1 = 0,$
$1/r_2 = -ab/B^3,$ **n** inward, $\alpha = 1,$ $\beta = B - abz/B^2.$

17. $a^2 N_\phi = F_{00} \csc^2 \phi + F_\phi \cot \phi + F.$
$a^2 N_0 = F_{\phi\phi} + F.$
$a^2 N_{\phi\theta} = -F_{\phi 0} \csc \phi + F_\theta \csc \phi \cot \phi.$

18. $A = B = c\sqrt{\cosh^2 x - \cos^2 y}.$
$A^3 N_x = AF_{yy} + A_x F_x - A_y F_y.$
$A^3 N_y = AF_{xx} - A_x F_x + A_y F_y.$
$A^3 N_{xy} = -AF_{xy} + A_y F_x + A_x F_y.$

21. $\bar{u}_x + (\nu w/a) = 0,$ $w_{xxxx} + (Eh/a^2 D)w = 0.$
At $x = 0,$ $w_{xx} = -M_0/D,$ $w_{xxx} = -Q_0/D.$

22. $w = \dfrac{p(a^2 - r^2)}{64D} \left(\dfrac{5 + \nu}{1 + \nu} a^2 - r^2 \right).$

23. $w = \dfrac{P}{64D} (a^2 - r^2)^2.$

CHAPTER 6

1. $P_{cr} = ka.$
2. $P_{cr} = ka^3.$
3. $P_{cr} = k/a,$ $P_{cr} = 3k/2a.$
4. $F_{cr} = 2kc^3/a^2 b^2,$ $F_{cr} = 2kc/b^2,$ $c^2 = a^2 + b^2.$
5. $F_{cr} = 16k/5a,$ $F_{cr} = 3k/a,$ $F_{cr} = 8k/3a.$
6. $P_{cr} = 6(2EI/L^2 + k/L).$
7. $P_{cr} = 42EI/L^2.$
8. $W_{cr} = k/r,$ where r is the radius of the shell.

9. $\eta = A \sin kx + B \cos kx + Cx + D$
$kL = \tan kL$. Hence $P_{cr} = 2.04\pi^2 EI/L^2$.

10. $N_{cr} = (D/b^2)[6(1 - \nu) + (\pi^2 b^2/a^2)]$.

11. 0.2938.

12. $F_{cr} = 2.49 EI/L^2$.

13. $\alpha \sinh \alpha b \cos \beta b + \beta \cosh \alpha b \sin \beta b = 0$.

$$\alpha = \left[\frac{n^2\pi^2}{a^2} + \sqrt{(N/D)(n^2\pi^2/a^2)} \right]^{\frac{1}{2}}, \qquad \beta = \left[-\frac{n^2\pi^2}{a^2} + \sqrt{(N/D)(n^2\pi^2/a^2)} \right]^{\frac{1}{2}}.$$

14. $F_{cr} = \dfrac{12EI}{L^2}, \qquad \dfrac{F}{F_{cr}} = \dfrac{\theta^3}{3(\sin \theta - \theta \cos \theta)}.$

If $\theta = 30°$, $\quad F/F_{cr} = 1.03$.

15. $P_{cr} = \dfrac{\pi^2 EI}{a^2} + \dfrac{\pi^2 bD}{3a^2} + \dfrac{2D(1 - \nu)}{b}$.

16. $L = \pi/2 \sqrt[3]{4\pi EI/\gamma(\pi^2 - 4)}$ or $L^3 = 8.3EI/\gamma$.

17. $D\nabla^4\zeta + N_x\zeta_{xx} + N_y\zeta_{yy} = 0$.

18. $M_{cr} = \dfrac{\pi Eht^3}{6(1 - \nu^2)L} \sqrt{(1 - \nu)/2 + \pi^2 h^2/48L^2}$.

CHAPTER 7

1. $T = \frac{1}{2}m\left[\dot{x}^2 - \sqrt{2}a\dot{x}\dot{\theta} \sin \left(\theta + \frac{\pi}{4} \right) + \frac{2}{3}a^2\dot{\theta}^2 \right]$.

2. $T = \frac{7}{10} m(\dot{x} + r\dot{\theta})^2 + \frac{1}{2}(I + mx^2)\dot{\theta}^2$.

4. $\ddot{\theta} + (k/ml^2)\dot{\theta}^3 + (g/l) \sin \theta = 0$.

5. $x = \dfrac{g(m_1 - m_2)t^2}{2(m_1 + m_2 + I/r^2)} + V_0 t + x_0$.

6. $(I + m_1r^2 + m_2r^2 + m_3r^2 + m_4r^2)\ddot{\theta} + r^2(m_3 - m_2)\ddot{\phi}$
$+ gr(m_1 - m_2 - m_3 - m_4) = 0$.
$r^2(m_3 - m_2)\ddot{\theta} + (I + r^2m_2 + r^2m_3)\ddot{\phi} + gr(m_2 - m_3) = 0$.

7. $\ddot{\phi} = 0, \qquad \ddot{r} - r\dot{\theta}^2 + \dfrac{F}{m} = 0, \qquad r^2\dot{\theta} = $ constant.

8. $I\ddot{\theta} + [m/(m + M)]\ddot{x} \cos \theta + g \sin \theta = 0$.
$\dot{x} + \frac{5}{7}I\dot{\theta} \cos \theta = $ constant.

9. $\ddot{r} - r\dot{\theta}^2 + (k/m)(r - l) - g \cos \theta = 0$.
$r\ddot{\theta} + 2\dot{r}\dot{\theta} + g \sin \theta = 0$.

11. $\ddot{\theta} - (r\omega^2/l) \sin (\omega t - \theta) + (g/l) \sin \theta = 0$.

12. $x = x_0 - \frac{5}{7}s_0 \sin \omega t$.

13. $\ddot{r} - r(\dot{\phi} + \omega)^2 \cos^2 \alpha + k/r^2 = 0$.
$r^2(\dot{\phi} + \omega) = $ constant.
Hence, $\ddot{r} - C^2/r^3 + k/r^2 = 0, \qquad r = (C^2/K)/[1 - e \cos (\phi - \beta)]$.

15. $\ddot{\theta}_1 - (GJ/aI_1)(\theta_2 - \theta_1) = M/I_1$.
$\ddot{\theta}_2 + (GJ/aI_2)(\theta_2 - \theta_1) = 0$.
Hence $I_1\ddot{\theta}_1 + I_2\ddot{\theta}_2 = M$.

16. $\ddot{x} - \omega^2 x - \dot{\omega}y - 2\omega\dot{y} = F_x/m$.
$\ddot{y} - \omega^2 y + \dot{\omega}x + 2\omega\dot{x} = F_y/m$.

17. $\ddot{r} - r\alpha^2 t^2 = -F/m$.

20. Equation of axis is $y = z = \sqrt{3}x/(5 - 2\sqrt{3})$.

21. $\theta = 72°\ 39'$, $\phi = 141°\ 20'$, $\psi = 69°\ 34'$.

24. $M_1 = A\dot{\omega}\cos\alpha - (B - C)\omega^2\cos\beta\cos\gamma,\ \cdots$.

25. $R_1 = 16.16$ lb, $R_2 = 26.92$ lb.

26. 2.55 lb, in a direction perpendicular to the plane of the figure.

28. $A\ddot{\theta} - (A - C)\dot{\phi}^2\sin\theta\cos\theta + C\omega\dot{\phi}\sin\theta - mgh\sin\theta = 0$.
$\dot{\phi}(A\sin^2\theta + C\cos^2\theta) + C\omega\cos\theta = $ constant. If $\theta = 90°$, $\dot{\phi} = mgh/C\omega$.

29. $(A + A')\ddot{\theta} + (-A + C + C' - B')\dot{\phi}^2\sin\theta\cos\theta + C\dot{\phi}\dot{\psi}\sin\theta$
    $+ k(\theta - \pi/2) = 0$.
    $\dot{\phi}[(A + B')\sin^2\theta + (C + C')\cos^2\theta + I] + C\dot{\psi}\cos\theta$
    $+ \dot{\theta}\dot{\phi}[2(A + B')\sin\theta\cos\theta - 2(C + C')\sin\theta\cos\theta] - C\dot{\psi}\dot{\theta}\sin\theta + k\phi = 0$.
    $\dot{\psi} + \dot{\phi}\cos\theta = $ constant; hence, $\omega_3 = $ constant.
    If $\theta \approx \pi/2$ and $\dot{\phi} \approx 0$, $(A + A')\ddot{\theta} + k\theta = k\pi/2$,
    $(A + B' + I)\ddot{\phi} + k\phi = 0$, $\psi = $ constant.

30. $dx - a\cos\phi\,d\theta - a\sin\theta\sin\phi\,d\psi = 0$.
    $dy - a\sin\phi\,d\theta + a\sin\theta\cos\phi\,d\psi = 0$.

CHAPTER 8

1. $m\ddot{x} + kx = F(t)$.

2. $m\ddot{x} + (48EI/l^3)x = 0$.

3. $I\ddot{\theta} + c\dot{\theta} + k\theta = 0$.

4. $\omega_1 = \sqrt{[(3 + \sqrt{5})/2]k/m}$, $\omega_2 = \sqrt{[(3 - \sqrt{5})/2]k/m}$.
   $z_1/z_2 = (\sqrt{5} + 1)/2$, $-(\sqrt{5} - 1)/2$.

5. $\omega_1 = 22.0$, $\omega_2 = 5.5$, $u/v = -1.40$, 0.116.

6. $m\omega^2/k = a$, $a_1 = 0.198$, $a_2 = 1.555$, $a_3 = 3.247$.
   If $a = a_1$, $z_2/z_1 = 1.802$, $z_3/z_1 = 2.247$.
   If $a = a_2$, $z_2/z_1 = 0.445$, $z_3/z_1 = -0.802$.
   If $a = a_3$, $z_2/z_1 = -1.247$, $z_3/z_1 = 0.555$.

9. $\ddot{a}_n + \omega_n^2 a_n = 0$, $\omega_n = (\pi^2 n^2/l^2)\sqrt{EI/\rho}$.

10. $m\ddot{x}_1 + 2kx_1 - kx_2 - kx_3 = 0,\ \cdots$.
    $m\ddot{y}_1 + 2ky_1 - ky_2 - ky_3 = 0,\ \cdots$.
    $\omega = \sqrt{3k/m}$.

11. $y_{xx} = (\rho/N)y_{tt}$; $\rho = $ mass per unit length.

12. $\nabla^2 w = (\rho/N)w_{tt}$; $\rho = $ mass per unit area.

13. $(d^2/dr^2)(r\, Dw_{rr} + v\, Dw_r) - (d/dr)[(1/r)\, Dw_r + v\, Dw_{rr}] + prw_{tt} = 0,$
    $g'' + \omega^2 g = 0.$
    $(d^2/dr^2)(r\, Df_{rr} + v\, Df_r) - (d/dr)[(1/r)\, Df_r + v\, Df_{rr}] - p\omega^2 rf = 0.$

14. $w_{rrrr} + (2/r)w_{rrr} - (1/r^2)w_{rr} + (1/r^3)w_r + (p/D)w_{tt} = p/D.$

15. $w_{rr} + (1/r)w_r = (p/N)w_{tt}, \qquad w = f(r)\,g(t), \qquad f'' + \dfrac{1}{r}f' + (p\omega^2/N)f = 0.$

    $\omega_1 = 2.405\sqrt{N/pa^2}, \qquad \omega_2 = 5.520\sqrt{N/pa^2}, \qquad \omega_3 = 8.654\sqrt{N/pa^2}.$

16. $\tan \beta l = \tanh \beta l, \qquad \beta^4 = p\omega^2/EI.$

17. $\cos \beta l \cosh \beta l = -1, \qquad \beta^4 = p\omega^2/EI, \qquad \beta_1 l = 1.875,$

    $$f(x) = A\left[(\sinh \beta x - \sin \beta x) - \frac{\sinh \beta l + \sin \beta l}{\cosh \beta l + \cos \beta l}(\cosh \beta x - \cos \beta x) \right].$$

18. $\omega_{mn} = \pi^2(m^2/a^2 + n^2/b^2)\sqrt{D/p}, \qquad p = $ mass per unit area.
    $f(x, y) = A_{mn}\sin(m\pi x/a)\sin(n\pi y/b).$

19. $\theta_{tt} = c^2\theta_{xx}, \qquad c^2 = GJ/pI, \qquad \theta = f(x)\,g(t).$
    $\omega_n = n\pi c/l, \qquad f(x) = A\sin(n\pi x/l).$

20. $pu_{tt} = (\partial/\partial x)(EAu_x); \qquad p = $ mass per unit length.

21. $u_{\theta\theta\theta\theta} + 2u_{\theta\theta} + (1 + 1/Z)u + \bar{v}_\theta/Z + (a^2 p/EZ)u_{tt} = 0.$
    $\bar{v}_{\theta\theta} + u_\theta - (pa^2/E)\bar{v}_{tt} = 0.$
    $u = \phi(\theta)f(t), \qquad \bar{v} = \psi(\theta)f(t), \qquad f'' + \omega^2 f = 0.$
    $\phi'''' + 2\phi'' + (1 + 1/Z - a^2 p\omega^2/EZ)\phi + \psi'/Z = 0.$
    $\psi'' + \phi' + (a^2\omega^2 p/E)\psi = 0.$

22. $\dfrac{c_0^2 u_{xx}}{(1 + u_x)^{\gamma+1}} = u_{tt}, \qquad c_0^2 = \dfrac{\gamma p_0}{p_0}.$

23. 39.2 ft/sec, $\qquad 219$ ft.

24. $(\lambda + 2G)(u_{rr} + (1/r)u_r - (1/r^2)u + w_{zr}) + G(u_{zz} - w_{zr}) = pu_{tt}.$
    $(\lambda + 2G)(u_{zr} + (1/r)u_z + w_{zz}) - (G/r)(u_z - w_r) - G(u_{zr} - w_{rr}) = pw_{tt}.$

APPENDIX

1. (a) Indefinite. (b) Positive semidefinite. (c) Positive definite. (d) Positive semidefinite. (e) Indefinite.

2. (a) Not factorable. (b) $Q = (7a + b - 2c - 4d)(a + 3b - 5c + d).$
   (c) $Q = (x + iy)(x - iy).$ (d) Not factorable.
   (e) $Q = (2x + y)(3x - 5y + 7z).$

3. Set $a = \frac{1}{8}(25 - \sqrt{513}), \qquad b = \frac{1}{8}(25 + \sqrt{513}).$
   Q is positive definite if $x < a$, positive semidefinite if $x = a$, indefinite if $a < x < b$, negative semidefinite if $x = b$, and negative definite if $x > b$.

(5)

	x	y	z
ξ	$-1/\sqrt{21}$	$4/\sqrt{21}$	$2/\sqrt{21}$
η	$2/\sqrt{6}$	$1/\sqrt{6}$	$-1/\sqrt{6}$
ζ	$-2/\sqrt{14}$	$1/\sqrt{14}$	$-3/\sqrt{14}$

$$Q = \xi^2 + \tfrac{5}{2}\eta^2 - \tfrac{5}{2}\zeta^2.$$

6. $Q = -7.582y_1^2 + 9.151y_2^2 + 3.431y_3^2$.

7. $(t_{ij}) = \begin{bmatrix} 2/\sqrt{5} & 1/\sqrt{5} \\ 1/\sqrt{5} & -2/\sqrt{5} \end{bmatrix}$.

$Q = 2\xi_1^2 - 3\xi_2^2, \qquad P = \eta_1^2 + \eta_2^2$.

8. $(t_{ij}) = \begin{bmatrix} 1/\sqrt{5} & -3/\sqrt{5} \\ 1/\sqrt{5} & 2/\sqrt{5} \end{bmatrix}$.

$Q = \xi_1^2 + 11\xi_2^2, \qquad P = \eta_1^2 + \eta_2^2$.

9. $Q = -0.664\xi_1^2 + 0.095\xi_2^2 + 7.926\xi_3^2$.

$P = \eta_1^2 + \eta_2^2 + \eta_3^2$.

Bibliography

The following incomplete bibliography lists some books and articles that are commendable for their presentations of special topics or for general developments in mechanics. The well-known treatises by Whittaker and Appell are classics in the mathematical aspects of the subject. The works of Routh, Osgood, Goldstein, and · many others are also noteworthy. The introductory book by Planck provides an excellent physical insight into the foundations of mechanics. Lanczos' work on variational principles of mechanics combines an advanced treatment with a lucid and penetrating philosophy. The treatise by Courant and Hilbert provides an excellent mathematical background for the variational theory of mechanics. Variational methods of approximation are treated in the works of Kantorovich and Krylov, Courant and Hilbert, Collatz, Biezeno, and Grammel, and Temple and Bickley. The recent historical work by Dugas gives a comprehensive survey of the development of ideas in mechanics.

Among the engineering works on mechanics, the treatise by Biezeno and Grammel is noteworthy. It contains a good presentation of variational principles in elasticity theory. This topic is also treated admirably in the book by Sokolnikoff. General principles and engineering applications in the mechanics of rigid and elastic bodies are developed precisely and concisely in the recent work by Parkus. Applications of energy principles in analyses of engineering structures are emphasized in the works of Argyris, Hoff, Phillips, and Charlton. Variational principles play a significant part in modern plasticity theory; the work of Prager and Hodge treats this topic.

BOOKS

1. Appell, P., *Traité de mécanique rationnelle*, vols. 1 (1941), 2 (1953), 3 (1952), 3rd ed., Gauthier Villars, Paris.
2. Argyris, J. H., and S. Kelsey, *Energy Theorems and Structural Analysis*, Butterworth., London, 1960.
3. Biezeno C., and R. Grammel, *Engineering Dynamics*, vols. 1 and 2, Blackie and Son, London, 1956.

4. Bleich, F., *Buckling Strength of Metal Structures*, McGraw-Hill, New York, 1952.
5. Bôcher, M., *Introduction to Higher Algebra*, Macmillan, New York, 1938.
6. Bolza, O., *Lectures on the Calculus of Variations*, Stechert-Hafner, New York, 1931.
7. Bowman, F., *Introduction to Bessel Functions*, Dover, New York, 1958.
8. Bullen, K. E., *Theory of Seismology*, 2nd ed., Cambridge University Press, 1959.
9. Carslaw, H. S., *Fourier's Series and Integrals*, 3rd ed., Dover, New York, 1930.
10. Charlton, T. M., *Energy Principles in Applied Statics*, Blackie and Son, London, 1959.
11. Collatz, L., *Eigenwertprobleme und ihre numerische Behandlung*, Chelsea, New York, 1948.
12. Courant, R., and K. Friedrichs, *Supersonic Flow and Shock Waves*, Interscience, New York, 1948.
13. Courant, R., and D. Hilbert, *Methods of Mathematical Physics*, vol. 1, Interscience, New York, 1953.
14. Courant, R., and D. Hilbert, *Methoden der mathematischen Physik* vol. 2, Interscience, New York, 1937.
15. Crandall, S. H., *Engineering Analysis*, McGraw-Hill, New York, 1956.
16. Dana, E. S. and W. E. Ford, *A Textbook of Mineralogy*, 4th ed., Wiley, New York, 1932.
17. De Bruijn, N. G., *Asymptotic Methods in Analysis*, North Holland Publishing Co., Amsterdam, 1958.
18. Deimel, R. F., *Mechanics of the Gyroscope*, Dover, New York, 1950.
19. Den Hartog, J. P., *Mechanical Vibrations*, 4th ed., McGraw-Hill, New York, 1956.
20. Dickson, L. E., *Elementary Theory of Equations*, Wiley, New York, 1914.
21. Dugas, R., *A History of Mechanics*, Central Book, New York, 1955.
22. Ewing, W. M., W. S. Jardetzky, and F. Press, *Elastic Waves in Layered Media*, McGraw-Hill, New York, 1957.
23. Faddeeva, V. N., *Computational Methods of Linear Algebra*, Dover, New York, 1959.
24. Fermi, E., *Thermodynamics*, Dover, New York, 1956.
25. Flügge, W., *Stresses in Shells*, Springer, Berlin, 1960.
26. Fox, C., *An Introduction to the Calculus of Variations*, Oxford University Press, London, 1950.
27. Goldstein, H., *Classical Mechanics*, Addison-Wesley, Cambridge, Mass., 1950.
28. Goursat, E. J. B., *A Course in Mathematical Analysis*, Ginn, Boston, 1904.
29. Green, A. E., and W. Zerna, *Theoretical Elasticity*, Oxford, Clarendon Press, 1954.
30. Hamel, G., *Mechanik der Kontinua*, Teubner, Stuttgart, 1956.
31. Hart, I. B., *The Mechanical Investigations of Leonardo da Vinci*, Open Court, Chicago, 1925.
32. Hausdorff, F., *Set Theory*, Chelsea, New York, 1957.
33. Hertz, H., *Principles of Mechanics*, Macmillan, London, 1899.
34. Hoff, N. J., *The Analysis of Structures*, Wiley, New York, 1956.
35. Ince, E. L., *Ordinary Differential Equations*, Dover, New York, 1944.
36. Jeffery, R. L., *The Theory of Functions of a Real Variable*, University of Toronto Press, Toronto, 1953.
37. Jones, B. W., *The Arithmetic Theory of Quadratic Forms*, Carus Math. Monograph, No. 10, Wiley, New York, 1950.

38. Kantorovich, L., and V. Krylov, *Näherungsmethoden der höheren Analysis*, VEB Deutscher Verlag der Wissenschaften, Berlin. 1956.
39. Kaplan, W., *Advanced Calculus*, Addison-Wesley, Cambridge, Mass., 1952.
40. Von Kármán, T., and M. Biot, *Mathematical Methods in Engineering*, McGraw-Hill, New York, 1940.
41. Kellogg, O. D., *Foundations of Potential Theory*, Dover, New York, 1953.
42. Kelvin, Lord, and P. G. Tait, *Natural Philosophy*, Part 2, Cambridge University Press, 1883.
43. Kimball, W. S., *Calculus of Variations*, Butterworth, London, 1952.
44. Klein, F., and A. Sommerfeld, *Über die Theorie des Kreisels*, Teubner, Leipzig, 1910.
45. Knopp, K., *Theory of Functions*, Part 2, Dover, New York, 1947.
46. Lagrange, J. L., *Mécanique analytique*, Paris, 1788.
47. Lanczos, C., *Applied Analysis*, Prentice-Hall, Englewood Cliffs, N.J. 1956.
48. Lanczos, C., *The Variational Principles of Mechanics*, University of Toronto Press, Toronto, 1949.
49. Langhaar, H. L., and A. P. Boresi, *Engineering Mechanics*, McGraw-Hill, New York, 1959.
50. Lindsay, R. B., *Mechanical Radiation*, McGraw-Hill, New York, 1960.
51. Love, A. E. H., *The Mathematical Theory of Elasticity*, 4th ed., Cambridge University Press, 1934.
52. McLachlan, N. W., *Theory of Vibrations*, Dover, New York, 1951.
53. Melan, E., and Parkus, H., *Wärmespannungen*, Springer, Vienna, 1953.
54. Milne-Thomson, L. M., *Theoretical Hydrodynamics*, Macmillan, London, 1938.
55. Morse, P. M., and H. Feshbach, *Methods of Theoretical Physics*, vols. 1 and 2, McGraw-Hill, New York, 1953.
56. Murnaghan, F. D., *Finite Deformation of an Elastic Solid*, Wiley, New York, 1951.
57. Muskhelishvili, N. I., *Some Basic Problems in the Mathematical Theory of Elasticity*, Noordhoff, Groningen, Holland, 1953.
58. Myklestad, N. O., *Fundamentals of Vibration Analysis*, McGraw-Hill, New York, 1956.
59. Nadai, A., *Elastische Platten*, Springer, Berlin, 1922.
60. Nadai, A., *Theory of Flow and Fracture of Solids*, 2nd ed., McGraw-Hill, New York, 1950.
61. Novozhilov, V. V., *Foundations of the Nonlinear Theory of Elasticity*, Graylock, Rochester, N.Y., 1953.
62. Osgood, W. F., *Mechanics*, Macmillan, New York, 1937.
63. Parcel, J. I., and G. A. Maney, *Statically Indeterminate Stresses*, 2nd ed, Wiley, New York, 1936.
64. Parkus, H., *Mechanik der festen Körper*, Springer, Vienna, 1959.
65. Phillips, A., *Introduction to Plasticity*, Ronald, New York, 1956.
66. Pippard, A. J. S., *Strain Energy Methods of Stress Analysis*, Longmans, Green, London, 1928.
67. Planck, M., *General Mechanics*, Macmillan, London, 1933.
68. Planck, M., *The Mechanics of Deformable Bodies*, Macmillan, London, 1932.
69. Planck, M., *Theory of Heat*, Macmillan, London, 1932.
70. Prager, W., and P. G. Hodge, *Theory of Perfectly Plastic Solids*, Wiley, New York, 1951.
71. Rayleigh, Lord, *Theory of Sound*, Dover, New York, 1945.
72. Routh, E. J., *Advanced Dynamics of a System of Rigid Bodies*, 6th ed., Dover, New York, 1905.

73. Schleusner, A., *Strenge Theorie der Knickung und Biegung*, Teubner, Leipzig, 1937.
74. Seely F. B., and J. O. Smith, *Advanced Mechanics of Materials*, 2nd ed., Wiley, New York, 1952.
75. Sokolnikoff, I., *Mathematical Theory of Elasticity*, 2nd ed., McGraw-Hill, New York, 1956.
76. Sokolovsky, V. V., *Theorie der Plastizität*, VEB Verlag Technik, Berlin, 1955.
77. Sommerfeld, A., *Mechanics of Deformable Bodies*, vol. 2 of Lectures on Theoretical Physics, Academic, New York, 1950.
78. Sternberg, W., and T. Smith, *The Theory of Potential and Spherical Harmonics*, University of Toronto Press, Toronto, 1946.
79. Struik, D. J., *Differential Geometry*, Addison-Wesley, Cambridge, Mass., 1950.
80. Synge, J. L., *The Hypercircle in Mathematical Physics*, Cambridge University Press, 1957.
81. Synge, J, L., and B. A. Griffith, *Principles of Mechanics*, 3rd ed, McGraw-Hill, New York, 1959.
82. Temple, G., and W. Bickley, *Rayleigh's Principle and Its Application to Engineering*, Dover, New York, 1956.
83. Timoshenko, S., *Vibration Problems in Engineering*, 3rd ed., Van Nostrand Princeton, N.J. 1955.
84. Timoshenko, S. and J. M. Gere, Theory of Elastic Stability, 2nd ed, McGraw-Hill, New York, 1961.
85. Timoshenko, S., and J. N. Goodier, *Theory of Elasticity*, 2nd ed., McGraw-Hill, New York, 1951.
86. Timoshenko, S., and S. Woinowsky-Krieger, *Theory of Plates and Shells*, 2nd ed., McGraw-Hill, New York, 1959.
87. Tong, K. N., *Theory of Mechanical Vibrations*, Wiley, New York, 1960.
88. Van den Broek, J. A., *The Elastic Energy Theory*, Wiley, New York, 1931.
89. Weinstock, R., *Calculus of Variations*, McGraw-Hill, New York, 1952.
90. Whittaker, E. T., *Analytical Dynamics*, Dover, New York, 1944.

ARTICLES

91. Benjamin, R. J., Analysis of Oval Rings by Fourier Series, *J. Aeronaut. Sci.*, **19**, 9, September 1952.
92. Bleich, F., and H. Bleich, Bending, Torsion, and Buckling of Bars Composed of Thin Walls, *Prelim. Pub., 2nd Congr. Internat. Ass'n Bridge & Structural Engrs.*, English ed., 871, Berlin, 1936.
93. Boresi, A. P., A Refinement of the Theory of Buckling of Rings under Uniform Pressure, *J. Appl. Mechanics*, **22**, 1, March 1955.
94. Brown, C. L., The Treatment of Discontinuities in Beam Deflection Problems, *Quart. appl. Math.*, **1**, 4, January 1944.
95. Brown, E. H., The Energy Theorems of Structural Analysis, Engineering: 1-Definitions and Fundamentals, 305–308, March 11, 1955; 2-Derivation and Discussion, 339–342, March 18, 1955; 3-Worked Examples, 400–403, April 1, 1955.
96. Brush, D. O., Strain-Energy Expression in Nonlinear Shell Analysis, *J. Aeronaut. Space Sciences*, **27**, 7, 555–556, July 1960.
97. Bryan, G. H., On the Stability of Elastic Systems, *Proc. Cambridge Phil. Soc.*, **VI**, 199–210, 1886–1889.
98. Bryan, G. H., On the Stability of a Plane Plate under Thrusts in Its Own Plane, with Applications to the "Buckling" of the Sides of a Ship, *Proc. London Math. Soc.*, **XXII**, 54–67, 1891.

99. Castigliano, A., Theorie de l'equilibre des systemes elastiques et ses applications, Turin, 1879.

100. Charlton, T. M., Some Notes on the Analysis of Redundant Systems by Means of the Conception of Conservation of Energy, *J. Franklin Inst.*, **250**, 6, December 1950.

101. Charlton T. M. The Analysis of Statically Indeterminate Structures by the Complementary Energy Method, *Engineering*, **174**, 4521, September 19, 1952.

102. Chien, W. Z., The Intrinsic Theory of Thin Shells and Plates, *Quart. Appl. Math.*, **1**, 4, January 1944.

103. Donnell, L. H., Stability of Thin-Walled Tubes under Torsion, NACA TR 479, 1933.

104. Felgar, R. P., Jr., Formulas for Integrals Containing Characteristic Functions of a Vibrating Beam, Circular No. 14, Bureau of Engineering Research, University of Texas, Austin, 1950.

105. Fourier, J., Memoire sur la statique contenant la demonstration du principe des vitesses virtuelles et la théorie des moments, Oeuvres de Fourier, **2**, Gauthier Villars, Paris, 1890.

106. Friedrichs, K. O., On the Minimum Buckling Load for Spherical Shells, Theodore von Kármán Anniversary Volume, Calif. Inst. of Technology, 1941.

107. Friedrichs, K. O., and J. J. Stoker, Buckling of Circular Plates Beyond the Critical Thrust, *J. Appl. Mechanics*, **9**, A-7, 1942.

108. Goldberg, J. E., On the Application of Trigonometric Series to the Twisting of I-Type Beams, *Proc. First U.S. Nat. Congress Appl. Mech.*, ASME., New York, 1952.

109. Goodier, J. N., Some Observations on Elastic Stability, *Proc. First U.S. Nat. Congress Appl. Mechanics*, ASME, New York, 1952.

110. Goodier, J. N., Discussion of paper by E. Reissner, *J. Appl. Mechanics*, **13**, 3, A-251, 1946.

111. Goodier, J. N., Torsional and Flexural Buckling of Bars of Thin-Walled Open Section under Compressive and Bending Loads, *Trans. ASME*, **64**, A-103, 1942.

112. Goodier, J. N., Flexural-Torsional Buckling of Bars of Open Sections Under Bending, Eccentric Thrust, or Torsional Loads, Cornell Univ., Eng. Exper. Sta. Bulletin 28, 1942.

113. Goodier, J. N., The Buckling of Compressed Bars by Torsion and Flexure, Cornell Univ. Eng. Exper. Sta. Bulletin No. 27, Dec., 1941.

114. Gossard, M., P. Seide, and W. Roberts, Thermal Buckling of Plates, NACA TN 2771, 1952.

115. Hamilton, W. R., The Mathematical Papers of Sir W. R. Hamilton, **II**, *Dynamics*, Cambridge University Press, 1940.

116. Hoff, N. J., A Strain Energy Derivation of the Torsional-Flexural Buckling Loads of Straight Columns of Thin-Walled Open Sections, *Quart. appl. Math.*, **1**, 4, 341–345, January 1944.

117. Kappus, R., Drillknicken zentrisch gedrückter Stäbe mit offenem Profil im elastischen Bereich, *Luftfahrtforschung*, **14**, 44, 1937. Also: Twisting Failure of Centrally Loaded Open-Section Columns in the Elastic Range, NACA TM 851, 1938.

118. Von Kármán, T., and H. S. Tsien, The Buckling of Spherical Shells by External Pressure, *J. Aeronaut. Sci.*, **7**, 2, December 1939.

119. Langhaar, H. L., General Theory of Buckling, *Appl. Mechanics, Revs*, **11**, 11, November 1958.

120. Langhaar, H. L., The Principle of Complementary Energy in Nonlinear Elasticity Theory, *J. Franklin Inst.*, **256**, 3, September 1953.

121. Langhaar, H. L., Note on the Energy of Bending of Plates, *J. Appl. Mechanics*, **19**, 2, June 1952.

122. Langhaar, H. L., On the Torsional-Flexural Buckling of Columns, *J. Franklin Inst.*, **255**, 2, February 1953.

123. Langhaar, H. L., Lateral Buckling of Asymmetrical Beams, *J. Appl. Mechanics*, **22**, 3, September 1955.

124. Langhaar, H. L., and A. P. Boresi, Strain Energy and Equilibrium of a Shell Subjected to Arbitrary Temperature Distribution, *Proc. Third U.S. Nat. Congress Appl. Mechanics, ASME*, New York, 1958.

125. Langhaar, H. L., A. P. Boresi, and D. R. Carver, Energy Theory of Buckling of Circular Elastic Rings and Arches, *Proc. Second U.S. Nat. Congress Appl. Mech., ASME, New York*, 1954.

126. Langhaar, H. L., and M. C. Stippes, Three-Dimensional Stress Functions, *J. Franklin Inst.*, **258**, 5, 371–382, November 1954.

127. Lundquist, E., and C. Fligg, A Theory for Primary Failure of Straight Centrally Loaded Columns, NACA Report No. 582, 1937.

128. Maxwell, J. C., On the Calculation of the Equilibrium and Stiffness of Frames, *The Scientific Papers of J. C. Maxwell*, Dover, New York, 1890.

129. Naghdi, P. M., On the Theory of Thin Elastic Shells, *Quart. appl. Math.*, **XIV**, 4, 369–380, January 1957.

130. Parkus, H., Über eine Erweiterung des Hamiltonschen Prinzipes auf thermoelastische Vorgänge, *Federhofer-Girkmann Festschrift*, Vienna, 1950.

131. Poincaré, H., Sur l'Equilibre d'une masse fluide animée d'un mouvement de rotation, *Acta Math.*, **VII**, 259–380, 1883.

132. Poritsky, H., Topics in Gyroscopic Motion, *J. Appl. Mechanics*, **20**, 1, March 1953.

133. Radok, J. R. M., and A. Heller, Die exakte Lösung der Integralgleichungen gewisser Schwingungsprobleme, *Z. angew. Math. Physik*, **V**, Facs. 1, 50–66, 1954.

134. Rankine, W. J. M., On the General Law of Transformation of Energy, *Proc. Phil. Soc., Glasgow*, **III**, V, 1853.

135. Reiss, E. L., H. J. Greenberg, and H. B. Keller, Nonlinear Deflections of Shallow Spherical Shells, *J. Aeronaut. Sci.*, **24**, 7, 533–543, July 1957.

136. Reissner, E., The Effect of Transverse Shear Deformation on the Bending of Elastic Plates, *J. Appl. Mechanics*, **12**, 2, June 1945.

137. Reissner, E., On the Bending of Elastic Plates, *Quart. appl. Math.*, **V**, 1, April 1947.

138. Reissner, E., On a Variational Theorem in Elasticity, *J. Math. and Phys.*, **29**, 90–95, 1950.

139. Reissner, E., On Non-Uniform Torsion of Cylindrical Rods, *J. Math. and Phys.*, **31**, 214–221, 1952.

140. Reissner, E., On a Variational Theorem for Finite Elastic Deformations, *J. Math. and Phys.* **32**, 129–135, 1953.

141. Reissner, E., On Variational Principles in Elasticity, *Proc. Symposia Appl. Math.*, **8**, 1–6, 1958.

142. Ritz, W., Ueber eine neue Methode zur Lösung gewisser Variationsprobleme der mathematischen Physik, *J. reine angew. Math.*, **135**, 1–61, 1909.

143. Ruffner, B. F., The Use of Fourier Series in the Solution of Beam Problems, Bull. Ser. No. 18, Eng. Exper. Sta., Oregon State College, 1944.

144. Seide, P., A Donnell Type Theory for Asymmetrical Bending and Buckling of Thin Conical Shells, *J. Appl. Mechanics*, **24**, 4, 547–552, December 1957.

145. Temple, G., The General Theory of Relaxation Methods Applied to Linear Systems, *Proc. Roy. Soc.* (London), **A, 169,** December–March 1938–39.
146. Trefftz, E., Zur Theorie der Stabilität des elastischen Gleichgewichts, *Z. angew. Math. Mech.*, **13,** 160–165, 1933.
147. Trefftz, E., Konvergenz und Fehlerschätzung beim Ritzschen Verfahren, *Math. Ann.*, **100,** 503–521, 1928.
148. Vlasov, V. S., Basic Differential Equations in the General Theory of Elastic Shells, NACA TM 1241, February 1951.
149. Wagner, H., Verdrehung und Knickung von offenen Profilen, 25th Anniversary Volume of the Technische Hochschule, Danzig, 1929. Also: Torsion and Buckling of Open Sections, NACA TM 807, 1936.
150. Westergaard, H. M., On the Method of Complementary Energy, *Trans. ASCE,* **107,** 765–803, 1942.
151. Young, D., and R. P. Felgar, Jr., Tables of Characteristic Functions Representing Normal Modes of Vibration of a Beam, Engineering Research Series No. 44, University of Texas, Austin, 1949.
152. Ziegler, H., On the Concept of Elastic Stability, Advances in Applied Mechanics, **IV,** Academic, New York, 1956.
153. Ziegler, H., Die Stabilitätskriterien der Elastomechanik, *Ingen. Arch.*, **20,** 49–56, 1952.

Index

347